In Memoriam

This volume is the last major work of Jonathan B. Tucker, an internationally respected expert on chemical and biological weapons issues, biotechnology, infectious disease, arms control, and nonproliferation, who died at the end of July 2011.

Jonathan Brin Tucker was born in Boston on August 2, 1954. After earning a biology degree (*cum laude*) from Yale University in three years, he went on to receive a master's degree from the University of Pennsylvania and a doctorate in political science from Massachusetts Institute of Technology.

Tucker's hybrid education was an ideal launch pad for a career at the nexus of science and policy, working at the Congressional Office of Technology Assessment, the U.S. Arms Control and Disarmament Agency, the Monterey Institute's James Martin Center for Nonproliferation Studies, Darmstadt University of Technology in Germany, and the Federation of American Scientists. Tucker interspersed these positions with visiting fellowships at renowned research institutes in America and abroad. Along the way, Tucker was also a member of the U.S. delegation to the preparatory commission for the Chemical Weapons Convention and did a stint in 1995 as a United Nations weapons inspector in Iraq. Such experiences seasoned his research and set the stage for his staffing of two blue-ribbon panels, the Presidential Advisory Committee on Gulf War Veterans' Illnesses and the Commission on the Prevention of WMD Proliferation and Terrorism.

Having been an editor at *Scientific American* and *High Technology*, Tucker went on to publish in many professional journals; to edit two books in addition to this one, *Germany in Transition: A Unified Nation's Search for Identity* (1999) and *Toxic Terror: Assessing Terrorist Use of Chemical and Biological Weapons* (MIT Press, 2000); and to pen *Ellie: A Child's Fight against Leukemia* (1982), *Scourge: The Once and Future Threat of Smallpox* (2001), and *War of Nerves: Chemical Warfare from World War I to Al-Qaeda* (2006). Tucker's track record of authoritative scholarship garnered him invitations to testify before Congress and made him sought after by journalists.

Tucker's hallmark was balanced, meticulous research and analysis presented in crisp, comprehensible prose. Generous to a fault, Tucker gave his time and intellect to the arms control and disarmament community, serving on the boards of the Arms Control Association, the *Bulletin of the Atomic Scientists*, and the *Nonproliferation Review*. He was the embodiment of the gentleman scholar; and his quiet demeanor offset his tenacity as a researcher and his ability to innovate approaches to reduce the unconventional weapons threats that he studied.

Tucker's contributions to curbing the dangers of chemical and biological weapons have made the world a safer place, and his loss is felt greatly by his many friends and colleagues around the globe.

Hussein Alramini

Brian Balmer

Kirk C. Bansak

Nancy Connell

Malcolm R. Dando

Gerald L. Epstein

Gail Javitt

Catherine Jefferson

Alexander Kelle

Lori P. Knowles

Filippa Lentzos

Caitríona McLeish

Matthew Metz

Nishal Mohan

Jonathan D. Moreno

Anya Prince

Pamela Silver

Amy E. Smithson

Ralf Trapp

Mark Wheelis

Raymond A. Zilinskas

Contents

Innovation, Dual Use, and Security

Innovation, Dual Use, and Security

Managing the Risks of Emerging Biological and Chemical Technologies

edited by Jonathan B. Tucker
foreword by Richard Danzig

The MIT Press
Cambridge, Massachusetts
London, England

MIT Press books may be purchased at special quantity discounts for business or sales promotional use. For information, please email special_sales@mitpress.mit.edu or write to Special Sales Department, The MIT Press, 55 Hayward Street, Cambridge, MA 02142.

This book was set in Sabon by Toppan Best-set Premedia Limited. Printed and bound in the United States of America. Printed on recycled paper.

Library of Congress Cataloging-in-Publication Data

Innovation, dual use, and security: managing the risks of emerging biological and chemical technologies / edited by Jonathan B. Tucker; foreword by Richard Danzig.
    p.   cm.
Includes bibliographical references and index.
ISBN 978-0-262-01717-6 (hardcover: alk. paper)—ISBN 978-0-262-51696-9 (pbk.: alk. paper)
1. Biotechnology—Social aspects. 2. Nanotechnology—Social aspects. 3. Biological weapons.
4. Chemical agents (Munitions). I. Tucker, Jonathan B.
TP248.23.I56   2012
660.6'3—dc23
2011036228

10   9   8   7   6   5   4   3   2   1

Foreword

In his autobiography *Disturbing the Universe* (1979), the eminent physicist Freeman Dyson recalls a Victorian story he read as a child called "The Magic City." The tale's author, Edith Nesbit, posits a hero who has the ability to invent any technology he desires, but whatever he creates can never be discarded. He must live with its consequences for the rest of his life.

Dyson invoked this story to convey his dismay about the nuclear weapons he had helped to invent. He recognized that the technologies of war can be turned against their makers. In modern terms, they proliferate. Moreover, as they proliferate, they often decline in cost, increase in accessibility, and multiply in killing power.

That two atomic bombs killed more than a hundred thousand people in Japan in 1945 provides a compelling case for trying to control nuclear weapons. But one cannot help but think that these weapons also attract so much of our attention because the problems they pose can be resolved, at least conceptually. The logistics of nuclear weapons are in our favor. Atmospheric and underground tests can be detected and can therefore be banned. Weapons-grade nuclear material is difficult to create: the production of plutonium or highly enriched uranium and its fabrication into weapons require time, considerable funding, specialized equipment, and uncommon skills. All these activities have distinct signatures and can thus be controlled, at least in theory. To be sure, it takes extraordinary political will, a complex regime of international inspections, and elaborate enforcement mechanisms to keep the nuclear genie in the bottle. But for two-thirds of a century, we have done it, albeit imperfectly.

For biological and chemical weapons, in contrast, the logistics are not in our favor. These weapons can be developed and produced with low visibility and cost, and they require skills and equipment that are rapidly simplifying and proliferating. Most critically, our efforts to control the development of biological and chemical arms are confounded by the phenomenon of "dual use." Whereas nuclear-weapon technology can be differentiated from the peaceful uses of nuclear energy, the techniques and equipment for making biological and chemical weapons are fundamentally the same as those that sustain biological and chemical science and industry.

This book describes and evaluates our limited tools to disentangle the harmful applications of these technologies from the beneficial ones. It usefully differentiates between hard law (legally enforceable measures), soft law (formally agreed norms, but without enforcement mechanisms), and informal measures. It describes and recommends techniques such as export restrictions (adopted by a minority of nations), access controls on stocks of dangerous pathogens, international treaties (some of which, like the Biological Weapons Convention, are hortatory but lack verification and enforcement mechanisms), threat-reduction programs to convert state weapons facilities to commercial uses and employ former weaponeers in peaceful research, and efforts to educate scientists about the risk of misuse.

The contributors to this book usefully map and explore parts of a large and growing problem, but much remains unclear and uncontrolled. Our current failure is one of concept as well as political will: we have some tactics, but we do not yet have a plausible strategy for asserting control over protean, proliferating, and powerful biological and chemical technologies and the new types of weapons they can spawn. Put metaphorically, the nuclear weapons genie stands alone, whereas the biological and chemical weapons genies are Siamese twins with one set of vital organs. We cannot kill the evil twin because we have let the good twin out into the world.

This book tells us where we stand with dual-use biological and chemical technologies, how we got here, and, as best it can, where we are likely to go. Since that is a bad place, the authors struggle with how we might change course. Unfortunately, this is no fairy tale and not a book for children. But we need some books for adults, too.

Richard Danzig
Washington, D.C.

1

Introduction

Jonathan B. Tucker

Rapid technical innovation in fields such as biotechnology, nanotechnology, and neuropharmacology promises great benefits for human health and welfare but could potentially be exploited for the development and production of biological or chemical weapons.[1] These technologies are said to pose a dual-use dilemma because it is difficult to prevent their misuse without forgoing beneficial applications.[2] Indeed, many of the emerging technologies with the potential to do the most good are also capable of the greatest harm.

Recent developments in the life sciences have raised the political salience and urgency of the dual-use problem. In 1999 a report by the British Medical Association titled *Biotechnology, Weapons, and Humanity* observed: "We are concerned that the emerging sciences of genetic engineering and biotechnology may be developed for malign purposes. The social and ethical safeguards which may prevent the escalation of conflict and weapons development therefore need to be discussed urgently."[3] In 2003 an expert panel, convened under the auspices of the U.S. National Academy of Sciences and chaired by biology professor Gerald Fink of the Massachusetts Institute of Technology, produced a landmark study titled *Biotechnology Research in an Age of Terrorism*. This report identified seven types of experiments in the fields of microbiology and molecular biology that entail dual-use risks and hence warrant a security review before being approved and funded.[4] Responding to a recommendation in the Fink committee's report, the U.S. government established a federal advisory committee called the National Science Advisory Board for Biosecurity (NSABB), with the mandate to develop policy recommendations for the oversight of "dual-use research of concern" in the life sciences.[5]

Other prominent organizations have also issued warnings about the potential misuse of biotechnology. In 2004 the World Health Organization warned that "every major new technology of the past has come to be intensively exploited, not only for peaceful purposes but also for hostile ones. Prevention of the hostile exploitation of biotechnology therefore rises above the security interests of individual states and poses a challenge to humanity generally."[6] In the same year, the International

Committee of the Red Cross (ICRC) launched an appeal urging governments, the scientific and medical communities, the military, industry, and civil society "to strengthen their commitment to the international humanitarian law norms which prohibit the hostile uses of biological agents and to work together to subject potentially dangerous biotechnology to effective controls."[7]

In light of these security concerns, this book has two objectives. The first is analytical: to examine the characteristics of emerging dual-use technologies in the biological and chemical fields. The second is practical: to help policy makers devise effective governance strategies that can manage the risks of these innovations while preserving their very real benefits.

Defining Dual Use

The term "dual use" has multiple meanings. In the context of defense procurement, it refers to technologies that have both civilian and military applications.[8] Policy makers have at times encouraged the transfer of civilian technologies to the defense sector to reduce the cost of conventional weapon systems. In a different context, however, "dual use" refers to materials, hardware, and knowledge that have peaceful applications but could also be exploited for the illicit production of nuclear, chemical, or biological weapons. For example, certain ingredients in the production of inks, pesticides, and fire retardants are also precursors, or intermediates, in the production of chemical warfare agents. Similarly, a stainless-steel fermentation tank can be used either to produce a protective vaccine or to grow virulent anthrax bacteria as a biological weapon.

When thinking about dual-use risks in the life sciences, it is instructive to consider how biotechnology differs from nuclear technology. Methods for the enrichment of uranium or the extraction and recycling of plutonium from spent nuclear fuel are considered to be dual use because they can be employed either to produce fuel rods for generating electricity or to construct nuclear weapons. Nevertheless weapons-grade fissile materials have several characteristics that make them amenable to physical protection, control, and accounting. Highly enriched uranium and plutonium are both difficult and costly to produce, do not exist in nature in a concentrated form suitable for weapons use, have few civilian applications, and emit ionizing radiation that can be detected at a distance.

In contrast, pathogenic bacteria and viruses are available from natural sources, are self-replicating and hence cannot be inventoried and tracked reliably in a quantitative manner, have numerous legitimate applications in biomedical research and the pharmaceutical biotechnology industry, are present in many types of facilities (e.g., hospitals and universities), and are impossible to detect at a distance with available technologies. Because of these differences, the process of acquiring biological weapons entails fewer technical hurdles and has a lower chance of discovery than the construction of a nuclear device. In addition, whereas dual-use nuclear technologies are advancing slowly—the basic methods of uranium enrichment and plutonium separation have not changed significantly in several decades— many areas of biotechnology are progressing at an exponential rate, and the time lag from scientific discovery to technological application is extremely short.

Of course, nearly every technology has some potential for misuse; a hammer can serve as either a tool or a murder weapon. Given the pervasiveness of the dual-use phenomenon, devising an operational definition requires striking a reasonable balance. Defining dual use too narrowly would fail to capture some potential threats, while defining the term too broadly would constrain relatively innocuous technologies. To limit the scope of the problem, this book does not seek to cover the entire universe of emerging biological and chemical technologies that might be misused for hostile purposes. Instead the analysis focuses more narrowly on the subset of emerging technologies that are potential game changers because they could result in harmful consequences far more serious than is possible with existing technologies. For a dual-use innovation to cause real concern, in other words, it must offer a significant qualitative or quantitative increase in destructive capacity over what is currently available.

The Dual-Use Landscape, Past and Present

Since 9/11 and the anthrax letter attacks, the potential terrorist use of emerging technologies for the development and production of biological and chemical weapons has become a major focus of government concern. The broader phenomenon of dual-use technologies, however, has a much longer history. In the twentieth century, the two world wars saw the intensive exploitation of chemistry and physics for military purposes, including the development of high explosives, chemical nerve agents, radar, ballistic missiles, and the atomic bomb.[9] Although biology played a much smaller role in these conflicts, it did not escape application as an instrument of warfare. During World War I, German saboteurs drew on the bacteriological discoveries of Louis Pasteur and Robert Koch to carry out covert operations in which they used anthrax and glanders bacteria to sicken Allied horses, a key component of military logistics at the time.[10]

Before and during World War II, the United States, the Soviet Union, Britain, France, Germany, Japan, and other countries harnessed the scientific advances in microbiology for the development of offensive biological warfare capabilities.[11] The only country known to have used biological weapons during this period was imperial Japan. Between 1932 and 1945, Japanese military scientists developed a variety of biological warfare agents, tested them on human prisoners, and delivered them on an experimental basis against eleven Chinese cities.[12] Although these attacks killed many thousands of civilians, they proved to be of minimal strategic significance.

The biotechnology revolution began in the early 1970s, two decades after James Watson and Francis Crick published their seminal paper of 1953, describing the double-helical structure of the DNA molecule and suggesting how it encoded genetic information. In 1973, Stanley Cohen of Stanford University and Herbert Boyer of the University of California at San Francisco invented a methodology for cutting and splicing segments of DNA from different sources, making it possible to transfer genes between different species. Practical applications of this technology, known as recombinant DNA or genetic engineering, included the production of human insulin in bacteria and gave rise to the modern biotechnology industry.

Although the first biotech firms were spun off from large research universities in the Boston and San Francisco areas, the industry has since expanded globally. Several factors have fueled this international growth, including economic globalization and the increasing use of international subcontracting and cooperation agreements.[13] In addition, a number of Asian countries, such as China, India, Malaysia, and Singapore, have championed biotechnology as a key element of their economic development plans.

The Rise of Synthetic Genomics

The progressive refinement since the 1980s of automated DNA synthesis machines, which can produce any desired sequence from off-the-shelf chemicals in the laboratory, has made it possible to construct entire genes and microbial genomes from scratch. This powerful new technology is known as synthetic genomics. Instead of using traditional recombinant DNA techniques to isolate individual genes from one species and splice them into the genome of another, scientists can now design a genetic sequence on a computer and convert it directly into a physical strand of DNA coding for a useful product or function. A global industry has emerged to produce custom DNA molecules to order for scientific and pharmaceutical-industry clients, who have increasingly outsourced their DNA-synthesis requirements. Commercial suppliers of synthetic DNA not only exist in the United States, Europe, and Japan but are also present in the developing world, including countries in Asia, South America, and the Middle East.

Synthetic genomics is a clear example of a dual-use technology. In 2002, scientists at the State University of New York at Stony Brook re-created the poliovirus by converting its published RNA sequence to DNA (which is easier to synthesize), ordering short sections of the DNA sequence from a commercial supplier, and assembling these fragments into the full viral genome, which, when placed in a cell-free extract, spontaneously directed the production of infectious virus particles.[14] This demonstration prompted fears that terrorists might exploit the same technology to re-create more deadly viral agents.[15] Since 2002, legitimate scientists have also synthesized a bat virus closely related to the causative agent of SARS, as well as the formerly extinct "Spanish" strain of influenza virus, which was responsible for the global pandemic in 1918 and 1919 that killed as many as 50 million people worldwide. (The scientific rationale for re-creating the Spanish influenza virus in 2005 was to gain insights into the genetic factors that made it particularly virulent so as to guide the development of antiviral drugs that would be effective against future pandemic strains of the disease.)

The technology underlying genome synthesis has continued to advance at a remarkable pace. In May 2010, the J. Craig Venter Institute announced the synthesis of the first self-replicating bacterial genome, consisting of more than a million DNA units.[16] At least in principle, it is now possible for skilled scientists to reconstruct any virus for which an accurate genetic sequence exists, including variola virus, the causative agent of smallpox, which was eradicated from nature in the late 1970s and exists only in a few highly secure repositories.

In addition to the powerful technology of genome synthesis, rapid advances in mapping the human genome (genomics), studying the structure and function of the myriad proteins in living organisms (proteomics), and analyzing the complex biochemical circuits that regulate cellular metabolism (systems biology) are yielding a profound new understanding of life at the molecular level. At the same time, the growing convergence of biological and chemical production methods has made it possible to produce specialty chemicals and drugs in engineered bacteria and to synthesize biological molecules such as peptides (short protein chains) by strictly chemical means.[17] The dynamic field of nanotechnology is also giving rise to engineered nanoparticles that can ferry drugs through the bloodstream to specific tissues while evading the host immune response.

Although the dramatic advances in biological science and technology promise valuable new drugs and therapies, they also have a potential dark side. During the 1980s, the Soviet biological warfare program drew on recombinant DNA technology to develop genetically engineered pathogens with greater virulence, stability, and antibiotic resistance.[18] Today there is growing concern that outlaw states or terrorist organizations could misuse synthetic-genomics technology to circumvent controls on access to viral pathogens of biological-weapons concern. Other emerging biotechnologies might be harnessed to develop "enhanced" biological or chemical warfare agents that are more lethal, controllable, or resistant to existing medical countermeasures.[19]

Dual-use risks have also emerged unexpectedly from basic and applied research in the life sciences. In 2000, for example, a group of Australian researchers who were developing a contraceptive vaccine to control mouse populations found that inserting a single gene for an immune regulatory protein called interleukin-4 into mousepox virus rendered the normally mild pathogen highly lethal in mice, even in animals that had been vaccinated against mousepox.[20] This surprising discovery had clear dual-use implications because mousepox virus is closely related to variola virus (the causative agent of smallpox) and to monkeypox virus, both of which can infect humans. Accordingly, that the interleukin-4 gene increased the virulence of mousepox suggested that a similar genetic manipulation of variola or monkeypox might render them resistant to the standard protective vaccine.[21] After debating whether or not to publish their findings, the Australian researchers finally did so in the *Journal of Virology* in early 2001. The troubling security implications of their paper whipped up a storm of controversy about whether certain types of scientific information are too sensitive to release into the public domain.[22]

Technical Hurdles to Bioterrorism

Despite growing concerns about dual-use biotechnologies such as synthetic genomics, some analysts contend that the risks of misuse by terrorists are not as great as they appear because acquiring a biological-warfare capability is a complex, multistep process. Even if a would-be bioterrorist were able to harness whole-genome synthesis to construct a highly pathogenic virus from scratch, building an effective weapon would still require overcoming several more technical obstacles.[23] First, the synthetic virus would have to be cultivated in sufficient quantity, but viruses are significantly harder to grow than bacteria because they replicate only in living cells. One low-tech option would be to cultivate the virus in fertilized eggs, but to avoid contamination, it would be necessary to use pharmaceutical-grade eggs that would have to be specially ordered—not easy to do without attracting attention. Cultivating a highly pathogenic virus would also be extremely hazardous to the perpetrators.

Disseminating biological agents effectively involves additional technical hurdles. Whereas persistent chemical-warfare agents, such as sulfur mustard and VX nerve gas, are readily absorbed through the intact skin, bacteria and viruses cannot enter the body by that route unless the skin has been broken. As a result, microbial agents must usually be ingested or inhaled to cause infection. To expose large numbers of people through the gastrointestinal tract, it might be possible to contaminate food or drinking water supplies, yet neither of these scenarios would be easy to accomplish. Although large urban reservoirs are usually unguarded, unless the terrorists dumped in a massive quantity of biological agent, the dilution factor would be so great that no healthy person drinking the water would receive an infectious dose.[24] Modern sanitary techniques such as chlorination and filtration are also designed to kill pathogens from natural sources and would probably be equally effective against a deliberately released agent.

Accordingly, the only way to inflict mass casualties with a biological agent is by disseminating it as an aerosol: an invisible cloud of microscopic droplets or particles that remains suspended in the air for long periods and can be inhaled by large numbers of people downwind, causing infection through the lungs. A concentrated aerosol of an infectious agent, released into the atmosphere of a densely populated urban area, could potentially infect many thousands of victims. After an incubation period of a few days (depending on the type of agent and the inhaled dose), the exposed population would experience an outbreak of incapacitating or fatal illness.

Nevertheless the aerosol delivery of a biological agent is technically challenging. To infect through the lungs, the infectious particles must be between one and five microns (millionths of a meter) in diameter. Generating an aerosol cloud with the particle size and concentration needed to cover a large geographic area would also require a sophisticated delivery system, such as an airborne sprayer with specialized nozzles. There is also a trade-off between the ease of production and the effectiveness of dissemination. The simplest way to produce microbial agents is in the form of a wet slurry, but when a liquid agent is sprayed into the air, it forms heavy droplets that fall to the ground, so that only a small percentage of the agent is aerosolized. In contrast, if the microbes are dried to a solid cake and then milled into a fine powder, they become far easier to aerosolize, but the drying and milling process is technically challenging.

Even if aerosolization can be achieved, the effective delivery of biological agents in the open air would depend on the prevailing atmospheric and wind conditions, creating additional uncertainties. Only under highly stable atmospheric conditions, such as an inversion, will an aerosol cloud stay close to the ground where it can be inhaled rather than being rapidly dispersed. Moreover, most microorganisms are sensitive to ultraviolet radiation and cannot survive more than thirty minutes in bright sunlight, limiting effective delivery to nighttime attacks. The one major exception to this rule is the anthrax bacterium, which can form spores with a tough protective coating that will survive for several hours in sunlight. Terrorists could, of course, stage a biological attack inside an enclosed space, such as a building, subway station, shopping mall, or sports arena, but even there the technical hurdles would be by no means trivial.

Finally, the outcome of a biological attack would depend on whether the agent is contagious from person to person, the basic health of the people who are exposed, and the speed with which the public health authorities detect the outbreak and move to contain it. A prompt response with effective medical countermeasures, such as the administration of antiviral drugs combined with vaccination, might significantly blunt the impact of an attack. In addition, simple, proven methods, such as the isolation and home care of infected individuals, the wearing of face masks, frequent hand washing, and the avoidance of hospitals where transmission rates are high, have been effective in the past at slowing the spread of epidemic diseases.

In sum, the technical challenges involved in carrying out a mass-casualty biological attack are formidable. Perhaps for this reason, only two bioterrorist attacks in the United States are known to have caused actual casualties. Both of these incidents involved the use of crude delivery methods: the deliberate contamination of restaurant salad bars with *Salmonella* bacteria by the Rajneeshee cult in Oregon in 1984, and the mailing of powdered anthrax spores through the U.S. postal system in 2001. Such low-tech attacks, which are likely to remain the most common form of bioterrorism, are potentially capable of inflicting at most tens to hundreds of fatalities— within the destructive range of high-explosive bombs, but not the mass death predicted by worst-case planning scenarios.[25]

Nevertheless growing access to certain emerging biotechnologies could potentially enable nonstate actors to overcome the existing technical hurdles and carry out more devastating attacks, a possibility that warrants a closer look at the associated risks. In addition, past state-level programs, such as those of the United States and the Soviet Union, have already demonstrated the ability to overcome the technical barriers involved in the weaponization and large-scale production of biological agents.

The Changing Nature of Security Threats

In parallel with the revolution in biology and chemistry, the nature of armed conflict has undergone a major transformation in recent decades. Since the end of the Cold War, the terrifying specter of vast land and sea battles between the two opposing military blocs, possibly culminating in a mutually annihilating nuclear exchange, has receded into history. Yet instead of ushering in a new era of peace, the opening years of the twenty-first century have been characterized by a variety of low-intensity conflicts, including global terrorist campaigns, ethnic and civil wars, insurgencies and counterinsurgencies, and paramilitary operations such as peacekeeping, drug wars, and counterterrorism.

This sea change in the nature of armed conflict could create new incentives and opportunities for the hostile exploitation of emerging biological and chemical technologies.[26] Indeed, one consequence of the focus on counterterrorism and counterinsurgency campaigns, in which combatants and noncombatants are often intermingled, has been a growing interest on the part of several states in developing so-called nonlethal or less-than-lethal chemical agents. Whereas riot-control agents (RCAs) such as tear gas have temporary irritating effects on the eyes and skin that dissipate rapidly after the exposure ends, incapacitating drugs (such as the synthetic-opiate anesthetic fentanyl) have persistent effects on the central nervous system and induce a state of disorientation, unconsciousness, euphoria, or depression that lasts for several hours.

Some countries have also explored the possibility of developing novel incapacitating agents based on natural body substances called bioregulators, many of which are short protein chains called peptides. From 1974 to 1989, the Soviet Union pursued a top-secret program code-named Bonfire, which involved the development of lethal and incapacitating agents based on peptide bioregulators.[27] The U.S. Department of Defense has also funded research on so-called calmatives, including some peptides that affect the brain.[28] In addition to Russia and the United States, several other countries have reportedly done development work on chemical incapacitants.[29]

Since 9/11 and the anthrax letter attacks in the United States, there has also been growing concern that terrorist organizations such as al-Qaeda will resort to biological or chemical terrorism. Fortunately, the number of groups that possess both the motivation and the technical capability to do so remains quite small. Most terrorist organizations are conservative in their choice of weapons and tactics, relying on standard guns and explosives and innovating only when forced to do so by the introduction of new countermeasures, such as improved aviation security. Although al-Qaeda is an exception to this rule, having openly declared its ambition to acquire weapons of mass destruction, the organization's chemical and biological warfare capabilities remain rudimentary.[30]

To date, the only terrorist group to have moved fairly high up the technological learning curve for developing unconventional weapons is Aum Shinrikyo, a doomsday cult in Japan that sought to stage biological and chemical attacks in Tokyo to trigger massive social disruption and seize control of the national government. In the early 1990s, Aum recruited biologists and chemists from Japanese universities into its unconventional weapons programs. Using vast sums of money acquired through a range of legitimate and criminal enterprises, cult leader Shoko Asahara put his scientists to work developing and manufacturing anthrax bacteria, botulinum toxin, and sarin nerve agent.

Because Aum inadvertently acquired a harmless vaccine strain of the anthrax bacterium and failed to produce botulinum toxin in significant quantities, its biological attacks caused no injuries or deaths. Numerous technical problems also prevented the cult from producing a large stockpile of sarin and carrying out mass-casualty chemical attacks. Ultimately Aum released limited quantities of sarin on two occasions, in Matsumoto in June 1994 and on the Tokyo subway in March 1995; these two incidents claimed a total of nineteen lives and injured hundreds more.[31] The sarin attack on the Tokyo subway was prepared in haste over a weekend because of an impending police raid on the cult's headquarters. If Aum's chemists had been given more time to prepare a purer solution of sarin and weaponize it more effectively, the incident might have caused a significantly higher number of casualties.

Harm versus Misuse

When discussing dual-use technologies, it is important to distinguish between harm and misuse. Harm encompasses a broad range of negative consequences, including fatal and nonfatal casualties, permanent disability, psychological trauma, social disruption, economic damage, and the incitement of fear. Whereas the potential to cause harm is an inherent characteristic of a dual-use technology or material, misuse is a function of the intent of the individual actor and the prevailing social norms. From a legal standpoint, misuse is an action that violates an existing national or international statute. International humanitarian law, for example, prohibits the use of certain types of weapons because they are indiscriminate and hence likely to kill civilians or because they have effects on the human body that cause unnecessary suffering.

The legal definition of misuse has changed over the course of history in response to the evolution of international law, which follows and embodies changing norms of state behavior. As a result, the relationship between harm and misuse today differs from how it was conceived in the past. During World War I, for example, the Germans believed that the use of biological weapons to kill or injure humans was immoral, but that biological attacks against horses and other draft animals were entirely legitimate. Today, however, any use of biological weapons, even one restricted to animals or crops, would be considered a flagrant violation of international law and a misuse of biology. In another example, although the 1925 Geneva Protocol banned the *use* in war of infectious-disease agents, it was still legal for states to develop, produce, and stockpile biological weapons until the entry into force of the Biological Weapons Convention (BWC) in 1975. Similarly, before the Chemical Weapons Convention (CWC) entered into force in 1997, states were allowed under international law to acquire and stockpile chemical arms.

Over time, certain categories of weapons, such as chemical and biological arms and more recently antipersonnel land mines, have come to be seen as morally and legally unacceptable by a large majority of the international community. Although most arms control accords are binding only on the states that sign and ratify them voluntarily, the 1925 Geneva Protocol has acquired the status of customary international law, meaning that it applies to all states. Meanwhile another category of armament, incendiary weapons, exists in a legal gray zone in which use in the vicinity of civilians has been banned by an international treaty, the Convention on

Certain Conventional Weapons (CCW), but use against military targets is still permitted.

Misuse Scenarios

Four scenarios can be envisioned for the misuse of emerging biological and chemical technologies. First, dual-use technologies could *facilitate or accelerate the production of standard biological or chemical warfare agents*. An example of this scenario is the application of whole-genome synthesis to construct dangerous viruses from scratch, circumventing access controls on pathogens of bioterrorism concern.

Second, dual-use technologies could help to *identify or develop novel biological or chemical warfare agents* that do not already exist. For example, it may become possible to synthesize artificial pathogens or toxins that are resistant to standard medical countermeasures.

Third, technologies that increase the understanding of human biology at a systems level may *result in new knowledge that could be misused* to develop a new or enhanced means for harming the human body. For example, advances in neuroscience and psychopharmacology could lead to the development of drugs that affect human memory, cognition, or emotion in specific ways.

Finally, dual-use biological and chemical technologies may *lead to harmful applications that undermine international legal norms*. For example, the CWC bans any use of toxic chemicals in warfare, yet Article II, paragraph 9 (d) permits the development, production, and use of chemical agents for "law enforcement including domestic riot control."[32] Because the treaty text does not specify the types and quantities of toxic chemicals that may be used for this purpose, the law-enforcement exemption creates a potential loophole. If interpreted broadly to cover chemicals that are substantially more potent than tear gas, the exemption could lead to the development and deployment of new types of incapacitating agents. Such a capability would seriously weaken the CWC because the chemicals intended for law-enforcement applications would be weaponized and thus hard to distinguish from a military capability.[33] As the British chemical weapons analyst Julian Perry Robinson has warned, "A regime that allows weaponization of one form of toxicity but not another cannot, under the circumstances, be stable."[34]

The hostile applications of chemical and biological agents might also move beyond warfare and terrorism to include systematic violations of human rights and humanitarian law. Examples include the use of incapacitating or calmative drugs to assist with coercive interrogation or to repress political dissent. Rulers of autocratic or totalitarian states might order the use of such agents against their own civilian populations to control unrest or put down a popular uprising. It might also be possible to engineer biological agents that can harm certain ethnic or racial groups selectively, based on their genetic makeup. Although genetic warfare is not a practical option today, information derived from ongoing research into the human genome might eventually be exploited for this purpose.

Structure of the Book

Proposals to ban dual-use biological or chemical technologies outright are not realistic because doing so would mean forgoing their benefits for human health, agriculture, energy, and economic development. A better approach is to devise policies that help to prevent the misuse of such technologies for harmful purposes while permitting their legitimate applications. Absent a systematic approach to risk assessment and governance, however, dual-use technologies have continued to proliferate in an uncontrolled manner, increasing the danger that they could fall into the hands of states, groups, and individuals with malign intent. Devising a practical decision-making framework for helping to manage the dual-use dilemma is the primary aim of this book.

In addition to this introduction, the book is organized into four parts. Part I, "Assessing and Managing Dual-Use Risks," contains three chapters. "Review of the Literature on Dual Use" summarizes the academic literature on technological risk assessment and governance measures as the starting point for developing a new risk-management methodology. "Current Dual-Use Governance Measures" surveys the existing approaches to technology governance in the United States, the European Union, Israel, and Singapore. "The Decision Framework" presents a system for designing tailored governance strategies to manage the risks of individual technologies. This framework focuses on two key dimensions of a dual-use technology—its risk of misuse and susceptibility to governance—and ranks each according to several parameters. The rankings of risk and governability then serve as the basis for developing packages of governance measures that can be implemented at the individual, institutional, national, and international levels to minimize the risk of misuse.

Part II, "Contemporary Case Studies," applies the decision framework to a set of fourteen dual-use technologies in the biological and chemical fields that either have emerged in recent years or are still emerging. Because many of these technologies will continue to evolve, the case studies should be considered "snapshots" at a given point in time, and the proposed governance measures will need to be reviewed periodically to adapt them to changing circumstances.

Part III, "Historical Case Studies," examines two past cases in which civilian technologies were transferred to military applications: the conversion of a failed pesticide called Amiton into the V-series nerve agents in Britain during the 1950s, and the development of LSD as an incapacitating chemical weapon, interrogation aid, and tool for covert operations by the U.S. Army and the CIA during the Cold War. These two historical cases yield valuable insights into the social processes that lead to the misuse of civilian technologies for harmful purposes.

Part IV, "Findings and Conclusions," performs a comparative analysis of the fourteen contemporary case studies, draws lessons from the two historical cases that are applicable to today's policy challenges, reviews the usefulness of the decision framework, and discusses how it might be institutionalized and implemented at both the national and international levels.

Acknowledgments

Generous funding for this research project was provided by the British Foreign and Commonwealth Office's Strategic Counter-Proliferation Programme Fund and the U.S. Defense Threat Reduction Agency's Advanced Systems and Concepts Office (ASCO). Thanks are also due to Professor William C. Potter, director of the James Martin Center for Nonproliferation Studies (CNS) at the Monterey Institute of International Studies, and Leonard S. Spector, head of the CNS office in Washington, D.C., for their encouragement and support, and to CNS research associates Kirk Bansak and Michael Tu for outstanding organizational and editorial assistance. Kirk Bansak in particular played an invaluable role in the project, serving as a sounding board for ideas, providing detailed comments on the chapters, and preparing figures and tables.

Gregory Koblentz, Filippa Lentzos, and three anonymous reviewers provided valuable comments on some or all of the book manuscript. Finally, Edith Bursac, manager of conference and event services at CNS, did a superb job of planning the two authors' workshops, which were held in Washington in September 2009 and at Oxford University in March 2010.

Notes

1. James B. Petro, Theodore R. Plasse, and Jack A. McNulty, "Biotechnology: Impact on Biological Warfare and Biodefense," *Biosecurity and Bioterrorism* 1, no. 3 (September 2003): 161–168; Eileen R. Choffnes, Stanley M. Lemon, and David A. Relman, "A Brave New World in the Life Sciences," *Bulletin of the Atomic Scientists* 62, no. 5 (September–October 2006): 28–29; Ronald M. Atlas and Malcolm Dando, "The Dual-Use Dilemma for the Life Sciences: Perspectives, Conundrums, and Global Solutions," *Biosecurity and Bioterrorism* 4, no. 3 (2006): 276–286.

2. Parliamentary Office of Science and Technology, "The Dual-Use Dilemma," *Postnote*, no. 340 (July 2009): 1.

3. British Medical Association, *Biotechnology, Weapons, and Humanity* (Amsterdam: Harwood Academic Publishers, 1999), 1.

4. National Research Council, *Biotechnology Research in an Age of Terrorism* (Washington, DC: National Academies Press, 2004).

5. The NSABB defines "dual-use research of concern" (DURC) as "research that, based on current understanding, can be reasonably anticipated to provide knowledge, products, or technologies that could be directly misapplied by others to pose a threat to public health and safety, agricultural crops and other plants, animals, the environment, or materiel." NSABB, *Proposed Framework for the Oversight of Dual-Use Life Sciences Research: Strategies for Minimizing the Potential Misuse of Research Information* (June 2007), http://oba.od.nih.gov/biosecurity/biosecurity_documents.html.

6. World Health Organization, *Public Health Response to Biological and Chemical Weapons: WHO Guidance*, 2nd ed. (Geneva, Switzerland: World Health Organization, 2004), Executive Summary.

7. International Committee for the Red Cross, "Appeal on Biotechnology, Weapons, and Humanity," Geneva, Switzerland, September 25, 2002.

8. John A. Alic, Lewis M. Branscomb, Harvey Brooks, Ashton B. Carter, and Gerald Epstein, *Beyond Spinoff: Military and Commercial Technologies in a Changing World* (Boston: Harvard Business School Press, 1992).

9. William H. McNeill, *The Pursuit of Power: Technology, Armed Force, and Society since A.D. 1000* (Chicago: University of Chicago Press, 1982).

10. Mark Wheelis, "Biological Sabotage in World War I," in *Biological and Toxin Weapons: Research, Development, and Use from the Middle Ages to 1945*, ed. Erhard Geissler and John Ellis van Courtland Moon, SIPRI Chemical and Biological Warfare Studies no. 18 (Oxford, UK: Oxford University Press for the Stockholm International Peace Research Institute, 1999), 35–62.

11. Malcolm Dando, "The Impact of the Development of Modern Biology and Medicine on the Evolution of Modern Biological Warfare Programmes in the Twentieth Century," *Defense Analysis* 15, no. 1 (1999): 51–65.

12. Sheldon H. Harris, *Factories of Death: Japanese Biological Warfare, 1932–45, and the American Cover-Up*, 2nd ed. (New York: Routledge, 2002).

13. Christopher Chyba and Alex Greninger, "Biotechnology and Bioterrorism: An Unprecedented World," *Survival* 46 (2004), 143–162.

14. Jeronimo Cello, Aniko V. Paul, and Eckard Wimmer, "Chemical Synthesis of Poliovirus cDNA: Generation of Infectious Virus in the Absence of Natural Template," *Science* 297, no. 5583 (August 9, 2002): 1016–1018.

15. Rick Weiss, "Mail-Order Molecules Brew a Terrorism Debate: Virus Created in Lab Raises Questions of Scrutiny for DNA Suppliers," *Washington Post*, July 17, 2002, A1.

16. Daniel G. Gibson, John I. Glass, Carole Lartigue, et al., "Creation of a Bacterial Cell Controlled by a Chemically Synthesized Genome," *Science* 329 (July 2, 2010): 52–56.

17. Jonathan B. Tucker, "The Convergence of Biology and Chemistry: Implications for Arms Control Verification," *Bulletin of the Atomic Scientists* 66, no. 6 (November–December 2010): 56–66; Ralf Trapp, "Advances in Science and Technology and the Chemical Weapons Convention," *Arms Control Today* 38, no. 2 (March 2008): 18–22.

18. John Hart, "The Soviet Biological Weapons Program," in *Deadly Cultures: Biological Weapons since 1945*, ed. Mark Wheelis, Lajos Rózsa, and Malcolm Dando (Cambridge, MA: Harvard University Press, 2006), 132–156.

19. Caitríonia McLeish and Paul Nightingale, "Biosecurity, Bioterrorism, and the Governance of Science: The Increasing Convergence of Science and Security Policy," *Research Policy* 36, no. 10 (December 2007): 1635–1654.

20. R. J. Jackson, A. J. Ramsay, C. D. Christensen, S. Beaton, D. F. Hall, and I. A. Ramshaw, "Expression of Mouse Interleukin-4 by a Recombinant Ectromelia Virus Suppresses Cyctolytic Lymphocyte Responses and Overcomes Genetic Resistance to Mousepox," *Journal of Virology* 75, no. 3 (2001): 1205–1210.

21. Michael J. Selgelid and Lorna Weir, "The Mousepox Experience," *EMBO Reports* 11, no. 1 (2010): 18–24.

22. National Science Advisory Board for Biosecurity, "Proposed Framework for the Oversight of Dual Use Life Sciences Research: Strategies for Minimizing the Potential Misuse of Research Information," June 2007, http://oba.od.nih.gov/biosecurity/pdf/Framework%20 for%20transmittal%200807_Sept07.pdf.

23. Rebecca L. Frerichs, Reynolds M. Salerno, Kathleen M. Vogel, et al., *Historical Precedence and Technical Requirements of Biological Weapons Use: A Threat Assessment*, SAND2004-1854 (Albuquerque, NM: Sandia National Laboratories, 2004). See also Raymond A. Zilinskas, "Technical Barriers to Successful Biological Attacks with Synthetic Organisms," in *From Understanding to Action: Community-Based Options for Improving Safety and Security in Synthetic Biology*, ed. Stephen M. Mauer, Keith V. Lucas, and Starr Terrell (Berkeley: University of California Press, 2006).

24. Jonathan B. Tucker, introduction to *Toxic Terror: Assessing Terrorist Use of Chemical and Biological Weapons*, ed. Jonathan B. Tucker (Cambridge, MA: MIT Press, 2000), 7.

25. Ibid., 6–9.

26. Mark Wheelis, "Will the New Biology Lead to New Weapons?" *Arms Control Today* 34 (July–August 2004): 6–13.

27. Ken Alibek with Stephen Handelman, *Biohazard* (New York: Random House, 1999), 154–155, 163–164.

28. Joan M. Lakoski, W. Bosseau Murray, and John M. Kenny, *The Advantages and Limitations of Calmatives for Use as a Non-lethal Technique* (College Park: Pennsylvania State University, College of Medicine and Applied Research Laboratory, October 3, 2000), 39–45.

29. According to a study by Michael Crowley of the University of Bradford (U.K.), other countries that have conducted research and development on chemical incapacitating agents include China, the Czech Republic, France, and the United Kingdom. See Michael J. A. Crowley, *Dangerous Ambiguities: Regulation of Riot Control Agents and Incapacitants under the Chemical Weapons Convention* (University of Bradford, 2009).

30. Scott Stewart, "The Jihadist CBRN Threat," *Stratfor Global Intelligence*, February 10, 2010, http://www.stratfor.com/weekly/20100210_jihadist_cbrn_threat?utm _source=SWeekly&utm_medium=email&utm_campaign=100210&utm_content=readmore &elq=b6bf5229ea8d488ba77fb8f36fc609d9.

31. David E. Kaplan, "Aum Shinrikyo (1995)," in *Toxic Terror: Assessing Terrorist Use of Chemical and Biological Weapons*, ed. Jonathan B. Tucker (Cambridge, MA: MIT Press, 2000), 207–226.

32. Alan M. Pearson, Marie Isabelle Chevrier, and Mark Wheelis, eds., *Incapacitating Biochemical Weapons: Promise or Peril?* (Lanham, MD: Lexington Books, 2007).

33. Mahdi Balali-Mood, Pieter S. Steyn, Leiv K. Sydnes, and Ralf Trapp, "Impact of Scientific Developments on the Chemical Weapons Convention (IUPAC Technical Report)," *Pure and Applied Chemistry* 80, no. 1 (2008): 185.

34. Julian P. Perry Robinson, "Ensuring that OPCW Implementation of the CWC Definition of Chemical Weapons Remains Fit for Purpose," discussion paper for the 54th Pugwash CBW Workshop, *The Second CWC Review Conference and After*, Noordwijk, the Netherlands, April 5–6, 2008.

I

Assessing and Managing Dual-Use Risks

2

Review of the Literature on Dual Use

Jonathan B. Tucker

A prerequisite for effective governance is the ability to assess the safety and security risks of a technology. Efforts to assess dual-use risks have traditionally revolved around the materials, methods, and products involved in misuse, and governance strategies have also taken an "artifact-centric" approach by seeking to control the availability of dual-use products and services. This traditional paradigm has serious limitations, however, for two reasons. First, whereas the traditional definition of technology emphasizes hardware, equipment, and tools, technology also encompasses people, processes, and know-how.[1] Second, dual-use biological and chemical technologies are increasingly diffuse, globalized, and multidisciplinary and are often based on intangible information rather than on specialized materials and equipment. Because of these factors, traditional technology-denial strategies such as export controls are often ineffective while also tending to stifle legitimate research and commerce. Accordingly, new approaches to dual-use technology governance are needed.

Assessing Uncertain Risks

Assessing the safety and security risks of an emerging technology at an early stage in its development is particularly challenging because of the lack of available hard information. Uncertainty is a perennial problem because emerging technologies often evolve in unexpected ways. A case in point is synthetic biology, which is the product of two different convergences: between biology and chemistry, and between engineering principles and living systems.[2]

Synthetic biology with standard parts is an emerging discipline that envisions the design and construction of genetic circuits and modules using a tool kit of DNA sequences called BioBricks. These pieces of DNA have known protein-coding or regulatory functions. Theoretically, BioBricks should behave in a predictable manner and have a standard interface, so that they can be joined together to produce functional genetic circuits in much the same way that transistors, capacitors, and resistors are assembled into electronic devices. A major goal of parts-based synthetic biology is to design and build genetic modules that endow microbes with useful functions not present in nature, such as the ability to convert sunlight into biofuels or to absorb greenhouse gases. To date, a few simple genetic circuits have been created, such as an oscillator that codes for the cyclic production of a green fluorescent protein, making the engineered bacteria blink slowly on and off like tiny Christmas tree lights. Yet even these simple constructs tend to be "noisy" and to manifest unexpected behaviors.

Given these uncertainties, it is difficult to ascertain at this early stage of development what questions to ask about the risks of synthetic biology, let alone what broader social values are at stake.[3] Denise Caruso argues that traditional approaches to risk assessment do not apply to synthetic biology, which has no historical precedent other than by analogy. To assess such unprecedented risks, she advocates an approach that combines data analysis with a deliberative process, drawing on a broad representation of relevant scientific expertise. The methodology involves the use of scenario narratives to develop risk models that are computable over time as hard scientific data become available. Caruso suggests that this approach can help government officials decide when existing regulations are suitable for emerging processes and products, and when they are inappropriate or insufficient.[4]

M. Granger Morgan calls for engaging the public in the process of risk assessment. "Laypeople have different, broader definitions of risk, which in important respects can be more rational than the narrow ones used by experts," he writes. "Furthermore, risk management is, fundamentally, a question of values. In a democratic society, there is no acceptable way to make these choices without involving the citizens who will be affected by them."[5] Accordingly, Morgan urges risk managers to allow ordinary citizens to become involved in risk assessment in a significant and constructive way by giving them access to information and enabling them to consult with experts. Along similar lines, Richard Sclove has argued that the United States should adopt the system of "participatory technology assessment" practiced in Denmark, where panels of ordinary citizens and subject-matter experts discuss the social implications of emerging technologies.[6]

Risk Communication

A separate literature from risk assessment deals with risk communication, which the National Research Council has defined as "an interactive process of exchange of information and opinion among individuals, groups, and institutions."[7] This field grew out of methods for estimating the risk to humans of exposure to toxic substances, as well as research on how individuals perceive risk.[8] In the mid-1980s, risk communication became a key component of risk management and community decision making in the fields of environmental and occupational health, particularly with respect to hazardous wastes, nuclear power plants, and toxic chemicals. Conflict resolution plays an important role in risk communication because community members, activists, government officials, scientists, and corporate executives often disagree about the nature, magnitude, and severity of the risk in question.[9]

Psychological research has identified a set of mental strategies or heuristics that people use to make sense of an uncertain world. These rules create permanent biases in risk perception that are resistant to change. Chauncey Starr observed in 1969, for example, that the public accepts risks from voluntary activities such as skiing that are roughly 1,000 times greater than what the public tolerates from involuntary hazards such as toxic pollution.[10] Paul Slovic also contends that the public's perception of risk is not based on statistics, such as expected numbers of deaths or injuries, but serves as a surrogate for other social and personal concerns. Thus unfamiliar or exotic risks elicit more calls for government regulation than do more mundane risks that actually cause a higher rate of death or injury. For example, the public tends to view the risks of nuclear power plants as unacceptably high because they are "unknown, dread, uncontrollable, inequitable, catastrophic, and likely to affect future generations."[11]

Building on Slovic's work, Jessica Stern argues that biological weapons fall into the category of "dreaded risks" because they possess characteristics such as involuntary exposure, invisibility, and indiscriminate effects, all of which elicit a disproportionate level of fear, disgust, and horror. As a result of this perceptual bias, politicians and the public tend to overestimate the danger of bioterrorism when compared to other, more likely risks.[12] In sum, risk communication requires active outreach and engagement with the scientific community and the broader public, including an understanding of the psychology of risk perception.

Assessing the Risk of Deliberate Misuse

Compared to health and safety risks, assessing the risk that an emerging technology will be misused for hostile purposes poses a far greater challenge because it depends as much on the intent and capabilities of the user as on the characteristics of the technology itself. Another problem is that because of the limited number of cases of biological and chemical terrorism in the historical record, little empirical evidence is available to provide a solid basis for dual-use risk assessment.[13] As Gregory Koblentz has observed, the small number of incidents suggests either that few terrorist groups are motivated to acquire and use biological or chemical agents, or that doing so is more difficult than is generally assumed, or both. Also unclear is the role of intangible factors, such as the internal dynamics of individual groups, for the ability of terrorists to build and deliver a biological or chemical weapon capable of causing mass casualties.[14]

Ideally, assessing the risk of deliberate misuse of an emerging technology should take into account the potential actors and their motivations, along with the likely targets and scale of an attack. In practice, however, such prospective assessments are extremely difficult to make in a rigorous manner. A further difficulty is that in contrast to an unthinking act of nature, an intelligent actor can adapt and modify his behavior to circumvent or neutralize defensive countermeasures.[15] Assessments of the risk of deliberate misuse must also take into account the vulnerability of potential targets, including—in the case of biological or chemical terrorism—the availability of effective medical countermeasures such as antidotes, vaccines, and therapeutic drugs. Some analysts have tried to operationalize the risk of unconventional terrorism by describing it as the product of the likelihood of an attack, the probability of its successful execution, and the losses that would result in terms of fatalities, injuries, and direct and indirect economic impacts.[16]

Despite the need for improved methods of risk assessment, it will always be difficult to predict whether or not a specific technology will be misused for harmful purposes. Indeed, the U.S. National Research Council panel that prepared the influential report *Biotechnology Research in an Age of Terrorism* examined several cases of "contentious" research in the life sciences, including the Australian mousepox experiment and the laboratory synthesis of poliovirus, but it did not identify a single case in which the research was so prone to misuse that it should not have been published.[17] Brian Rappert argues that instead of examining individual cases, a better approach is to examine the cumulative development of an emerging technology and assess the extent to which further incremental improvements would increase the potential for, or the consequences of, deliberate misuse.[18]

Explicit versus Tacit Knowledge

In assessing the risk that terrorist organizations could misuse technologies such as whole-genome synthesis to construct "select-agent" viruses from scratch, a relevant question is how they would acquire the technical skills needed to do so effectively. Skeptics point out that mastering whole-genome synthesis demands multiple sets of expertise and resources, including considerable know-how that cannot be written down but must be gained through years of hands-on experience in the laboratory. But other scholars disagree, arguing that genome synthesis is subject to a process of "deskilling," or a gradual decline in the amount of tacit knowledge required for its use, that will eventually make the technology accessible to nonexperts with malicious intent. This debate is of more than academic interest because it has important implications for assessing the security risks associated with the rapid pace of progress in the life sciences.

Scientists and sociologists of science distinguish between two types of technical knowledge: explicit and tacit. Explicit knowledge is information that can be codified, written down in the form of a recipe or laboratory protocol, and transferred from one individual to another by impersonal means, such as publication in a scientific journal. Tacit knowledge, in contrast, involves skills, know-how, and sensory cues that are vital to the successful use of a technology yet cannot be reduced to writing and must be acquired through hands-on practice and experience.[19] As a result, scientific procedures and techniques requiring tacit knowledge do not diffuse as rapidly as those that are readily codified.

Tacit knowledge can further be subdivided into two types. *Personal* tacit knowledge is held by individuals and can be conveyed from one person to another through a master-apprentice relationship (learning by example) or acquired by a lengthy process of trial-and-error problem solving (learning by doing). The amount of time required to gain such knowledge depends on the complexity of a task and the level of skill involved in its execution. Moreover, personal tacit knowledge tends to decay if it is not used on a regular basis and transmitted to the next generation of specialists.[20] *Communal* tacit knowledge is more complex because it is not held by a single individual but resides in an interdisciplinary team of specialists, each of whom has skills and experience relevant to a particular facet of a scientific protocol or technology. This social dimension makes communal tacit knowledge particularly difficult to transfer from one laboratory to another because doing so requires adapting a complex set of technical practices to a new context and replicating them in communal form.[21]

Field research by sociologists of science has shown that advanced biotechnologies, such as whole-genome synthesis, demand a high level of personal and communal tacit knowledge. In a study of the chemical synthesis of poliovirus in 2002, for example, Kathleen Vogel found that the researchers did not rely exclusively on written protocols but extensively used intuitive skills acquired through years of experience, particularly when preparing the cell-free extracts needed to translate the synthetic genome into infectious virus particles. If the cell-free extract was not prepared in the correct manner by relying on subtle tricks and sensory cues, it proved impossible to reproduce the published experiment.[22] Based on her empirical research, Vogel concludes that biotechnology is a "sociotechnical assemblage," meaning an activity whose technical and social dimensions are inextricably linked.[23]

Other case studies of technological innovation in the life sciences have confirmed the importance of the sociotechnical dimension, which includes tacit knowledge, teamwork, laboratory infrastructure, and organizational factors. Such factors help to explain the problems that scientists often encounter when trying to replicate a research protocol developed in another laboratory or when translating a scientific discovery from the research bench to commercial application.[24] Despite the ongoing revolution in the life sciences, these traditional bottlenecks persist.

Indeed, the synthesis of ever larger and more complex genomes has given rise to additional challenges. In 2010, for example, the synthesis of an artificial bacterial genome consisting of more than 1 million DNA units by a research team at the J. Craig Venter Institute required a unique working configuration of expertise and resources.[25] In an interview, Dr. Venter noted that at each stage in the process, a team of highly skilled and experienced molecular biologists had to develop new methodologies, which could be made to work only through a lengthy process of trial and error. One problem was that the long molecules of synthetic bacterial DNA were fragile and had to be stored in supercoiled form inside gel blocks and handled carefully to prevent them from breaking up. "As with all things in science," Venter explained, "it's the little tiny breakthroughs on a daily basis that make for the big breakthrough."[26]

In view of this evidence, Vogel and Sonia Ben Ouagrham-Gormley have argued that the technical and socio-organizational hurdles that must be overcome to apply whole-genome synthesis effectively pose a major obstacle to the ability of terrorist organizations to exploit the technology for harmful purposes.[27]

The Deskilling Agenda

Other scholars, such as Chris Chyba and Gerald Epstein, are less convinced. They note that emerging technologies often entail a process of deskilling that, over time, reduces the amount of tacit knowledge required for their use. For example, genetic-engineering techniques that a few decades ago were found only in cutting-edge laboratories have since become available in the form of commercial kits and services, making them accessible to individuals with limited scientific training and experience, such as college students. Accordingly, as dual-use technologies such as genome synthesis become increasingly automated, commercialized, and "black-boxed," they will become more accessible to terrorists and other malicious actors who possess only basic scientific skills.[28]

The debate over deskilling has focused both on the technology of whole-genome synthesis (synthetic genomics) and on the related but much broader field of synthetic biology. Despite the overlap between the two disciplines, they differ in important ways. Whereas synthetic genomics is an enabling technology, synthetic biology is an umbrella term that covers several distinct research programs. Two prominent and outspoken scientists, Tom Knight and Drew Endy, advocate a particular synthetic-biology paradigm that aims to facilitate the engineering of biology through the development of a Registry of Standard Biological Parts.

In principle, the use of standard genetic parts and modular design techniques should significantly reduce the need for tacit knowledge in the construction of synthetic organisms. As Gautam Mukunda, Kenneth Oye, and Scott Mohr have argued, "Deskilling and modularity . . . have the potential to . . . decrease the skill gradient separating elite practitioners from non-experts."[29] Nevertheless not everyone in the synthetic biology community has bought into the standardized-parts approach, and some believe that it is destined to fail—or, at the least, not live up to its ambitious claim of providing a simple and predictable way to design and build artificial genomes. One problem is that as the size of synthetic constructs increases, nonlinear interactions among the constituent genetic elements may become increasingly difficult to predict or control, resulting in unexpected behaviors and other emergent properties. At least in theory, such constructs could pose safety hazards that are impossible to predict.[30] Thus, although certain aspects of parts-based synthetic biology may well become more accessible to nonexperts, the deskilling agenda is still far from becoming an operational reality.

The Democratization of Biology

Another key element in the agenda of parts-based synthetic biology, as conceived by Knight and Endy, is "open-access biology," or making the Registry of Standard Biological Parts and related know-how freely available to interested researchers without patents or other restrictions. An important vehicle for the open-access biology movement is the International Genetically Engineered Machine (iGEM) competition, which is held annually at MIT by the BioBricks Foundation.[31] The goals of iGEM are "to enable the systematic engineering of biology, to promote the open and transparent development of tools for engineering biology, and to help construct a society that can productively apply biological technology."[32] In addition, in May 2008 a group of amateur biologists in Cambridge, Massachusetts, launched an open-access initiative called DIYbio (do-it-yourself biology), with the goal of making biotechnology more accessible to nonspecialists, including the potential use of synthetic-biology techniques to carry out personal projects.[33] DIYbio has since expanded to several cities and today has more than 2,000 members.[34]

Some observers contend that the deskilling and open-access agendas promoted by iGEM and DIYbio will unleash a wave of innovation as a growing number of people from various walks of life acquire the ability to engineer biology for useful purposes. According to Gaymon Bennett and his colleagues, "The good news is that open access biology, to the extent that it works, may help actualize the long-promised biotechnical future: growth of green industry, production of cheaper drugs, development of new biofuels and the like."[35] Extrapolating from these trends a few decades into the future, Freeman Dyson published a controversial article in 2007 envisioning a world in which synthetic biology has been deskilled to the point that it is fully accessible to amateur scientists, hobbyists, and even children, who use it to play "biotech games, designed like computer games for children down to kindergarten age, but played with real eggs and seeds rather than with images on a screen. . . . The winner could be the kid whose seed grows the prickliest cactus, or the kid whose egg hatches the cutest dinosaur."[36]

Such predictions may or may not come true, depending on the degree to which synthetic biology is deskilled in the future. Skeptics note that iGEM teams have often had trouble creating or using biological parts that work reliably and predictably in different contexts. Moreover, certain techniques depend on an extensive technical infrastructure that is not available to amateurs. According to Andrew Ellington, "The difference between DIY Bio and Michael Dell putting computers together in his garage is the difference between the availability of the raw materials. There is no 'Radio Shack' for DNA parts, and even if there were, the infrastructure required to manipulate those parts is non-trivial for all but the richest amateur scientist."[37]

Reviewing the historical record, one finds that the scientific claims about deskilling made in the past have often failed to pan out. For example, Helen Anne Curry has studied the development of techniques for the genetic modification of plants from the 1930s through the 1950s. During that period, agricultural interests promised that new genetic technologies involving the use of radium, x-rays, and chemicals to generate mutations would enable the rapid breeding of new and useful plant varieties, and that these methods would soon become accessible to amateur gardeners. In fact, although plant-breeding techniques did result in new varieties of roses and orchids, the broader claims about deskilling never materialized.[38]

In addition to the potential benefits of deskilling and open access, a number of commentators have warned that the democratization of synthetic biology could increase the associated safety and security risks. One concern is that substantially expanding the pool of individuals who have access to synthetic-biology techniques would inevitably increase the likelihood of accidents, creating unprecedented hazards for the environment and public health.[39] Even Dyson's generally upbeat article acknowledges that the recreational use of synthetic biology "will be messy and possibly dangerous" and that "rules and regulations will be needed to make sure that our kids do not endanger themselves and others."[40]

Beyond the possible safety risks, Mukunda, Oye, and Mohr worry that the deskilling of synthetic biology would make this powerful technology accessible to individuals and groups who would use it deliberately to cause harm. "Synthetic biology," they write, "includes, as a principal part of its agenda, a sustained, well-funded assault on the necessity of tacit knowledge in bioengineering and thus on one of the most important current barriers to the production of biological weapons."[41] Indeed, the experience to date with "black-hatted" computer hackers who create software viruses and worms for criminal purposes, for espionage, or simply to demonstrate their technical prowess provides grounds for concern about the possible emergence of "biohackers" who engage in reckless or malicious experiments with synthetic organisms in basement laboratories.[42]

Resolving the Debate

To resolve the debate about the extent to which terrorists could exploit synthetic biology to cause harm, it is important to determine if deskilling affects aspects of the technology that require tacit knowledge, thereby removing a major obstacle to misuse. Preliminary evidence suggests that some biotechnologies are more amenable to deskilling than others. In particular, deskilling has already occurred with respect to several genetic-engineering techniques that have been around for more than twenty years, including transfection, gene cloning, and the polymerase chain reaction, which makes it possible to copy any particular gene sequence several million-fold. Although these methods require access to natural genetic material, the associated skill sets have diffused widely across the international virology community. In fact, some of these techniques have been deskilled to the point that they are now performed routinely by undergraduates and even advanced high school students and could therefore be appropriated fairly easily by terrorist groups.

Other techniques, in contrast, have proved resistant to deskilling for a variety of reasons, including the complexity of biological organisms, tacit knowledge, and other sociotechnical factors. For example, although scientists commonly use genetic-engineering kits containing all the materials and reagents needed to perform a particular laboratory procedure, studies have shown that these kits do not necessarily remove the need for tacit knowledge when applied in the context of a particular experiment.[43]

Moreover, analysts who contend that the deskilling of biotechnology will facilitate misuse often overstate the risk because they focus on one or two steps of what is actually a complex, multistep process. For example, practitioners of de novo viral synthesis note that the most challenging steps do not involve the synthesis of DNA fragments, which can be ordered from commercial suppliers, but rather encompass the assembly of the fragments into a functional genome. This task requires extensive familiarity with techniques such as DNA ligation and the expression of viral proteins. According to a report by the U.S. National Science Advisory Board for Biosecurity, "The technology for synthesizing DNA is readily accessible, straightforward and a fundamental tool used in current biological research. In contrast, the science of constructing and expressing viruses in the laboratory is more complex and somewhat of an art. It is the laboratory procedures downstream from the actual synthesis of DNA that are the limiting steps in recovering viruses from genetic material."[44]

Along similar lines, the virologist Jens H. Kuhn has provided a highly differentiated assessment of the technical challenges involved in de novo viral synthesis:

Many priority pathogens have simple and short genomes, which themselves are infectious. The methods to create these genomes are so standard that they are not even described in the method sections of publications anymore. This means that the tacit knowledge to apply these methods is widely spread and practically speaking "for hire." Other agents require more sophistication. The genomes of negative-stranded RNA viruses, for instance the Zaire ebola virus or 1918 H1N1 influenza A virus, are not infectious by themselves, but require the presence of viral helper proteins, which also have to be synthesized and present inside of a cell in the right numbers. It takes longer to create such reverse genetic systems, and a limited number of people have the skill to succeed, but the methods themselves are not any different from those routinely used by thousands of scientists and taught to students within months. Also, scientists have already created and published many RNA viruses of concern, using individual sequences obtained separately—not synthesized. Other agents, such as the variola [smallpox] virus and bacteria, are not within reach of individuals, because the methods to synthesize them are not as widely distributed or not yet developed.[45]

It is also important to note that creating an effective biological weapon involves far more than simply acquiring a virulent pathogen, whether by isolating it from nature or by synthesizing it from scratch. Tacit knowledge also plays an important role in production and weaponization, which include the following steps: (1) growing the bacterium or virus in the needed quantity, (2) formulating the agent with chemical additives to enhance its stability and shelf life, (3) processing the agent into a concentrated slurry or a dry powder, and (4) devising a delivery system that can disseminate the agent as a fine-particle aerosol that infects through the lungs. These steps entail greater technical hurdles than the synthesis of the DNA fragments making up the viral genome. Indeed, the study of past state-level programs has shown that the acquisition of biological weapons requires an interdisciplinary team of scientists and engineers with expertise and tacit knowledge in the fields of microbiology, aerobiology, formulation, and delivery. States are generally more capable of organizing and sustaining such teams than are nonstate actors. According to Jens Kuhn, "The methods to stabilize, coat, store, and disperse a biological agent are highly complicated, known only to a few people, and rarely published. . . . [Even if terrorists manage to synthesize a viral agent successfully,] they will in all likelihood get stuck during the weaponization process."[46]

Along similar lines, Michael Levi of the Council on Foreign Relations has questioned the ability of terrorists to construct an improvised nuclear device from stolen fissile materials. He notes that the process of building such a device would involve a complex series of technical steps, all of which the terrorists would have to perform correctly to succeed.[47] In much the same way, when assessing the risk that terrorists could exploit whole-genome synthesis to create a biological weapon, one must break the problem down into its component parts and calculate the overall probability of success as the product of the probabilities of performing all the intermediate steps correctly, as well as the likelihood that the necessary enabling conditions will be in place.

The preliminary findings discussed in the previous paragraphs suggest that deskilling does not proceed in a uniform manner but affects some biotechnologies more than others. Whether commercial kits and how-to manuals will merely make it easier for experienced scientists to perform certain difficult or tedious operations more quickly and easily or whether deskilling will truly make advanced biotechnologies available to nonexperts is still an open question and will probably remain so for some time. It is also important to disaggregate the risk of misuse by distinguishing among potential actors that differ greatly in resources and technical capacity, including states that possess advanced biowarfare capabilities, terrorist organizations of varying size and sophistication, and individuals motivated by ideology or personal grievance.

Before making the grand claim that all biotechnologies will inevitably become deskilled and accessible to nonexperts, it will be necessary to move beyond assertions based on anecdotal evidence and conduct more empirical research on the nature of tacit knowledge and the process of deskilling. To shed light on this policy debate, it will be necessary to examine what specific conditions, skills, and socio-organizational contexts are required for advanced biotechnologies to work reliably; why certain tools, techniques, and practices of biotechnology become deskilled whereas others do not; and what conditions, both technical and social, facilitate or hamper the process of deskilling.

Answering these questions will require disaggregating specific biotechnologies into their component parts and assessing the importance of tacit knowledge and other sociotechnical factors for each of these elements. Possible methodological approaches include historical analysis based on archival research, in-depth interviews with practicing scientists, and ethnographic observation of laboratory work. Such research should permit a more nuanced assessment of the security risks associated with synthetic biology and other emerging biotechnologies and help policy makers identify the areas of greatest concern.

Dual Use as a Societal Process

Whereas determinist theories view dual use as an inherent property of certain technologies, another school of analysis contends that the *social context* plays a dominant role in shaping how a technology is developed and used. Theories of the "social construction of technology," pioneered by Trevor Pinch and Wiebe Bijker, posit that science and technology are not objective and value neutral but instead reflect the political agendas of those who practice them.[48] As a result, technological development is an "open process that can produce different outcomes depending on the social circumstances of development."[49] When viewed through this theoretical lens, the transfer of a technology from the civilian to the military sector, or from legitimate civilian use to misuse for hostile purposes, involves reinterpreting a technological artifact as an instrument for causing harm and developing it further within that context.

In general terms, social constructivist theory holds that a new technology inspires multiple competing interpretations, each of which implies a different potential path of development.[50] Interest groups and institutions coalesce around these different interpretations, which are then contested and negotiated through a social process that reflects the power relationships of the players and the rules governing their interactions. The dominant interpretation that emerges from this negotiation in turn defines and legitimates the technology's subsequent development and use. Once a technological artifact has been created that fits the specifications of the dominant interpretation, closure has been achieved, and the development process comes to an end.

A familiar historical example that illustrates the lack of inevitability in technology development and the importance of contending social interpretations is the QWERTY keyboard, which is now used in all typewriters and computers throughout the English-speaking world. The QWERTY keyboard is not the most efficient layout for typing English. In fact, it was introduced on manual typewriters to make typists *less* efficient so that they would not type too fast and jam the keys. By the time that more efficient keyboard layouts were proposed, however, managers and office staff had already invested time and money in training personnel to type with the QWERTY keyboard and thus had no incentive to switch to a different layout. Even today QWERTY keyboards remain predominant, although jamming is no longer a problem with electronic keyboards. This case suggests that technological development is far from inevitable and that what is interpreted as the best solution depends on one's perspectives and interests.[51]

Because dual-use technologies, by definition, inspire competing interpretations, the transfer from civil to military application involves a process in which social actors reinterpret the purpose of a technology from a peaceful to a hostile context. This reinterpretation is mediated by a socioscientific network whose structure and composition may vary from one technology to the next. For this reason, the social context of technology development not only is important for understanding the dual-use problem but also may reveal new avenues for policy intervention to reduce the risk of misuse.[52] For example, it may be possible to modify the structure or practices of the socioscientific network in such a way as to delegitimize the use of a technology for hostile purposes.

Approaches to Technology Governance

The growing prevalence in technology-policy literature of the word "governance" reflects a paradigm shift from an earlier focus on "governing," or top-down efforts by the state to regulate the behavior of people and institutions. Governance includes a range of approaches to the management of technology that are not limited to top-down, command-and-control regulation. According to Jan Kooiman, responsibility for the oversight of new technologies is no longer based exclusively in the state but is increasingly shared with the private sector and nongovernmental organizations.[53]

R. A. W. Rhodes notes that the various social actors involved with emerging technologies exchange resources and negotiate shared purposes. Thus, whereas governing involves goal-directed interventions by the state, governance is the net result of a set of complex sociopolitical administrative interactions.[54] The key actors in this process are the scientists and engineers who develop a new technology, the policy makers and regulators who promote innovation and regulate its products, and the civil society groups that promote a new technology or express concerns about its risks. Gerry Stoker argues that when these interactions achieve a high level of mutual understanding and shared vision, they become a "self-governing network" in which a coalition of actors and institutions coordinate their resources, skills, and purposes.[55] For a self-governing network to be sustainable over time, it must be capable of evolving, learning, and adapting. The state may also attempt to steer the network indirectly through facilitation, accommodation, and bargaining, an approach that Stoker calls "managed governance."[56]

A good example of managed governance is the creation by the National Institutes of Health of the *NIH Guidelines for Research Involving Recombinant DNA Molecules*. The process began when concerns about the potential safety hazards of genetic engineering caused the leading scientists in the field to impose a voluntary research moratorium. The relevant scientific community then organized the 1975 Asilomar Conference in Pacific Grove, California, which developed a set of voluntary biosafety rules for the creation of recombinant DNA molecules. Subsequently, the NIH converted the Asilomar rules into a more formal set of biosafety guidelines for recipients of federal research grants. Over the next few decades, the *NIH Guidelines* went through a series of revisions in response to real-world experience with the technology. This process remains a landmark example of self-governance by the scientific community and of managed governance by the state.[57]

In general, strategies to manage the risk of emerging dual-use technologies must seek out a delicate balance. Although inadequate regulation can result in harm to human health, the environment, or national security and undermine public confidence, excessive regulation can smother a promising technology in the cradle and deprive society of its benefits. Effective governance of emerging technologies is particularly challenging because it involves assessing risks at an early stage of development, when uncertainty is high. Joyce Tait has called for "appropriate risk governance," by which she means policies that enable technological innovation, minimize risks to people and the environment, and balance the interests and values of the relevant stakeholders. To achieve these goals, governance should concentrate as much as possible on evidence of harm, accommodate the values and interests of all affected social groups, and maximize the scope for choice among a range of technology options.[58]

Anticipatory Governance

David Guston and Daniel Sarewitz address the problem of uncertainty when assessing the risks of emerging technologies by calling for a system of "anticipatory governance" as an integral part of the research and development (R&D) process. In their view, the key to dealing with knowledge gaps when assessing technological risk is to create a process that is "continuously reflexive, so that the attributions of and relations between co-evolving components of the system become apparent, and informed incremental response is feasible."[59] Such a capability requires building "real-time technology assessment" into the R&D cycle, encouraging communication among the potential stakeholders, and allowing for the modification of development paths and outcomes in response to ongoing risk analysis.

Daniel Barben and his colleagues disaggregate the concept of anticipatory governance into three components: foresight, engagement, and integration. Foresight involves anticipating the implications of an emerging technology through methods such as forecasting, scenario development, and predictive modeling. Engagement is public involvement that goes beyond opinion polls to include substantive upstream consultations with a variety of stakeholders, using vehicles such as museum exhibits, public forums, Internet sites, and citizens' panels. Integration involves encouraging natural scientists to engage with societal issues as an integral part of their research.[60] Barben and his colleagues conclude that anticipatory governance "comprises the ability of a variety of lay and expert stakeholders, both individually and through an array of feedback mechanisms, to collectively imagine, critique, and thereby shape the issues presented by emerging technologies before they become reified in particular ways."[61]

Daan Schuurbiers and Erik Fisher argue that the research and development phase provides a largely overlooked opportunity for addressing social concerns before the results are translated into products or services. "Shaping technological trajectories," they write, "will, at some point, include shaping the very research processes that help to characterize them."[62] One approach, which Schuurbiers and Fisher call "midstream modulation," involves evaluating and adjusting research decisions in the light of societal factors such as environmental, health, and safety considerations. This method challenges scientists to consider issues of social responsibility when designing and carrying out their research activities. To permit such policy interventions at the laboratory scale without compromising research integrity or hampering scientific productivity, Schuurbiers and Fisher call for new types of collaborative interaction between social scientists and natural scientists and engineers.

Adaptive Governance

Another generic problem faced by technology governance is to ensure that a regulatory mechanism, once established, is capable of responding to technological change. The challenge for policy makers is to create a system of ongoing technology assessment and flexible or "adaptive" governance. This approach generally requires an iterative cycle of data gathering, evaluation, and rule modification as a technology evolves and the scientific understanding of its risks matures.[63] Unfortunately, institutional and legal hurdles, such as the complex requirements of the U.S. Administrative Procedures Act, may impede the establishment of adaptive governance mechanisms.[64]

One possible solution to this problem is to incorporate multiple options or contingencies into a regulation, thereby providing a degree of flexibility in its implementation. For example, both the Environmental Protection Agency (EPA) and the Food and Drug Administration (FDA) have designed regulations so that they can be updated and corrected as new information becomes available. The EPA's National Ambient Air Quality Standards program has revised the air-quality standards for particulate matter several times after systematic review of the latest information on health effects. Similarly, the FDA's post-marketing surveillance program tracks the adverse health effects of new drugs based on information collected after they have been approved and marketed. Both sets of regulations are designed to accommodate new scientific findings and other knowledge into an iterative decision-making process.[65] The problem with applying this approach to emerging technologies, however, is that it is often difficult to predict how a new technology will evolve and hence what specific regulatory options will be needed in the future. An alternative approach is to create an expedited process for making technical amendments, so that regulations can be modified or expanded rapidly in response to the emergence of new risks and benefits.[66]

Types of Governance Measures

Current approaches to dual-use technology governance comprise a broad spectrum of measures, including "hard-law" measures (mandatory, statute based), "soft-law" measures (voluntary, nonbinding), and informal measures (based on moral suasion). (See figure 2.1.)

Examples of hard-law measures are licensing, certification, civil liability, insurance, indemnification, testing, labeling, and mandatory oversight. Formal regulations can range from minimalist (e.g., the requirement to report a new technology before marketing it) to extremely stringent (e.g., the FDA's rules for the premarket

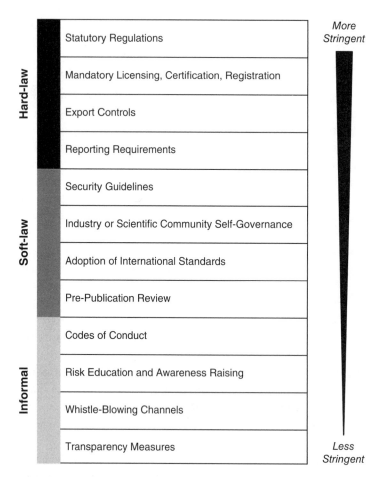

Figure 2.1
Illustrative spectrum of governance measures.

testing and approval of pharmaceuticals). Soft-law mechanisms include voluntary guidelines and industry best practices, and informal measures include education and awareness-raising programs, codes of conduct, and transparency measures. Integrating these approaches in a complementary, synergistic way can give rise to a "web of prevention" that functions at multiple levels, from the individual (e.g., codes of conduct) to the international (e.g., multilateral treaties).[67]

For the governance of emerging dual-use technologies, policy analysts differ over the merits of formal, top-down regulation versus self-regulation by the relevant professional communities. A strictly top-down approach brings with it the danger of excessive government control and limited responsiveness to public concerns.

Conversely, an undue emphasis on bottom-up approaches may enable certain stakeholder groups to "hijack" the governance process in pursuit of their parochial objectives. In the case of synthetic biology, for example, Stephen Maurer and Laurie Zoloth argue that voluntary guidelines offer several advantages: they can be developed in months rather than years, and consensus-based solutions are less disruptive and more likely to be respected by the community of practitioners.[68] A cooperative rather than coercive approach to governance, they argue, is particularly important in a fast-moving field like synthetic biology, where formal regulation cannot keep pace with technological change.

Other advocates of self-regulation contend that government intervention can have harmful unintended consequences. Robert Carlson argues that top-down regulation of synthetic biology would lead to a black market in synthetic DNA that would be more difficult to monitor and control than the current unfettered free market. "Our best potential defense against biological threats," he writes, "is to create and maintain open networks of researchers at every level, thereby magnifying the number of eyes and ears keeping track of what is going on in the world."[69] Nevertheless the transparency of private-sector activities cannot be ensured, because companies seek to protect trade secrets to retain a competitive advantage.

Those favoring the legally binding regulation of emerging technologies such as synthetic biology argue that voluntary governance measures are inadequate because there is no guarantee that all the relevant players will participate. Indeed, a system lacking formal sanctions for noncompliance would tend to attract cheaters and free riders. Another problem with self-governance is that because scientists have a strong professional interest in promoting and publishing their research, they may deliberately overlook or downplay its security implications.[70] Jan van Aken observes, for example, that "scientists hesitate to place any restrictions on each other's work and regard oversight mechanisms largely as a bureaucratic burden."[71] While this statement may be exaggerated, it is also true that without active buy-in and participation by the affected professional community, self-regulation cannot be effective.

It is widely assumed that scientists and companies oppose formal regulation of any kind because it entails burdensome paperwork, unpaid mandates, and rigid performance standards. In fact, however, certain rules that apply to all members of a given discipline or industry can be generally beneficial. Regulation not only creates a predictable framework for research and development but ensures a level playing field for scientific and commercial competition. Formal regulations may also build public confidence in an emerging technology, creating a more favorable political environment for the continued funding of development and its eventual adoption. Because scientists and companies are aware of these advantages, they do not tend to be as antiregulation as is generally assumed. Instead of opposing regulation in principle, they object to poorly informed or technically incompetent regulation.

Hybrid Approaches to Governance

The choice between top-down regulation and community self-governance is rarely clear-cut, and in many cases the best solution is a hybrid of the two approaches. In the case of synthetic biology, for example, Gabrielle Samuel and her colleagues advocate a combination of institutional and governmental regulation to achieve an optimal balance between academic freedom and public safety.[72] Filippa Lentzos calls the debate over top-down versus bottom-up governance a false dichotomy and instead favors a multiplicity of approaches.[73] She identifies three different modes of regulation, which she terms *coercive*, *normative*, and *mimetic*. The coercive mode involves statutory regulations based on the authority of the state and accompanied by penalties for noncompliance. The normative mode is less formal and involves conceptions of what is socially desirable, along with behavioral standards such as codes of conduct and professional self-regulation. Finally, the mimetic mode involves the emulation of successful practices and models of behavior through peer observation and mentoring.[74]

Lentzos argues that for technology governance to be effective, the coercive mode must be combined with the normative and mimetic modes. The relative importance of the three modes varies as a technology moves through the development pipeline. During the early stages of R&D, the normative and mimetic modes tend to predominate, but as a new technology approaches the market, the importance of coercive regulation increases.[75]

Gregory Mandel notes the dilemma that although the early stages of R&D provide a window of opportunity to introduce new regulations before vested interests and sunk costs have reinforced the status quo, intervention at an early stage is also problematic because political interest is low and little is known about the risks and benefits of the technology.[76] As Gerald Epstein observes, the best time to introduce governance measures is during the brief period—if it exists at all—between "too early to tell" and "too late to change."[77]

Informal measures implemented by members of the life sciences community, including universities, medical and veterinary schools, trade associations, and biotechnology and pharmaceutical companies, also have an important role to play in managing the dual-use risks of biotechnology. Such measures include education and awareness-raising campaigns and professional codes of conduct. According to the 2008 report of the Commission on the Prevention of Weapons of Mass Destruction Proliferation and Terrorism, it is important to "sensitize researchers to biosecurity issues and concerns . . . [so that they] design and conduct experiments in a way that minimizes safety and security risks."[78] To this end, a number of international organizations have launched initiatives to foster a culture of biosecurity awareness, including the International Criminal Police Organization (Interpol) and the Organisation for Economic Cooperation and Development (OECD).

The Web of Prevention

Because the problem of dual-use technologies is multifaceted, mitigating the risk of misuse requires a variety of mutually reinforcing governance measures. In the early 1990s, Graham Pearson, then director general of the United Kingdom's Chemical and Biological Defence Establishment at Porton Down, introduced the concept of a "web of deterrence" for countering the threat of biological warfare. His proposed web had four key elements: verifiable biological arms control, export monitoring and control, defensive and protective measures, and national and international responses to the acquisition or use of biological weapons. As the primary security challenges have shifted from the Cold War confrontation to the threat of rogue states and nonstate actors, other scholars have reconceptualized the web of deterrence as a "web of prevention" that places greater emphasis on international and national controls on the storage, transfer, and use of potentially dangerous pathogens.[79]

Given the impossibility of a silver-bullet solution to the problem of biological weapons, the concept of a web of governance measures continues to hold currency in policy debates. Subsumed under this approach are initiatives, measures, and activities ranging from awareness raising and scientific codes of conduct to export controls and the oversight of dual-use technologies. A drawback of the "web of prevention" concept, however, is that it obscures the prioritization of issues and the need to identify key actors and intervention points.[80]

International Governance Regimes

Mihail Roco writes that although organizations and measures for technology governance are often "stovepiped" by area of jurisdiction, product, and process, the governance of emerging technologies that have transborder and global implications requires an integrated approach that includes both anticipatory and corrective measures.[81] For example, commercial gene-synthesis providers exist not only in the United States, Europe, and Japan but also in China, India, and other emerging economies. For this reason, applying a regulatory framework selectively to a few countries would have limited security benefits and could even be counterproductive by driving illicit users to patronize unregulated suppliers.

As biotechnology continues to globalize, governance measures must be harmonized internationally through outreach to the affected industry in different parts of the world or the negotiation of international agreements under the auspices of the United Nations or some other multilateral body.[82] Chyba argues that although international treaties such as the Biological Weapons Convention and the Chemical Weapons Convention seek to prevent the misuse for hostile purposes of biology and chemistry, respectively, their effectiveness is impeded by "the mismatch between the rapid pace of technological change and the relative sluggishness of multilateral negotiation and verification, as well as the questionable suitability of monitoring and inspections with a widely available, small-scale technology."[83] Moreover, because formal treaties take a great deal of time, effort, and political will to negotiate, states normally reserve them for the most serious and urgent threats to international health and security, such as global climate change, nuclear weapons proliferation, and persistent chemical pollutants.

Gary Marchant and Douglas Sylvester suggest a number of alternatives to formal treaties for regulating dual-use technologies at the international level, including forums for transnational dialogue and information sharing, civil-society-based monitoring, international consensus standards, and confidence-building measures (meaning incremental steps to build trust in the context of an enduring dispute).[84] International governance mechanisms have also been established in a few specialized areas of dual-use science and technology, such as the scientific committee created by the World Health Organization to oversee research with live variola virus, the causative agent of smallpox.[85]

As dual-use chemical and biological technologies become increasingly globalized, traditional nonproliferation tools based on technology denial, such as export controls, are declining in effectiveness. Elizabeth Turpen notes that the expansion of international trade has outpaced the ability of the United States and other like-minded countries to control access to dual-use materials and equipment. As an alternative, she calls for "a new grand bargain whereby the developing world gains access to critical technologies while being fully vested in a comprehensive nonproliferation and global control regime." While state-level governance is crucial, Turpen writes, "industry and other nongovernment organizations must increasingly work in concert with governments to meet the burgeoning proliferation challenges."[86]

Another approach to international technology governance is the creation of informal "global issue networks." According to scholars who have studied this process, such as Anne-Marie Slaughter, Jean-François Rischard, and Caroline Wagner, global issue networks link scientists in various parts of the world both to one another and to representatives of governments, nongovernmental organizations, and the private sector. By helping to generate an international consensus around specific proposals, global issue networks can foster "rapid norm production" and pressure states to behave responsibly.[87] In the area of dual-use technology, it may be possible to create a global network of informed scientists, connected through the Internet, who will observe when a new technology is being used for harmful purposes and report their concerns to the appropriate national authorities.[88]

Conclusions

This review of the scholarly literature on the risk assessment and governance of emerging dual-use technologies suggests that a one-size-fits-all approach to managing the risk of misuse is neither possible nor appropriate. Methods for assessing the safety and security risks of emerging technologies should be both flexible and capable of integrating new information as the development process unfolds. The most effective way to achieve this objective is to incorporate an iterative process of technology assessment into the research and development cycle itself. Once the risks of an emerging dual-use technology have been identified, it will be necessary to identify a tailored package of governance measures—made up of hard-law, soft-law, and informal elements—to ensure a reasonable balance of risks and benefits and their equitable distribution across the various stakeholders.[89] To provide some additional background, chapter 3 describes the relevant governance mechanisms already in existence and evaluates their suitability for managing the risks of emerging dual-use technologies. Chapter 4 then presents a decision framework for assessing dual-use risk and devising governance strategies in a consistent manner across a wide variety of technologies.

Notes

1. Keith Grint and Steve Woolgar, *The Machine at Work: Technology, Work, and Organization* (Cambridge: Polity Press, 1997).

2. I am grateful to Lori Knowles for this observation.

3. National Research Council, Committee on Risk Characterization, *Understanding Risk: Informing Decisions in a Democratic Society* (Washington, DC: National Academy Press, 1996).

4. Denise Caruso, *Synthetic Biology: An Overview and Recommendations for Anticipating and Addressing Emerging Risks* (Washington, DC: Center for American Progress, November 2008), 10.

5. M. Granger Morgan, "Risk Analysis and Management," *Scientific American* (July 1993): 32–41.

6. Richard Sclove, *Reinventing Technology Assessment: A 21st Century Model* (Washington, DC: Woodrow Wilson International Center for Scholars, April 2010).

7. National Research Council, Committee on Risk Perception and Communication, *Improving Risk Communication* (Washington, DC: National Academy Press, 1989), 2.

8. U.S. Public Health Service, "Risk Communication: Working with Individuals and Communities to Weigh the Odds," *Prevention Report*, February–March 1995.

9. Ibid.

10. Chauncey Starr, "Social Benefit versus Technological Risk: What Is Our Society Willing to Pay for Safety?" *Science* 165, no. 3899 (September 19, 1969): 1232–1238.

11. Paul Slovic, "Perception of Risk," *Science* 236 (April 17, 1987): 280–285.

12. Jessica Stern, "Dreaded Risk and the Control of Biological Weapons," *International Security* 27, no. 3 (winter 2002–3): 89–123.

13. For case studies of historical incidents of bioterrorism, see Jonathan B. Tucker, ed., *Toxic Terror: Assessing Terrorist Use of Chemical and Biological Weapons* (Cambridge, MA: MIT Press, 2000).

14. Gregory D. Koblentz, "Biosecurity Reconsidered: Calibrating Biological Threats and Responses," *International Security* 34, no. 4 (spring 2010): 131.

15. British Royal Society, *New Approaches to Biological Risk Assessment* (London: British Royal Society, 2009), 11.

16. Barry Charles Ezell, Steven P. Bennett, Detlof von Winterfeldt, John Sokolowski, and Andrew J. Collins, "Probabilistic Risk Analysis and Terrorism Risk," *Risk Analysis* 30, no. 4 (2010): 577.

17. National Research Council, *Biotechnology Research in an Age of Terrorism* (Washington, DC: National Academies Press, 2004), 24–29.

18. Brian Rappert, "The Benefits, Risks, and Threats of Biotechnology," *Science and Public Policy* 35, no. 1 (February 2008): 1–6.

19. Michael Polanyi, *Personal Knowledge* (London: Routledge and Kegan Paul, 1958). Polanyi was a chemist writing about his observations of technical practice.

20. H. M. Collins, "Tacit Knowledge, Trust, and the Q of Sapphire," *Social Studies of Science* 31, no. 1 (2001): 71–85. See also Kathleen M. Vogel, "Bioweapons Proliferation: Where Science Studies and Public Policy Collide," *Social Studies of Science* 36, no. 5 (October 2006): 659–690.

21. Donald MacKenzie and Graham Spinardi, "Tacit Knowledge, Weapons Design, and the Uninvention of Nuclear Weapons," *American Journal of Sociology* 101, no. 1 (1995): 44–99. See also Laura McNamara, "Ways of Knowing about Weapons: The Cold War's End at the Los Alamos National Laboratory," Ph.D. diss. (Albuquerque: University of New Mexico, 2001).

22. Kathleen M. Vogel, "Framing Biosecurity: An Alternative to the Biotech Revolution Model?" *Science and Public Policy* 35, no. 1 (2008): 45–54.

23. Kathleen M. Vogel, "Biodefense: Considering the Sociotechnical Dimension," in *Biosecurity Interventions: Global Health and Security in Question*, ed. Andrew Lakoff and Stephen J. Collier (New York: Columbia University Press, 2008), 240–241.

24. Paul Nightingale and Paul Martin, "The Myth of the Biotech Revolution," *Trends in Biotechnology* 22, no. 11 (November 2004): 564–569.

25. Daniel G. Gibson, John I. Glass, Carole Lartigue, et al., "Creation of a Bacterial Cell Controlled by a Chemically Synthesized Genome," *Science* 329 (July 2, 2010): 52–56.

26. Andrew Marshall, "The Sorcerer of Synthetic Genomes," *Nature Biotechnology* 27, no. 12 (December 2009): 1122.

27. Sonia Ben Ouagrham-Gormley and Kathleen M. Vogel, "The Social Context Shaping Bioweapons (Non)proliferation," *Biosecurity and Bioterrorism* 8, no. 1 (2010): 15.

28. Christopher F. Chyba, "Biotechnology and the Challenge to Arms Control," *Arms Control Today* 36, no. 8 (October 2006): 11–17.

29. Gautam Mukunda, Kenneth A. Oye, and Scott C. Mohr. "What Rough Beast? Synthetic Biology, Uncertainty, and the Future of Biosecurity," *Politics and the Life Sciences* 28, no. 2 (September 2009): 14–15.

30. Jonathan B. Tucker and Raymond A. Zilinskas, "The Promise and Perils of Synthetic Biology," *New Atlantis*, no. 12 (spring 2006): 25–45.

31. BioBricks Foundation Web site, http://biobricks.org.

32. iGEM Web site, http://ung.igem.org/Main_Page.

33. Marcus Wohlsen, "Amateurs Are Trying Genetic Engineering at Home," Associated Press, December 25, 2008; Carolyn Y. Johnson, "As Synthetic Biology Becomes Affordable, Amateur Labs Thrive," *Boston Globe*, September 16, 2008; Phil McKenna, "Rise of the Garage Genome Hackers," *New Scientist*, no. 2689 (January 7, 2009): 20–21.

34. DIYbio Web site, http://diybio.org.

35. Gaymon Bennett, Nils Gilman, Anthony Stavrianakis, and Paul Rabinow, "From Synthetic Biology to Biohacking: Are We Prepared?" *Nature Biotechnology* 27, no. 12 (December 2009): 1109.

36. Freeman Dyson, "Our Biotech Future," *New York Review of Books* 54, no. 12 (July 19, 2007), http://www.nybooks.com/articles/archives/2007/jul/19/our-biotech-future.

37. Andrew Ellington, "Opinion: On Regulation," *Scientist*, March 3, 2011, http://www.the-scientist.com/news/display/58035.

38. Helen Anne Curry, "The Evolution Engineers: Plant Breeding with Chemicals in the Laboratory and at Home, 1937–1950," Science and Technology Studies Colloquium Series, Cornell University, February 7, 2011.

39. Markus Schmidt, "Diffusion of Synthetic Biology: A Challenge to Biosafety," *Systems and Synthetic Biology* (2008), DOI 10.1007/s11693-008-9018-2.

40. Dyson, "Our Biotech Future."

41. Mukunda, Oye, and Mohr, "What Rough Beast," 15.

42. "Hacking Goes Squishy," *Economist*, September 5, 2009, 30–31.

43. Michael Lynch, "Protocols, Practices, and the Reproduction of Technique in Molecular Biology," *British Journal of Sociology* 53, no. 2 (June 2002): 203–220.

44. National Science Advisory Board for Biosecurity, *Addressing Biosecurity Concerns Related to the Synthesis of Select Agents* (Bethesda, MD: National Institutes of Health, 2006), 4.

45. Jens H. Kuhn, "Defining the Terrorist Risk," in the roundtable discussion forum "Is the Availability of Genetic Information Dangerous?" *Bulletin of the Atomic Scientists*, February 28, 2008, http://www.thebulletin.org/web-edition/roundtables/is-the-availability-genetic-information-dangerous#rt623.

46. Ibid.

47. Michael A. Levi, *On Nuclear Terrorism* (Cambridge, MA: Harvard University Press, 2009).

48. Trevor J. Pinch and Wiebe E. Bijker, "The Social Construction of Facts and Artifacts, or How the Sociology of Science and the Sociology of Technology Might Benefit Each Other," in *The Social Construction of Technological Systems: New Directions in the Sociology and History of Technology*, ed. Wiebe E. Bijker, Thomas P. Hughes, and Trevor J. Pinch

(Cambridge, MA: MIT Press, 1987), 17–50; Susan Leigh Star, *Ecologies of Knowledge: Work and Politics in Science and Technology* (Albany: State University of New York Press, 1995), 3. Other seminal works on the social construction of technology include Daryl Chubin and Ellen Chen, *Science off the Pedestal: Social Perspectives on Science and Technology* (Belmont, CA: Wadsworth, 1989); Bruno Latour and Steve Woolgar, *Laboratory Life: The Social Construction of Scientific Facts* (Beverly Hills, CA: Sage, 1979); Karin Knorr-Cetina, *The Manufacture of Knowledge: An Essay on the Constructivist and Contextual Nature of Science* (Oxford: Pergamon Press, 1981); and Andrew Pickering, ed., *Science as Practice and Culture* (Chicago: University of Chicago Press, 1992).

49. Hans K. Klein and Daniel Lee Kleinman, "The Social Construction of Technology: Structural Considerations," in *Science, Technology, and Human Values* 27, no. 1 (winter 2002): 29.

50. For a full treatment of the social construction of technology as traced through three case studies, see Wiebe E. Bijker, *Of Bicycles, Bakelites, and Bulbs: Toward a Theory of Technological Change* (Cambridge, MA: MIT Press, 1995). For critiques of the theory, see Stuart Russell, "The Social Construction of Artifacts: A Response to Pinch and Bijker," *Social Studies of Science* 16 (May 1986): 331–345; and Langdon Winner, "Upon Opening the Black Box and Finding It Empty: Social Constructivism and the Philosophy of Technology," *Science, Technology, and Human Values* 16, no. 3 (summer 1993): 362–378.

51. Robert Pool, *Beyond Engineering: How Society Shapes Technology* (Oxford: Oxford University Press, 1997).

52. Jennifer Croissant and Sal Restivo, "Science, Social Problems, and Progressive Thought: Essays on the Tyranny of Science," in Star, *Ecologies of Knowledge*, 57.

53. Jan Kooiman, ed., *Modern Governance* (London: Sage, 1993).

54. R. A. W. Rhodes, "The New Governance: Governing without Government," *Political Studies* 44, no. 4 (September 1996): 657, 660.

55. Gerry Stoker, "Governance as Theory: Five Propositions," *International Social Science Journal* 50, no. 155 (1998): 17–23.

56. Ibid., 23–24.

57. Marcia Barinaga, "Asilomar Revisited: Lessons for Today?" *Science* 287, no. 5458 (March 3, 2000): 1584–1585; Gregory A. Petsko, "An Asilomar Moment," *Genome Biology* 3, no. 10 (September 25, 2002), DOI 10.1186/gb-2002-3-10-comment1014.

58. Joyce Tait, "Systemic Interactions in Life Science Innovation," *Technology Analysis and Strategic Management* 19, no. 3 (May 2007): 257–277.

59. David H. Guston and Daniel Sarewitz, "Real-Time Technology Assessment," *Technology in Society* 24 (2002): 100.

60. Daniel Barben, Erik Fisher, Cynthia Selin, and David H. Guston, "Anticipatory Governance of Nanotechnology: Foresight, Engagement, and Integration," in *The Handbook of Science and Technology Studies*, 3rd ed., ed. Edward J. Hackett, Olga Amsterdamska, Michael Lynch, and Judy Wajcman (Cambridge, MA: MIT Press, 2007), 979–1000.

61. Ibid., 992–993.

62. Daan Schuurbiers and Erik Fisher, "Lab-scale intervention," *EMBO Reports* 10, no. 5 (2009): 424–427.

63. Gregory Mandel, "Nanotechnology Governance," *Alabama Law Review* 59 (2008): 1379.

64. The U.S. Administrative Procedures Act (APA) mandates a complex and lengthy process for revising final rules after they have been promulgated. First, agencies must publish in the *Federal Register* a notice of proposed rule making that references the legal authority under which the rule is proposed and a description of the topics and issues to be addressed by the proposed rule. Second, agencies must give the public an opportunity to submit comments on the proposed rule and must address all significant comments. If the affected parties believe that a federal regulatory agency has made an unlawful decision owing to procedural or substantive error, they may seek judicial review of the decision under the APA.

65. Lawrence E. McCray, Kenneth A. Oye, and Arthur C. Petersen, "Planned Adaptation in Risk Regulation: An Initial Survey of U.S. Environmental, Health, and Safety Regulation," *Technological Forecasting and Social Change* 77, no. 6 (July 2010): 951–959.

66. Mandel, "Nanotechnology Governance," 1379.

67. Jez Littlewood, "Managing the Biological Weapons Problem: From the Individual to the International," commissioned paper no. 14, WMD (Blix) Commission, August 2004, http://www.blixassociates.com/wp-content/uploads/2011/03/no14.pdf.

68. Stephen M. Maurer and Laurie Zoloth, "Synthesizing Biosecurity," *Bulletin of the Atomic Scientists* 63, no. 6 (November–December 2007): 16–18.

69. Robert Carlson, "The Pace and Proliferation of Biological Technologies," *Biosecurity and Bioterrorism* 1, no. 3 (September 2003): 203–214.

70. Parliamentary Office of Science and Technology, "The Dual-Use Dilemma," 3.

71. Jan van Aken, "When Risk Outweighs Benefit," *EMBO Reports* 7, no. S1 (2006): S13.

72. Gabrielle N. Samuel, Michael J. Selgelid, and Ian Kerridge, "Managing the Unimaginable," *EMBO Reports* 10, no. 1 (2009): 9.

73. Filippa Lentzos, "Countering Misuse of Life Sciences through Regulatory Multiplicity," *Science and Public Policy* 35, no. 1 (February 2008): 62.

74. Ibid.

75. Ibid., 64.

76. Mandel, "Nanotechnology Governance," 1378.

77. Gerald L. Epstein, "The Challenges of Developing Synthetic Pathogens," *Bulletin of the Atomic Scientists*, May 19, 2008, http://www.thebulletin.org/web-edition/features/the-challenges-of-developing-synthetic-pathogens.

78. Commission on the Prevention of Weapons of Mass Destruction Proliferation and Terrorism, *World at Risk* (New York: Vintage Books, 2008).

79. Daniel Feakes, Brian Rappert, and Caitríona McLeish, "Introduction: A Web of Prevention?" in *A Web of Prevention: Biological Weapons, Life Sciences, and the Governance of Research*, ed. Brian Rappert and Caitríona McLeish (London: Earthscan, 2007), 1–13.

80. Filippa Corneliussen, "Adequate Regulation, a Stop-Gap Measure, or Part of a Package?" *EMBO Reports* 7 (2006): 50–54.

81. Mihail Roco, "Possibilities for Global Governance of Converging Technologies," *Journal of Nanoparticle Research* 10, no. 1 (January 2008): 11–29.

82. Kendall Hoyt and Stephen G. Brooks, "A Double-Edged Sword: Globalization and Biosecurity," *International Security* 28, no. 3 (winter 2003–4): 123–148.

83. Christopher F. Chyba, "Biotechnology and the Challenge to Arms Control," *Arms Control Today* 36, no. 8 (October 2006): 11–17.

84. Gary E. Marchant and Douglas J. Sylvester, "Transnational Models for Regulation of Nanotechnology," *Journal of Law, Medicine, and Ethics* (winter 2006): 715–723.

85. Jonathan B. Tucker, "Preventing the Misuse of Biology: Lessons from the Oversight of Smallpox Virus Research," *International Security* 31, no. 2 (fall 2006): 116–150.

86. Elizabeth Turpen, "Achieving Nonproliferation Goals: Moving from Denial to Technology Governance," *Policy Analysis Brief* (Muscatine, IA: Stanley Foundation, June 2009).

87. Caroline S. Wagner, *The New Invisible College: Science for Development* (Washington, DC: Brookings Institution Press, 2008); Anne-Marie Slaughter, *A New World Order* (Princeton, NJ: Princeton University Press, 2005); Jean-François Rischard, "Global Issue Networks," *Washington Quarterly* 26, no. 1 (winter 2002–3): 17.

88. National Research Council, *Globalization, Biosecurity, and the Future of the Life Sciences*, 251–256.

89. Parliamentary Office of Science and Technology, "The Dual-Use Dilemma," 2.

3

Current Dual-Use Governance Measures

Lori P. Knowles

The effective governance of dual-use technologies requires a multifaceted approach that includes three types of measures: hard law (treaties, statutes, and regulations), soft law (voluntary standards and guidelines), and informal measures (awareness raising, professional codes of conduct). These three types of governance measures are not mutually exclusive. For example, voluntary standards and guidelines can be bolstered by criminal or tort laws that impose penalties for breaches of legal standards or the harm caused by accidental or deliberate misuse.[1]

Although this chapter focuses mainly on specific dual-use governance measures, a host of other national laws and regulations relating to environmental protection, transportation, imports, labeling, confidentiality, and privacy may be relevant as well. In addition to national measures, international consortia such as the International Organization for Standardization publish industrial-security standards for fields ranging from freight security to cybersecurity. These standards have been widely adopted, becoming de facto international regulations, and have been translated into domestic regulations and standards that provide a backdrop to dual-use governance regimes.

Emerging biotechnologies require flexible governance strategies because they are advancing so fast that more rigid measures would rapidly become obsolete. Recent trends to enhance the governance of dual-use technologies include an increased emphasis on criminal law, efforts at international harmonization, oversight of dual-use scientific research, and efforts to create an ethics-based "culture of responsibility" in the life sciences.[2] These diverse approaches have created a "web of prevention" that is dynamic and adaptable, although gaps in governance remain.[3]

This chapter reviews some current technology governance measures at the national, regional, and international levels and assesses their effectiveness. The focus is primarily on managing the safety and security risks of biological (rather than chemical) science and technology, for two reasons. First, regulating the dual-use aspects of chemistry is more straightforward because chemical warfare agents, such as sulfur mustard or sarin, are manufactured compounds that do not exist in nature and have few if any peaceful applications; they are also derived from a limited set of precursor chemicals whose export and import can be controlled. Biological pathogens, in contrast, are available from natural sources and have legitimate uses in research, vaccine production, and drug development. Second, because of the anthrax letter attacks in 2001 and controversial cases of dual-use biological research such as the synthesis of poliovirus, policymakers have paid far more attention to biosecurity issues than to chemical security issues. Thus, although environmental laws cover the import, export, testing, and manufacture of chemicals,[4] the disposal of hazardous wastes, and accidental releases of toxic substances,[5] there has been relatively little focus on the dual-use aspects of chemical science and technology.

Arms Control and Disarmament Treaties

Legally binding arms control and disarmament treaties provide important instruments of hard law at the international level. Such regimes seek to prevent the development, production, acquisition, and use of certain categories of weapons and technologies. Although treaties have limitations, they play an essential role in codifying and harmonizing norms of state behavior, sometimes through customary international law.[6] Indeed, in cases where penal sanctions do not exist or cannot be enforced, international norms with respect to the handling and application of dual-use technologies may provide an effective deterrent to misuse. The Geneva Protocol, the Biological Weapons Convention, and the Chemical Weapons Convention constitute the backbone of international law with respect to the governance of dual-use technologies in the chemical and biological fields.

The Geneva Protocol of 1925

A foundational instrument of the international law of armed conflict is the Geneva Protocol of 1925, which bans the use of chemical and biological weapons in warfare.[7] Although the Hague Conventions of 1899 and 1907 had similar provisions, the Geneva Protocol was the first widely accepted prohibition on the military use of asphyxiating gases and bacteriological agents.[8] Still, the treaty had several weaknesses: it was limited to a ban on use in war, did not forbid the member states from developing and stockpiling chemical and biological weapons, and lacked verification and enforcement measures. Moreover, many countries that ratified the Geneva Protocol reserved the right to retaliate in kind if they were attacked, in effect limiting the ban to a no-first-use declaration.[9] Finally, because the treaty was structured as a contract among the parties, it did not bind the participating states with respect to nonparties. Today this structure is less of a weakness because many legal scholars believe that the Geneva Protocol has achieved the status of customary international law, so that it is binding on all nations whether or not they have formally ratified or acceded to it.

A major impediment to U.S. ratification of the Geneva Protocol was disagreement over whether it bans the use in war of nonlethal chemicals, such as riot-control agents and defoliants. Contrary to the large majority of member states, the United States does not consider riot-control agents (such as CS or CN tear gas) to be chemical weapons. Because of this controversy, Washington delayed ratifying the Geneva Protocol until 1975. In addition, when President Ford signed the instrument of ratification, he issued an executive order reserving the right of the United States to employ riot-control agents with presidential authorization "in defensive military modes to save lives," such as rescuing downed pilots behind enemy lines or when civilians are used to mask or screen attacks.[10]

The Biological Weapons Convention of 1972

During the Cold War, the Soviet Union and the United States pursued a biological arms race until 1969, when President Nixon decided to renounce the U.S. offensive biological weapons program and limit all further work to defensive research and development. This unilateral decision, along with Moscow's agreement in 1971 to pursue separate treaties to control biological and chemical arms, created a positive political climate for the negotiation of the Biological and Toxin Weapons Convention (BWC), which was concluded in 1972 and entered into force in March 1975. The BWC builds on the Geneva Protocol by prohibiting the development, production, possession, and transfer of biological weapons and requiring the destruction of all existing stockpiles and production facilities. Article IV of the BWC urges each member state to adopt implementing legislation making the treaty prohibitions binding on its citizens and imposing penal sanctions for violations.

The dual-use nature of pathogens and toxins, however, precludes the BWC from imposing a comprehensive ban on all activities involving these materials.[11] Instead the treaty's definition of a biological weapon is based largely on intent. Article I prohibits "microbial or other biological agents, or toxins whatever their origin or method of production, of types and in quantities that have no justification for prophylactic, protective or other peaceful purposes." This definition, known as the General Purpose Criterion, goes to the heart of the dual-use problem. BWC compliance depends on how a member state uses a biological agent, in what quantities, and with what equipment. Yet because the treaty lacks formal declaration and inspection measures, there is no effective way to monitor, verify, or enforce compliance. Although Article VI of the BWC empowers states to report suspected treaty violations to the UN Security Council for investigation, this provision has never been used because the five permanent members of the Security Council have the power to block an inquiry into their own treaty compliance or that of an ally.

In the 1990s, international confidence in the BWC was severely shaken when defectors revealed that the Soviet Union had secretly maintained a vast biological warfare program in violation of its treaty commitments. In 1995 concern about the BWC's lack of verification measures led the member states to start negotiating a legally binding compliance protocol to augment the convention by enhancing transparency and deterring violations. The multilateral talks culminated in June 2001 in a chairman's text of the protocol, but the United States subsequently rejected the draft treaty on the grounds that it would be both ineffective at detecting violations and overly burdensome to the U.S. biodefense program and to the biotechnology and biopharmaceutical industries. Whether or not effective verification of the BWC is technically feasible is the crux of the continuing debate about how best to strengthen the convention.

Review conferences of the BWC are held every five years to survey the operation of the treaty and to assess the impact on the regime of advances in science and technology. From 2002 through 2010, BWC member states also pursued an "intersessional work program" consisting of annual meetings of experts and diplomats. These meetings focused on topics related to BWC implementation and the prevention of bioterrorism, such as securing dangerous pathogens and creating codes of conduct for life scientists. The combination of the five-year review conferences and the intersessional work program has helped to keep international attention focused on biological disarmament issues.

The Chemical Weapons Convention of 1993

A few years after the BWC was concluded, the UN Conference on Disarmament in Geneva began what proved to be a quarter century of negotiations on a separate treaty banning chemical arms. The Chemical Weapons Convention (CWC), which entered into force in April 1997, requires member states to declare and destroy all existing stockpiles of chemical weapons and prohibits any future development, production, transfer, and use of such arms. The CWC and the BWC share a number of similarities. First, to avoid being overtaken by technological change, the CWC also uses a broad, purpose-based definition of chemical weapons as "toxic chemicals and their precursors, except where intended for purposes not prohibited under this Convention, as long as the types and quantities are consistent with such purposes."[12] Nonprohibited uses of toxic chemicals include industrial, agricultural, research, medical, pharmaceutical, and other peaceful applications; the development of defensive countermeasures; and "law enforcement including domestic riot control." Second, the CWC requires member states to adopt domestic implementing legislation that makes the terms of the treaty binding on their citizens, both at home and abroad, and imposes penal sanctions for violations. Third, review conferences of the CWC are held roughly every five years.

In contrast to the BWC, however, the CWC has extensive verification measures to monitor compliance with its provisions, including the routine inspection of chemical industry plants that produce certain dual-use chemicals. To provide the basis for routine verification, an annex to the treaty includes three lists, or schedules, of toxic chemicals and precursors. The drafters of the CWC recognized that the schedules were not comprehensive and would require periodic updating as science and technology evolved. As of 2011, however, the schedules have not been amended since the treaty entered into force in 1997.

The international entity responsible for overseeing the implementation of the CWC is the Organization for the Prohibition of Chemical Weapons (OPCW), based in The Hague. The OPCW has three main subsidiary bodies: the Technical Secretariat, which conducts inspections and helps member states to meet their treaty obligations; the Conference of the State Parties, which meets annually to make policy decisions related to the convention; and the forty-one-country Executive Council, which is responsible for carrying out the decisions of the Conference. A Scientific Advisory Board (SAB) reporting to the director general monitors relevant scientific and technological developments.

UN Security Council Resolution 1540

UN Security Council Resolution 1540 of April 28, 2004, requires all UN member states to adopt and implement national legislation to prevent the proliferation of nuclear, chemical, and biological weapons and their means of delivery, particularly to terrorists. Countries must establish "appropriate [export] controls over related materials" and pass laws that prohibit efforts by "non-states to manufacture, acquire, possess, develop, transport, transfer or use nuclear, chemical or biological weapons and their means of delivery, in particular for terrorist purposes."[13] The legislative details are left to the individual states.

National Biosafety and Biosecurity Measures

Biosafety governance seeks to keep scientific personnel safe from accidental exposures to the hazardous biological agents they are working with and to prevent accidental releases of pathogens from the laboratory that could threaten public health and the environment. Biosecurity measures, in contrast, seek to prevent the deliberate theft, diversion, or malicious release of pathogens for hostile purposes.

A number of countries have introduced guidelines for laboratory biosafety and biosecurity (sometimes combined under the term "biorisk") that have also been the object of international harmonization efforts by groups such as the World Health Organization (WHO),[14] the European Committee for Standardization (CEN),[15] and the Organization for Economic Cooperation and Development (OECD).[16] In 2008, for example, CEN created a system of biorisk management for laboratories that handle dangerous pathogens, including a scheme to certify laboratory compliance with the CEN standards and applicable national regulations. Each country that adopts the standards is responsible for selecting its own certification method and agency. Although the harmonized rules may require the use of specific equipment, which can be burdensome and expensive, the existence of common standards facilitates technology transfer and collaboration among legitimate researchers in the participating countries. Over time, the harmonization of biosafety and biosecurity standards can reduce costs and increase efficiencies through mutual recognition and reciprocity agreements.

The following sections discuss biosafety and biosecurity governance in some representative countries: the United States, the member states of the European Union, Israel, and Singapore.

The United States

The United States has a large number of laws, regulations, and guidelines pertaining to the safe handling of biological pathogens. In addition to laws covering the transportation of human pathogens and plant and animal pests, the Occupational Safety and Health Act of 1970 provides health and safety standards for exposure to pathogens.[17] Since 1984 the U.S. Centers for Disease Control and Prevention (CDC) and the National Institutes of Health (NIH) have jointly published a manual titled *Biosafety in Microbiological and Biomedical Laboratories* (*BMBL*), which includes a graduated risk-assessment and containment model for work with dangerous pathogens. The levels of precaution range from Biosafety Level 1, for research on microbes not known to cause human disease, to Biosafety Level 4, for research on dangerous and exotic agents that pose a high risk of life-threatening infection and person-to-person spread and for which no vaccines or treatments are available. The four biosafety levels require increasingly stringent measures for the handling, containment, and disposal of biohazardous materials, as well as risk-management efforts involving equipment, facilities, and personnel.[18]

The *BMBL* is not legally binding on U.S. laboratories, serving instead as an advisory document that codifies best practices rather than prescriptive regulations. Laboratories are expected to adhere to the standards in the *BMBL* if they receive federal funds or choose to bind themselves voluntarily. Many commercial labs and private pharmaceutical firms also comply voluntarily with the *BMBL* because of liability concerns and the strict regulations associated with the licensing and marketing of new drugs and vaccines.[19] Nevertheless, because the *BMBL* sets out performance standards without enforcement mechanisms or a clearly prescribed method of implementation, various U.S. institutions implement the guidelines in different ways and thus achieve inconsistent levels of biosafety.

Partially filling this legal gap is tort law under actions for negligence, for which the *BMBL* provides standards of reasonable (i.e., nonnegligent) behavior.[20] In addition, regulatory liability may be assigned based on failure to follow standards that are referenced through regulation. For example, the Public Health Security and Bioterrorism Preparedness and Response Act of 2002 requires anyone who works with Select Agents (a list of microbial pathogens and toxins of bioterrorism concern) to follow the biosafety guidelines in the *BMBL*.[21] By referencing the *BMBL* guidelines, the Bioterrorism Act effectively makes them legally binding, so that failure to comply could result in a finding of civil or criminal liability.

Genetic engineering in the United States is governed by the *NIH Guidelines on Research Involving Recombinant DNA Molecules*.[22] These guidelines, first developed in 1976, specify safe laboratory practices and appropriate levels of physical and biological containment for basic and clinical research with recombinant DNA, including the creation and use of organisms containing foreign genes. The *NIH Guidelines* classify research with recombinant microbes into four risk categories based on the pathogenicity of the agent in healthy adult humans, accompanied by increasingly stringent safety and oversight precautions. In principle, failure to comply with the *NIH Guidelines* could lead to the revocation of federal funding for recombinant DNA research projects, although noncompliance is rarely punished.

Both the *NIH Guidelines* and the *BMBL* are revised periodically, and during this process cross references between the risk levels in the *NIH Guidelines* and the biosafety levels in the *BMBL* have been inserted. The *NIH Guidelines* have also been amended repeatedly in an attempt to keep them current with scientific advances. In 2009, in response to developments in synthetic genomics, the NIH published proposed changes extending the coverage of the guidelines to molecules constructed outside living cells by joining natural or synthetic DNA segments to DNA molecules that can replicate in a living cell.[23]

Like the *BMBL*, the *NIH Guidelines* apply to laboratories and institutions that receive federal funding for recombinant DNA research and to additional institutions that accept the rules voluntarily. Under the *NIH Guidelines*, proposed experiments with recombinant DNA must be reviewed at the local level by an Institutional Biosafety Committee (IBC), which assesses the potential harm that might occur to public health and the environment. Based on this risk assessment, the IBC determines the appropriate level of biocontainment; evaluates the adequacy of training, procedures, and facilities; and assesses the compliance of the investigator and the institution with certain elements of the *NIH Guidelines*, such as the reporting of adverse events.

Local IBCs review human gene transfers and the use of recombinant DNA in whole animals, and they also function as sentinels to identify emerging safety and policy issues for consideration by a federal-level body called the Recombinant DNA

Advisory Committee (RAC), which falls under the NIH. An IBC can refer research proposals that lack a clear precedent and have a high level of potential risk to the RAC, which then develops appropriate guidelines. In exceptional cases, the RAC may refer cases to the NIH director for adjudication. The RAC also serves as a forum for policy deliberations on scientific, ethical, and legal issues related to recombinant DNA research.

A separate set of local oversight bodies called Institutional Review Boards (IRBs) weigh the risks and benefits of research on human subjects and ensure that human volunteers give their fully informed consent to participate. Whereas IBCs are based on the *NIH Guidelines*, IRBs were established by statute and hence are mandatory for institutions that receive federal research funding.[24] Both IBCs and IRBs are staffed by volunteers and have been criticized for their excessive workload and lack of expertise in key areas. Indeed, the growing number of research protocols that raise complex biosafety and bioethical issues has strained the ability of some IBCs and IRBs to make careful, informed decisions.[25]

Nevertheless research oversight at the local level offers certain advantages over a centralized or national oversight system. In particular, local reviewers tend to have an institutional memory and personal familiarity with the community and key individuals, including those investigators who tend to understate the risks of their research. In addition to leveraging institutional expertise, local review committees are more economical and less bureaucratic than a national oversight system. These efficiencies result in a more streamlined review process for proposed experiments and impose fewer administrative burdens and time constraints on research.

A weakness of the current U.S. biosafety system is that apart from clear-cut cases in which a proposed experiment involves recombinant DNA or human subjects, the review of an experimental protocol from the standpoint of biosafety or bioethics may not occur unless the principal investigator or some other individual identifies an issue and initiates an external review. Another drawback of the system is that the source of research funding—private or public—determines whether the *NIH Guidelines* are binding or merely advisory. Whereas the *BMBL* and *NIH Guidelines* function as de facto research standards for all federally funded institutions that conduct research with biological pathogens or recombinant DNA, some private institutions have chosen not to follow these rules when conduct of the research would have benefited considerably from the application of safeguards.[26]

U.S. Biosecurity Governance
The United States leads the rest of the world with respect to the extent and detail of its biosecurity legislation.[27] Shortly after the terrorist attacks of September 11, 2001, and the subsequent anthrax mailings, Congress passed the Uniting and

Strengthening America by Providing Appropriate Tools Required to Intercept and Obstruct Terrorism Act of 2001 (USA PATRIOT Act), which prohibits "restricted persons" from shipping, possessing, or receiving Select Agents and Toxins.[28] The definition of "restricted persons" includes citizens of countries on the State Department's list of state sponsors of terrorism, individuals with a criminal background or a history of mental instability or drug abuse, and persons connected with organizations suspected of domestic or international terrorism. The USA PATRIOT Act also criminalizes the possession of Select Agents in types or quantities that cannot be justified for prophylactic, protective, or peaceful purposes, and makes it a federal crime for convicted felons, illegal aliens, and fugitives to possess or transport Select Agents in any quantity, for any reason.

The USA PATRIOT Act illustrates the increasing use of criminal law as a tool in combating biological weapons proliferation and terrorism.[29] The criminalization of certain activities involving biological and chemical agents empowers law enforcement officials to investigate the stipulated activities. This trend toward criminalization is also apparent in the international arena, as in the draft treaty proposed by the Harvard Sussex Program to criminalize the acquisition of chemical or biological weapons.[30]

The Department of Health and Human Services first established the Select Agent Regulations under the Antiterrorism and Effective Death Penalty Act of 1996,[31] which required that U.S. laboratories that transferred or received Select Agents register with the CDC and report all such transactions. This initial rule had a serious flaw, however, in that it overlooked facilities that merely possessed or worked with such agents without transferring them. Congress later closed this loophole through a provision in the Public Health Security and Bioterrorism Preparedness and Response Act of 2002, which required all institutions that possess, use, or transfer Select Agents affecting humans to register with and notify the CDC.[32] In addition, entities that work with plant or animal pathogens on the Select Agent List must notify the Animal and Plant Health Inspection Service (APHIS) of the U.S. Department of Agriculture.[33]

All institutions and persons registered under the Select Agent Regulations must undergo a "security risk assessment" by the Federal Bureau of Investigation that involves fingerprinting and screening against terrorist and other databases. This vetting process aims to identify "restricted persons" and others who are legally denied access to Select Agents. Registered institutions and personnel are also required to report any release, loss, theft, or accident involving Select Agents. The regulations require the U.S. government to review and update the Select Agent List every two years. Eventually the list may be replaced by a system for specifying microbial pathogens and toxins based on DNA sequence rather than microbial species, in an attempt to keep pace with rapid advances in biotechnology.[34]

Dual-Use Research of Concern

In recent years, biosecurity concerns have extended to certain areas of basic research in the life sciences that could yield dangerous bioengineered pathogens or knowledge with a potential for deliberate misuse. Because researchers, reviewers, funders, and publishers must be able to recognize "dual-use research of concern" (DURC) when they see it, a shared definition is essential. The National Science Advisory Board for Biosecurity (NSABB), a U.S. federal advisory committee, has defined DURC as "research that, based on current understanding, can be reasonably anticipated to provide knowledge, products, or technologies that could be directly misapplied by others to pose a threat to public health and safety, agriculture, plants, animals, the environment, or material."[35]

The NSABB has recommended that IBCs be made responsible for the oversight of dual-use research, in addition to their current role in ensuring the biosafety of recombinant DNA experiments.[36] To the extent that the IBC system relies on self-identification of risks by researchers, without more education about the nature of dual-use risks in life sciences research, its success as a dual-use oversight mechanism is questionable.[37] It is necessary to develop a set of criteria for identifying DURC that can be applied consistently across different institutions. The National Research Council, for example, has identified seven types of "experiments of concern" that may warrant dual-use review, such as those aimed at rendering vaccines ineffective, impairing the immune system, or making a pathogen more virulent.[38]

Another element of biosecurity governance concerns possible restrictions on the publication of sensitive information. In 2003 the editors of several major scientific journals issued a joint statement calling for the review of security-sensitive research papers submitted for publication, with the default position favoring public release.[39] In response to concerns about dual-use information, the NSABB suggested conducting a risk-benefit analysis before publishing articles that could lead to dual-use concerns. Based on biosecurity concerns, the editors can ask authors to modify an article, delay publication, or reject it entirely. In 2004 an expert panel of the U.S. National Academies considered restrictions on the publication of pathogen genomes but ultimately decided not to endorse the restrictions.[40] A major problem with pre-publication security reviews is that identifying a priori which research findings entail dual-use risks can be extremely difficult. Critics also argue that scientific freedom and access to information are crucial to technological innovation and that restricting publication would slow the development of medical countermeasures against biological threats.[41]

Export Controls

U.S. export regulations on dual-use items and materials have been promulgated pursuant to several laws, including the USA PATRIOT Act of 2001, the Homeland

Security Act of 2002,[42] and the Export Administration Act of 1979.[43] The Commerce Control List (CCL), established by part 738 of the Export Administration Regulations (EAR), sets out the combinations of dual-use goods and destinations for which an exporter must obtain a license from the Department of Commerce's Bureau of Industry and Security (BIS). The CCL also provides "reasons for control" for each item, ranging from counterterrorism and crime prevention to national security and regional stability.

The Department of Homeland Security's Customs and Border Protection (CBP) agency is responsible for ensuring that licensable exports from U.S. ports comply with nonproliferation export controls, but the system has gaps.[44] According to a report in 2005 by the department's inspector general, "CBP does not consistently enforce federal export licensing laws and regulations at all U.S. ports of exit. CBP's ability to effectively and efficiently control exports licensed by State and Commerce is limited by inadequate information and staff resources."[45] Greater industry awareness of dual-use export requirements, coupled with more advanced information-sharing systems, would facilitate enforcement efforts. A separate set of regulations governs transfers of sensitive information ("deemed exports") to foreign nationals within the United States.[46]

Dual-use export controls are designed to prevent countries seeking nuclear, biological, or chemical arms from obtaining the necessary materials and production equipment. The Australia Group (AG) is an informal group of like-minded states, formed in 1985, that harmonize their national export regulations on dual-use materials and technology related to chemical and biological weapons. Made up of more than forty participants, including the United States and most of its allies, the AG has developed common control lists of chemical weapons precursors, biological pathogens and toxins, and dual-use chemical and biological production equipment.[47] Advocates of this approach contend that harmonized dual-use export controls can help to slow proliferation while facilitating trade among legitimate users. (Complementing the work of the AG are three other supplier cartels: the Nuclear Suppliers Group, the Missile Technology Control Regime, and the Wassenaar Arrangement on Export Controls for Conventional Arms and Dual-Use Goods and Technologies.)

Although export controls are an important tool of dual-use governance, they suffer from a number of weaknesses. First, their effectiveness depends in part on exporters' knowledge of export requirements, such as when it is necessary to obtain an export license. Second, monitoring, enforcement, and targeted sanctions are key to the effectiveness of export controls and require considerable investment by governments. Third, export controls must be harmonized internationally, because the alternative is "an uneven patchwork of regulations, creating pockets of lax implementation or enforcement."[48]

The European Union

Despite European leadership in promoting the international norms embodied in the BWC and the CWC, efforts are needed to strengthen and reinforce those norms through biosecurity governance at the regional and national levels. In contrast to the United States, where biosafety and biosecurity have been developed on separate tracks, the European Union has pursued these two forms of governance in tandem. This approach has resulted in part from the history of attacks on laboratories by animal-rights extremists, which prompted the development of laboratory security measures in Europe long before they were adopted in North America.

In general, the EU has been less preoccupied than the United States with the threat of bioterrorism and has given priority to other biological risks, such as food safety.[49] European concerns in this area have focused on genetically modified organisms (GMOs) and crops; incidents of food contamination in Belgium, France, and the United Kingdom; and outbreaks of BSE (mad cow disease) and its human variant in the United Kingdom. This backdrop has made Europeans more skeptical about genetic engineering and assurances of its safety from developers and regulators. In contrast to the United States, EU regulations on recombinant DNA research are based in hard law and apply regardless of the source of funding. The value of this approach is that it provides more consistent oversight and greater assurance that all relevant laboratories will comply. Another factor influencing biosafety governance in Europe has been the EU's embrace of the "precautionary principle," which requires evidence that serious hazards will not materialize or can be controlled before a new technology is approved for broad public release.

A number of EU directives on biosafety provide guidelines for national implementation by member states. For example, EU Directive 2000/54/EC of September 18, 2000, sets out a legislative framework for protecting workers from risks related to occupational exposures to biological agents.[50] The directive includes a list of animal and human pathogens (but not GMOs), provides a model for risk assessment and biocontainment, and sets out employer obligations for worker safety and reporting. Individual EU countries have also adopted biosafety regulations that are consistent with this approach.

A series of terrorist attacks in Europe, including the transportation bombings in Madrid in 2004 and in London in 2005, caused the EU countries to become more concerned about biosecurity threats from nonstate actors.[51] In response to UN Security Council Resolution 1540, the EU adopted Regulation 428/2009 of August 2009.[52] This regulation creates lists of controlled goods that are subject to export restrictions and licensing, including dual-use biological and chemical materials and production equipment, and restricts the brokering and transit of such goods through EU territory. Member countries are legally bound to implement the regulation through national legislation, although they are free to adopt export controls that are stricter than the EU standard.

The EU customs security program, adopted in 2006, aims to create a harmonized, secure customs system to identify hazardous goods entering EU territory, thereby removing impediments to trade among member states.[53] Companies are required to submit information on exports before their arrival, and the EU has also established a secure electronic system for exchanging risk information. Although the tightening of European customs regulations promises to enhance security in a region with many borders and ports, the effective implementation of the new system requires the adoption of common standards and a substantial financial investment.

Israel

Israel conducts world-class biological research, but until recently the country had no biosecurity or dual-use risk governance measures either woven into its biosafety legislation or drafted specifically to address these risks. Moreover, Israel is not a party to the BWC because of its geopolitical situation in the Middle East. For many years, the military threats emanating from neighboring states focused Israel's efforts on biodefense, to the exclusion of governance mechanisms to manage the risks of biological agents, equipment, and know-how within its own borders.

Laboratory biosafety in Israel is regulated under the Safety Oversight Order for Medical, Biological, and Chemical Laboratories (2001). While no single oversight body is responsible for ensuring that laboratories adhere to these safety regulations, the Ministry of Industry, Labor, and Trade shares this responsibility with the Ministry of Health. Most Israeli research in the life sciences takes place in academic laboratories, which are subject to oversight by institutional biosafety committees and human and animal experimentation committees. At each institution, safety officials oversee research involving dangerous biological and chemical agents, human blood and tissue samples, and recombinant DNA. Laboratories are inspected regularly to ensure compliance with the biosafety regulations.

Israel is keenly aware of the need to keep dangerous substances, information, and equipment out of the hands of hostile actors. To that end, the Defense Export Control Law, a piece of dual-use export control legislation, came into effect in December 2007. This legislation created an Export Control Authority within the Israeli Ministry of Defense and also established control lists of human, animal, and plant pathogens and dual-use production equipment similar to those developed by the Australia Group. (Israel does not participate in the AG, nor is Israel a member of the CWC.)

U.S. and international initiatives in the field of dual-use governance increased Israel's awareness and commitment to developing a legal biosecurity framework. In

2005 the Israel Academy of Sciences and Humanities and the Israel National Security Council jointly established the Steering Committee on Issues in Biotechnological Research in an Age of Terrorism (COBRAT) to review Israel's biosecurity oversight practices. This committee published a report in 2008 that recommended several measures to ensure that dangerous organisms, information, and technologies do not fall into hostile hands, while permitting research and information exchange to flourish.[54] COBRAT's two main recommendations were to establish a regulatory system for research with dangerous biological agents and dual-use technologies and to educate the life sciences community about these risks.[55]

Similar to the NSABB in the United States, COBRAT proposed expanding the responsibilities of Israel's existing institutional biosafety committees to cover biosecurity and dual-use reviews. In response to the COBRAT recommendations, the Israeli parliament quickly adopted a national biosecurity law, the Regulation of Research into Biological Disease Agents Act of 2008 (Disease Agents Act). This act established a list of select viruses, bacteria, and toxins of bioterrorism concern, designated "Disease Agents"; defined certain types of dual-use research of concern; and created a national oversight authority, the Council for Dangerous Biological Agents, which regulates institutions that store or conduct research with Disease Agents. The Israeli minister of health and the director general of the Ministry of Health are jointly responsible for implementing the law.

With respect to research oversight, the Disease Agents Act applies to both private and public institutions and academic, industrial, and government laboratories that possess or work with Disease Agents. Institutions and individuals who wish to store or work with Disease Agents must be accredited by the Ministry of Health and obtain approval for proposed research from an institutional committee responsible for both biosafety and biosecurity issues. These committees assess the risk posed by a proposed experiment to the health, welfare, and security of the Israeli people. In determining whether the research should be permitted, the committee must assess its dual-use risks, for example, determining if it would increase the virulence or resistance to therapy of a Disease Agent or make it harder to detect or identify. The act prohibits research whose "sole purpose . . . is to cause or exacerbate a disease or illness or to impair the ability to prevent or treat it." Additionally, when research not involving designated Disease Agents produces unexpected results that could increase the virulence or contagiousness of a microbe in humans, the research must be stopped, and a request for approval submitted to the institutional committee.

As in other countries, Israel faces the challenge that researchers in the life sciences currently receive little or no formal education about dual-use risks and thus are generally unaware of the need for biosecurity measures. Accordingly, along with overseeing the implementation of biosecurity regulations at research institutions, the

Council for Dangerous Biological Agents seeks to raise the awareness of life scientists with respect to dual-use research and other biosecurity issues.

Singapore

In contrast to Israel's growing awareness of dual-use issues and swift legislative response, Singapore has a more traditional emphasis on biosafety and relatively little awareness of dual-use risks. A country with an advanced biotechnology industry, Singapore has ambitions to become a biomedical science hub of the Asia-Pacific region. It is no surprise, therefore, that Singapore has comprehensive biosafety legislation modeled on the WHO biosafety guidelines, as well as legislation governing the research, handling, and transport of GMOs.

In 2003 researchers in Singapore and other countries in the region contracted severe acute respiratory syndrome (SARS) while conducting laboratory research on the SARS virus after it had stopped circulating naturally.[56] This experience shaped the countries' priorities with respect to biosafety, biosecurity, and dual-use technologies. The laboratory-acquired SARS infections underlined the need for stringent oversight of biosafety procedures. Another consequence of the SARS experience has been that the safe handling of hazardous pathogens to protect public health and the environment is considered a critical component of biosecurity.

Singapore passed biosecurity legislation in 2006 in the form of the Biological Agents and Toxins Act.[57] This legislation specifies five schedules of biological agents and substances with different levels of control, based on risk assessment. The act provides mechanisms to track and inventory biological agents by controlling their import, possession, and transfer through a system of permits and approvals. Both the sender and the recipient of scheduled agents and toxins must be authorized to possess them. The Ministry of Health must certify all facilities that work with biological agents and toxins, and annual recertification is required.[58]

Recent studies suggest that life scientists in the Asia-Pacific region have a low awareness of dual-use issues. Instead concerns tend to center on preserving biodiversity, protecting public health, and maintaining agricultural security. One reason for this low awareness is a lack of academic courses that examine the dual-use risks associated with certain types of biotechnological research.[59]

Soft-Law and Informal Measures

One of the primary trends in the governance of dual-use technologies is the reinforcement of international treaties via domestic legislation that establishes biosafety and biosecurity measures at the national level. Another set of governance tools involves "soft-law" and informal measures that are not legally binding, such as professional guidelines, codes of ethics, and education and awareness raising. These measures focus on achieving self-governance by institutions and people with access to dual-use materials and equipment.[60] Soft-law measures are voluntary and lack strict enforcement mechanisms. They aim instead to create a culture of responsibility for dual-use research and to forestall more stringent government intervention.

For self-governance to succeed at controlling dual-use risks, there must be a community of practitioners with which individuals can self-identify. A profession is a legally mandated association whose members typically have a license and specialized knowledge or skills. The state grants each profession a monopoly over certain practices and powers of self-governance to align the expertise of the practitioners with the public good.[61] Because dual-use biological and chemical technologies are so diverse, however, the lack of a group identity rules out professionalization as a governance strategy in this area.

The commercial sector offers an alternative model of self-regulation centered on products and services rather than legally mandated professions. In the biotechnology industry, for example, companies involved in commercial gene synthesis have formed two associations, the International Association of Synthetic Biology and the International Gene Synthesis Consortium, which have prepared guidelines for their members to screen new customers and DNA synthesis orders to prevent the misuse of this technology for constructing dangerous pathogens from scratch. Participating gene synthesis companies realized that it was in their own long-term economic interest to help mitigate the security risks of gene synthesis technology. In response to this industry initiative, some of the major customers of synthetic DNA, such as large pharmaceutical companies, have pledged to purchase synthetic genes exclusively from firms that adhere to a code of responsible conduct, reinforcing the self-regulatory scheme.

The self-regulation by gene-synthesis companies provides a powerful model for contemporary governance that could theoretically be adapted to other emerging dual-use technologies. This model is anchored in market-based interests, however, and does not necessarily apply to the large pool of scientists and researchers who work in academia and the government. In these sectors, ethics education can help foster a culture of responsibility in the life sciences. The International Union of Microbiological Societies, following the lead of the American Society for Microbiology, has issued a code of ethics that prohibits its members from developing biological weapons.[62]

Nevertheless such codes have yet to be integrated into science education, professional development, or certification requirements. Indeed, it can be difficult to inculcate scientists with a sense of personal responsibility. Although medical students are bound by the Hippocratic oath to do no harm, science is often presented as a search for objective truth that is value neutral and unconstrained by ethical norms.

According to the InterAcademy Panel on International Issues, however, "Scientists have an obligation to do no harm. . . . They should therefore always bear in mind the potential consequences—possibly harmful—of their research and recognize that individual good conscience does not justify ignoring the possible misuse of their scientific endeavor."[63]

In 2009 the Federation of American Societies for Experimental Biology (FASEB) noted that "scientists who are educated about the potential dual-use nature of their research will be more mindful of the necessary security controls which strike the balance between preserving public trust and allowing highly beneficial research to continue."[64] Surveys indicate, however, that many researchers in the life sciences lack an awareness of dual-use concerns, including the risk of misuse associated with their own work.[65] Overcoming this deficit will require a commitment to ethics education for science and engineering students, as well as training in identifying and managing dual-use risks. The task is daunting because of the difficulty of defining dual-use research and the lack of experts in the field.[66]

To help contain dual-use risks, ethics education should be combined with mechanisms for reporting risks once they have been identified. Whenever a student or researcher suspects that a colleague is misusing a technology for harmful purposes or is willfully blind to the risks that his or her research might pose, a confidential channel should exist for relaying this information to the appropriate authorities so that they can act on it. Although medical students in the United States, Canada, Australia, and Europe are trained in professional ethics, it remains unclear how to motivate them to identify and report ethical breaches in view of the hierarchical culture that prevails in most laboratories and medical schools.

Critics of self-regulation call for reducing risk by bolstering soft-law and informal approaches with enforceable standards. EU member states, for example, do not consider self-governance and norm building to be viable alternatives to hard law.[67] In contrast, in the United States, historical deference to industry and the scientific community has created more space for self-regulation in lieu of binding legislation. Some preliminary studies suggest that formal oversight mechanisms may be less effective than the decision by individual scientists to forgo research with potential dual-use risks.[68]

Nevertheless the security risks of emerging technologies such as synthetic genomics may be sufficient to warrant a mixed-governance approach.[69] In such cases, there is a critical need for collaboration between the government and nongovernmental actors, including industry and scientific communities, as well as the ability to strike the appropriate balance between top-down and bottom-up regulatory tools. Mixed-governance strategies also provide greater flexibility to keep pace with the rapid evolution of emerging technologies.

Conclusions

The governance of dual-use biological and chemical technologies is grounded in international treaties such as the BWC and the CWC, which embody the international norm against the use of these technologies and materials for hostile purposes. Informal associations of like-minded states, such as the Australia Group, also bolster the nonproliferation regime by harmonizing national export controls on dual-use materials and technologies relevant to biological and chemical weapons. In recent years, high-profile terrorist attacks have shifted the focus of biosecurity activities from states to nonstate actors. National legislation to implement the BWC and the CWC, as well as UN Security Council Resolution 1540, has fostered efforts to harmonize and strengthen domestic safety and security regulations governing the use of biological pathogens, toxic chemicals, and GMOs. Although the EU and Singapore have focused primarily on biosafety measures and placed a lesser emphasis on biosecurity, the United States and Israel have created a dedicated set of laws and regulations to strengthen biosecurity.

Ongoing efforts to build a web of prevention through multiple, mutually reinforcing governance measures must include a greater awareness by industry and individual researchers of the dual-use risks associated with many emerging technologies and the recognition that they are the ultimate gatekeepers. If education is to become a powerful tool for technology governance, it must be coupled with the recognition that science is a morally bounded enterprise, and those who practice it have a responsibility to ensure that it is used for good, not ill.

Notes

1. Filippa Lentzos, "Countering Misuse of Life Sciences through Regulatory Multiplicity," *Science and Public Policy* 35, no. 1 (February 2008): 55–64.

2. Michael John Garcia, "Biological and Chemical Weapons: Criminal Sanctions and Federal Regulations," CRS Report for Congress (Washington, DC: Congressional Research Service, 2004).

3. Brian Rappert and Caitríona McLeish, eds., *Web of Prevention: Biological Weapons, Life Sciences, and the Governance of Research* (London: Earthscan, 2007).

4. Toxic Substances Control Act of 1976, 15 U.S.C. §2622 et seq.

5. Clean Air Act of 1970, 42 U.S.C. §7401 et seq.

6. Catherine Jefferson, "The Chemical and Biological Weapons Taboo: Nature, Norms, and International Law," D.Phil. diss., University of Sussex, 2009.

7. Protocol for the Prohibition of the Use in War of Asphyxiating, Poisonous, or Other Gases and of Bacteriological Methods of Warfare [Geneva Protocol], 1925.

8. Daniel H. Joyner, *International Law and the Proliferation of Weapons of Mass Destruction* (London: Oxford University Press, 2009), 88.

9. Nicholas Sims, "Legal Constraints on Biological Weapons," in *Deadly Cultures: Biological Weapons since 1945*, ed. Mark Wheelis, Lajos Rózsa, and Malcolm Dando (Cambridge, MA: Harvard University Press, 2006), 330.

10. President Gerald R. Ford, "Executive Order 11850—Renunciation of Certain Uses in War of Chemical Herbicides and Riot-Control Agents," April 8, 1975.

11. The BWC does not ban research, in part because of the difficulty of assessing whether or not it is being conducted for (illicit) offensive or peaceful purposes. Some commentators believe that research was excluded because a ban on offensive research would not be verifiable. See Nicholas Sims, "Banning Germ Weapons: Can the Treaty Be Strengthened?" *Armament and Disarmament Information Unit* 8, no. 5 (September–October 1986): 2–3.

12. Convention on the Prohibition of the Development, Production, Stockpiling, and Use of Chemical Weapons and on Their Destruction (CWC), September 3, 1992, United Nations Treaty Series (1974), 317.

13. UN Security Council Resolution 1540, April 28, 2004.

14. In 2004 the United States provided funding to the World Health Organization to develop guidelines for laboratory biosecurity. This effort led to the WHO manual *Biorisk Management: Laboratory Biosecurity Guidance*, WHO/CDS/EPR/2006.6 (Geneva, Switzerland: WHO, 2006).

15. European Committee for Standardization (CEN), "Laboratory Biorisk Management Standard," CEN Workshop Agreement 15793 (Brussels: CEN, 2008).

16. Organization for Economic Cooperation and Development, *OECD Best Practice Guidelines for Biological Resource Centers* (Paris: OECD, 2007).

17. Occupational Safety and Health Act of 1970, 29 U.S.C. §651 et seq.

18. U.S. Department of Health and Human Services, Centers for Disease Control and Prevention and National Institutes of Health, *Biosafety in Microbiological and Biomedical Laboratories*, 5th ed. (Washington, DC: U.S. Government Printing Office, 2007).

19. *Compliance through Science: U.S. Pharmaceutical Industry Experts on a Strengthened Bioweapons Nonproliferation Regime*, Henry L. Stimson Center Report No. 48, ed. Amy E. Smithson (September 2002), 45.

20. John H. Keene, "Ask the Experts—Non-compliant Biocontainment Facilities and Associated Liability," *Applied Biosafety Journal* 11, no. 2 (2006): 99–102.

21. Department of Health and Human Services, "Possession, Use, and Transfer of Select Agents and Toxins," 42 CFR Part 72 and 73.

22. National Institutes of Health, *NIH Guidelines for Research involving Recombinant DNA Molecules (NIH Guidelines)*, as amended, Federal Register 74, no.182, September 22, 2009.

23. National Institutes of Health, "Notice Pertinent to the September 2009 Revisions of the NIH Guidelines for Research Involving Recombinant DNA Molecules," http://www.ecu.edu/cs-dhs/prospectivehealth/upload/NIH_Gdlines_2002prn-1.pdf.

24. U.S. Department of Health and Human Services, "Protection of Human Subjects," 45 *Code of Federal Regulations* 46, revised July 14, 2009.

25. Tora K. Bikson, Ricky N. Blumenthal, Rick Eden, and Patrick P. Gunn, eds., "Ethical Principles in Social-Behavioral Research on Terrorism: Probing the Parameters," RAND Working Paper, WR-490-4-NSF/DOJ (January 2007), 119.

26. Lori P. Knowles, "The Governance of Reprogenetic Technologies: International Models," in *Reprogenetics: Law, Policy, and Ethical Issues*, ed. Lori P. Knowles and Gregory E. Kaebnick (Baltimore, MD: Johns Hopkins University Press, 2007), 127–129.

27. Jonathan B. Tucker, "Preventing the Misuse of Pathogens: The Need for Global Biosecurity Standards," *Arms Control Today* 33, no. 5 (June 2003): 3–10.

28. Uniting and Strengthening America by Providing Appropriate Tools Required to Intercept and Obstruct Terrorism Act of 2001 (USA PATRIOT Act), Pub. L. No. 107-56, October 12, 2001. These regulations strengthen a system of rules governing the importation and transport of pathogens: the CDC certifies facilities to receive and handle dangerous pathogens and administers regulations governing the importation of etiologic agents of human disease (as regulated in 42 CFR 72 and 42 CFR 71 and 71.54); APHIS oversees regulations regarding the importation of etiological agents of livestock, poultry, and other animal diseases and the federal plant pest regulations (respectively, see 9 CFR 92, 94, 95, 96, 122, and 130 and 7 CFR 330). The Commerce Department's export control rules are specified in 15 CFR 730–799, and the U.S. Postal Service's regulations on mailing etiologic agents are in 39 CFR 111. The regulations of the U.S. Department of Transportation are located in 19 CFR 171–178.

29. David P. Fidler and Lawrence O. Gostin, *Biosecurity in the Global Age: Biological Weapons, Public Health, and the Rule of Law* (Stanford, CA: Stanford University Press, 2008), 59–73.

30. Harvard Sussex Program on CBW Disarmament and Arms Limitation, "Draft Convention on the Prevention and Punishment of the Crime of Developing, Producing, Acquiring, Stockpiling, Retaining, Transferring, or Using Biological or Chemical Weapons," http://www.sussex.ac.uk/Units/spru/hsp/documents/Draft%20Convention%20Feb04.pdf.

31. Antiterrorism and Effective Death Penalty Act of 1996, Public Law No. 104-132, 110 Stat. 1214.

32. Public Health Security and Bioterrorism Preparedness and Response Act of 2002, 42 U.S.C. §262a.

33. Agricultural Bioterrorism Protection Act of 2002, 7 U.S.C. §8401.

34. National Research Council, *Sequence-Based Classification of Select Agents: A Brighter Line* (Washington, DC: National Academies Press, 2010).

35. National Science Advisory Board for Biosecurity, *Proposed Framework for the Oversight of Dual Use Life Sciences Research: Strategies for Minimizing the Potential Misuse of Research Information* (Bethesda, MD: National Institutes of Health, June 2007), 17.

36. U.S. Congress, Congressional Research Service, "Oversight of Dual-Use Biological Research: The National Science Advisory Board for Biosecurity," *CRS Report for Congress*, RL33342, April 27, 2007.

37. Brian Rappert, "Education as . . . ," in *Education and Ethics in the Life Sciences: Strengthening the Prohibition of Biological Weapons*, ed. Brian Rappert (Canberra, Australia: ANU E Press, 2010), 13–15.

38. National Research Council, Committee on Research Standards and Practices to Prevent the Destructive Application of Biotechnology, *Biotechnology Research in an Age of Terrorism* (Washington, DC: National Academies Press, 2004), 114–115.

39. Journal Editors and Authors Group, "Statement on Scientific Publication and National Security," *Science* 299 (February 21, 2003): 1149.

40. National Research Council, Committee on Genomics Databases for Bioterrorism Threat Agents, *Seeking Security: Pathogens, Open Access, and Genome Databases* (Washington, DC: National Academies Press, 2004).

41. U.S. National Academies, Committee on a New Government-University Partnership for Science and Security, *Science and Security in a Post 9/11 World: A Report Based on Regional Discussions between the Science and Security Communities* (Washington, DC: National Academies Press, 2007).

42. Homeland Security Act of 2002, Pub. L. 107-296, November 25, 2002.

43. The Export Administration Act of 1979 lapsed in August 2001 and has not been renewed by Congress. However, the Export Administration Regulations have remained in effect under Executive Order 13222, issued on August 17, 2001, pursuant to the International Emergency Economic Powers Act and extended annually by the president.

44. Authority of the Office of Export Enforcement, Bureau of Industry and Security, Customs Offices and Postmasters in Clearing Shipments, 15 C.F.R. Part 758.7.

45. Office of Inspector General, Department of Homeland Security, "Review of Controls over the Export of Chemical and Biological Commodities," OIG-05-21, June 2005, http://www.dhs.gov/xoig/assets/mgmtrpts/OIGr_05-21_Jun05.pdf.

46. *Export Administration Regulations*, §734.2(b)(2)(ii).

47. For information on the Australia Group, see http://www.australiagroup.net.

48. Jonathan B. Tucker, "Strategies to Prevent Bioterrorism: Biosecurity Policies in the United States and Germany," *Disarmament Diplomacy*, no. 84 (Spring 2007): 36–47.

49. See Alexander Kelle, "Synthetic Biology and Biosecurity Awareness in Europe," *Bradford Science and Technology Report*, no. 9 (November 2007): 9.

50. Directive 2000/54/EC of the European Parliament and of the Council of 18 September 2000, on the Protection of Workers from Risks Related to Exposure to Biological Agents at Work (seventh individual directive within the meaning of Article 16(1) of Directive 89/391/EEC), *Official Journal L* 262, October 17, 2000, 21–45.

51. Caitríona McLeish and Paul Nightingale, "Biosecurity, Bioterrorism, and the Governance of Science: The Increasing Convergence of Science and Security Policy," *Research Policy* 36 (2007): 1640.

52. Council Regulation (EC) No. 428/2009 of May 5, 2009, "Setting Up a Community Regime for the Control of Exports, Transfer, Brokering, and Transit of Dual-Use Items," *Official Journal L* 134/1, May 29, 2009.

53. Regulation (EC) No. 648/2005 of the European Parliament and the Council of 13 April 2005 amending Council Regulation (EEC) No. 2913/92 establishing the Community Customs Code.

54. Steering Committee on Issues in Biotechnological Research in an Age of Terrorism, Report by the Israeli Academy of Sciences and Humanities and the Israeli National Security Council, 2008.

55. David Friedman, Bracha Rager-Zisman, Eitan Bibi, et al., "The Bioterrorism Threat and Dual-Use Biotechnological Research: An Israeli Perspective," *Science and Engineering Ethics* 16, no. 1 (July 2008): 85–97.

56. Tin Tun, Kristen E. Sadler, and James P. Tam, "Biological Agents and Toxins Act: Development and Enforcement of Biosafety and Biosecurity in Singapore," *Applied Biosafety* 12, no. 1 (2007): 39–43.

57. Biological Agents and Toxins Act, 2006, C.24A.

58. Tun, Sadler, and Tam, "Biological Agents and Toxins Act."

59. Masamichi Minehata, "An Investigation of Biosecurity Education for Life Scientists in the Asia-Pacific Region," University of Exeter and University of Bradford, 2010.

60. InterAcademy Panel on International Issues, "Statement on Biosecurity," November 7, 2005.

61. Laura Weir and Michael J. Selgelid, "Professionalization as Governance Strategy for Synthetic Biology," *Systems and Synthetic Biology* 3 (2009): 91–97.

62. International Union of Microbiological Societies, "Code of Ethics against Misuse of Scientific Knowledge: Research and Resources," http://www.iums.org/about/Codeethics.html. American Society for Microbiology, "Code of Ethics" (revised 2005), http://www.asm.org/ccLibraryFiles/FILENAME/000000001596/ASMCodeofEthics05.pdf.

63. InterAcademy Panel on International Issues, "Statement on Biosecurity," November 7, 2005.

64. Federation of American Societies for Experimental Biology, "Statement on Dual Use Education," http://www.faseb.org/portals/0/pdfs/opa/2009/FASEB_Statement_on_Dual_Use_Education.pdf.

65. Malcolm R. Dando, "Dual-Use Education for Life Scientists," *Disarmament Forum*, no. 2 (2009): 41–44.

66. National Science Advisory Board for Biosecurity, *Strategic Plan for Outreach and Education on Dual Use Issues* (Washington, DC: NSABB, 2008).

67. Agomoni Ganguli-Mitra, Markus Schmidt, Helge Torgersen, et al., "Of Newtons and Heretics," *Nature Biotechnology* 27 (2009): 321–322.

68. Rappert, "Education as. . . ," 8–20.

69. For a discussion of governance options for synthetic genomics, including mixed approaches, see Michele Garfinkel, Drew Endy, Gerald L. Epstein, and Robert M. Friedman, *Synthetic Genomics: Options for Governance* (J. Craig Venter Institute, MIT, and CSIS, October 2007).

4

The Decision Framework

Jonathan B. Tucker

This chapter presents a decision framework that policy makers can use to assess the risk that individual emerging technologies will be misused for hostile purposes, and to develop tailored governance strategies. The model was developed through an iterative process that combined deductive reasoning with feedback from the analysis of empirical case studies. For reasons of clarity, the decision framework is first discussed in this chapter and then applied to fourteen emerging dual-use technologies in the following section.

The decision framework comprises three interconnected processes: (1) technology *monitoring* to detect emerging dual-use innovations with a potential risk of misuse, (2) technology *assessment* to determine the likelihood of misuse and the feasibility of regulation for individual technologies, and (3) the selection of *governance* measures based on the technology assessment and a cost–benefit analysis. These three processes and the methodology for applying the decision framework are explained in detail hereafter.

Technology Monitoring

A prerequisite for effective technology governance is a technology-watch program that monitors emerging technologies in the public and private sectors and identifies the potential for misuse. Some rudimentary mechanisms for technology monitoring already exist. The regular five-year review conferences of the Biological Weapons Convention (BWC) and the Chemical Weapons Convention (CWC) include surveys of advances in science and technology that have implications for the respective treaty regimes. Given the rapid pace of developments, however, the BWC expert Nicholas Sims contends that "five years is too long an interval" for the science and technology reviews and that they should be conducted every year in an expert forum.[1]

A few other organizations also conduct assessments of dual-use technologies in the biological and chemical fields. The Australia Group has established ad hoc committees to review certain technologies of concern, such as chemical microreactors and gene synthesis, to advise on whether these items should be added to the group's harmonized export-control lists. Independent scientific advisory bodies such as the National Research Council (an arm of the U.S. National Academies) and the British Royal Society also assess emerging technologies when asked to do so by their governments, but they do not have an ongoing mandate to perform this function. Finally, a small number of scholars in academia and the think-tank world monitor emerging technologies and assess their dual-use risks. Examples include the Program on Emerging Technologies (PoET) at the Massachusetts Institute of Technology; the Synthetic Biology Engineering Research Center (SynBERC) at the University of California, Berkeley; the Center for Nanotechnology in Society at Arizona State University; the Synthetic Biology Project at the Woodrow Wilson International Center for Scholars in Washington, D.C.; and the Action Group on Erosion, Technology, and Concentration (ETC Group) in Ottawa, Canada.

In the future, it would be desirable to establish an institutionalized technology-watch program that is permanent, centralized, and comprehensive in scope; that draws on a wide range of expertise from government, industry, and academia; and that reports to a key agency involved in regulatory decision making, such as the Office of Management and Budget, or an interagency group that exerts significant influence on U.S. government regulatory processes.

Technology Assessment

Once an emerging dual-use technology has been identified, it is important to assess its characteristics in a systematic manner. As illustrated in figure 4.1, the decision framework focuses on two key dimensions of a technology: its risk of misuse and its governability. Each of these dimensions is evaluated with a set of specific parameters.

Assessing the Risk of Misuse
The risk of misuse is assessed on the basis of four parameters: accessibility, ease of misuse, magnitude of potential harm resulting from misuse, and imminence of potential misuse.

Accessibility This parameter measures how easy it is to acquire the technology. The first step in the misuse of a technology is obtaining the hardware, software, and intangible information that enable its use. These components and information may be commercially available, proprietary if developed in the private sector, or restricted because of classification or some other reason. The accessibility parameter also takes into account the amount of money needed to purchase the technology and whether this level of expenditure is within the means of an individual, a group, or nation-state. (Of course, a scientist working in a research

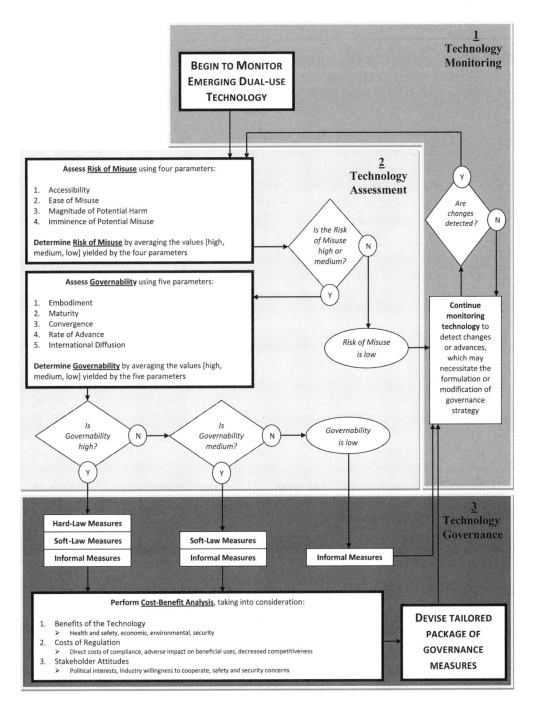

Figure 4.1
Decision framework for technology governance.

laboratory could potentially exploit available equipment at minimal personal expense.) Another component of accessibility is the dependence of a given dual-use technology on other technologies, such as those associated with agent delivery. Thus, in some cases, the acquisition of a technology also requires access to its upstream technologies, and this requirement is also included in the accessibility parameter.

Ease of misuse   This parameter considers the level of expertise and tacit knowledge required to master the technology, as well as the extent to which the technology is becoming deskilled and hence available to individuals with less formal expertise and hands-on experience. Ease of misuse accounts for the fact that simply acquiring a technology is not sufficient for its misuse; an actor must also possess the appropriate types and level of expertise.

Magnitude of potential harm resulting from misuse   This parameter is a function of both the technology itself and the vulnerability of the likely targets. The magnitude of potential harm includes the approximate number of deaths and injuries that could result from malign use; the economic costs associated with an incident, including mitigation and cleanup; and the societal effects of an attack, such as disruption, terror, and loss of faith in government. In particular, this parameter considers the extent to which a technology offers a qualitative or quantitative enhancement in the ability to do harm over existing technologies.

Imminence of potential misuse   This parameter indicates how soon a malicious actor seeking to cause harm could exploit the technology based on its current level of maturity. In general, during the early phase of research and development, a technology remains inchoate, and the risk of misuse is low. As the technology matures, however, it eventually reaches a stage at which its potential for misuse is more manifest, thereby increasing the risk. For example, DNA synthesis technology has now advanced to the point where it has become possible to synthesize from scratch any pathogenic virus whose genetic sequence is known.

Each of the four parameters describing the risk of misuse is ranked on a three-level ordinal scale (high, medium, and low). The four scores are then averaged to provide a rough estimate of the level of risk associated with a given technology. This method provides a good indication of whether or not the technology in question warrants the application of governance measures. In general, an aggregate score of high means that the technology has both an imminent risk of misuse and a significant potential for large-scale harm, whereas a score of low means that the risk posed by the technology either lies far in the future or is not significant in terms of the potential for large-scale harm, or both. Technologies with an aggregate score of medium fall somewhere between the high and low extremes.

In general, to meet the criteria for an aggregate risk score of high or medium, an emerging technology must also have characteristics that provide a significant qualitative or quantitative increase in capacity to cause harm over existing dual-use technologies. For example, the ability to synthesize the entire genomes of dangerous viral pathogens from scratch entails a new and salient risk of misuse that goes beyond what is already possible with conventional recombinant DNA technology, and thus warrants some type of additional governance. The level of concern is particularly great with respect to the possible re-creation of extinct pathogens (such as the Spanish influenza virus) or eradicated pathogens (such as the smallpox virus) that no longer exist in nature and are restricted to a few high-security labs.

An example of an emerging dual-use technology that lies at the low end of the risk spectrum is gene therapy. The ease of misuse of this technology is low because using viral vectors to deliver harmful genes to a target population would require expertise and tacit knowledge that do not currently exist. Although the accessibility and the magnitude of potential harm of gene-therapy technology are medium, the imminence of potential misuse is low because the technology is not advanced enough to be effectively weaponized.

The decision framework forwards technologies that have risk scores of high and medium to the next step of technology assessment to determine what types of governance measures should be applied. If the risk score of an emerging technology is low, it does not currently warrant the development of governance measures. In such cases, the decision framework will conclude the analysis without proceeding to the next step. Nevertheless it is prudent to continue monitoring the technology in case it acquires new capabilities in the future that could significantly increase its risk of misuse. (This return to the technology-monitoring process is represented by the feedback loop in figure 4.1.)

An important issue in assessing dual-use risk is the role of intent, which varies according to the potential user. Some evidence suggests, for example, that state actors may be attempting to militarize certain emerging technologies. Nevertheless, making "intent to misuse" a standard element of the decision framework would require speculation about human motives that is unsupported by the available evidence, particularly with respect to nonstate actors—unless, of course, one has access to secret intelligence. In most cases, the analysis of intent is possible only in retrospect, through historical studies in which the analyst has access to declassified documents or can interview key decision makers years after the fact.

Because of the difficulty of predicting that someone will be motivated to misuse a particular technology for harmful purposes within a specified time frame, the decision framework simply presumes the existence of malign intent on the part of certain potential actors, such as outlaw states or terrorists, and views them as continually probing for enhanced weapons capabilities and weak spots in the governance structure. If hostile intent is treated as a constant, then what varies from case to case are the accessibility of the technology to various actors, the extent to which misuse is technically feasible, the magnitude of the potential harm resulting from misuse, and the imminence of potential misuse based on the state of the technology. In sum, although the intent of an actor is critical to whether or not a technology is actually misused, the decision framework does not include intent as a variable because it is practically impossible to predict.

Assessing Governability

If the risk-assessment process described in the previous section determines that an emerging technology has a high or medium risk of misuse, the decision framework proceeds to the next step: determining the extent to which the technology is susceptible to various types of governance measures. The dimension of governability is measured according to five parameters: embodiment, maturity, convergence, rate of advance, and international diffusion.

Embodiment   Some technologies consist primarily of hardware, others are based largely on intangible information, and still others are a hybrid of the two. Innovations embodied in the form of hardware are relatively easy to govern through measures such as registration and export controls. In contrast, technologies embodied in the form of intangible information are less governable because data can be transmitted easily and often undetectably, particularly in electronic form. Formal regulation is particularly challenging with respect to so-called enabling technologies, which are based largely on intangible information and are widely employed in research laboratories around the world.

Maturity   The second parameter affecting governability is the maturity of a technology, namely, its position in the development pipeline extending from basic research to commercialization. Examples of different levels of maturity include early research and development, advanced development and prototyping, early marketing, and widespread commercial availability.

Convergence   This parameter refers to the number of different disciplines that are brought together to create a new device or technology.[2] So-called NBIC technologies, for example, combine elements of nanotechnology, biotechnology, information technology, and cognitive neuroscience.[3] Similarly, nanobiotechnology involves the convergence of nanotechnology and biotechnology, and synthetic biology combines elements of nanoscale biology, bioinformatics, and engineering into a new discipline for the design and construction of biological parts and devices that can perform useful tasks. Because each discipline has its own professional community, culture, lexicon, and level of awareness of dual-use issues,

highly convergent technologies that draw on multiple disciplines are more difficult to govern than technologies derived from only one or two disciplines. Nevertheless even a highly convergent technology may have critical elements that
provide effective targets for policy intervention, thereby increasing governability.
Synthetic biology, for example, relies heavily on automated DNA synthesis, a
technology that is the focus of both national and international governance
measures.

Rate of advance   This parameter refers to whether the effectiveness of a technology (as measured by its reliability, speed, throughput, accuracy, or cost) is increasing linearly, increasing exponentially, plateauing, or declining over time. Some
technologies progress slowly and incrementally until they reach a threshold of
speed, throughput, or capacity at which their dual-use potential becomes manifest. Others may advance rapidly and attain more immediate dual-use potential.
In general, the faster a technology is advancing, the harder it is for governance
measures to keep pace.

International diffusion   Emerging technologies vary greatly in the extent to
which they are available in international markets. Some technologies are limited
to one or a few countries or may be secret or under patent protection, while other
technologies are widely available. The annual International Genetically Engineered Machines (iGEM) competition held at the Massachusetts Institute of
Technology has accelerated the diffusion of synthetic biology by attracting the
participation of student teams from around the world. In general, the smaller the
number of countries that have access to a technology, the easier it is to govern
because of the reduced need for coordination and harmonization. In contrast, the
governance of a widely diffused technology requires the international harmonization of regulations or guidelines, which can be a difficult task.

As with the four parameters defining the risk of misuse, the five parameters
defining governability are scored on a simple ordinal scale (high, medium, and low)
and then averaged to give a rough assessment of a technology's susceptibility to
governance. Determining these scores is less clear-cut than with the risk of misuse,
however, because the five parameters do not correlate linearly with governability.
Instead three of the variables (convergence, rate of advance, and international diffusion) are inversely related to governability, while the other two variables (embodiment and maturity) have a more complex relationship to governability. In the case
of embodiment, for example, the governability score of hardware-based technologies
is high, that of information-based technologies is low, and that of hybrid technologies is medium.

With respect to maturity, the sweet spot of maximum governability occurs during
the brief period between advanced development and early commercialization, when

Table 4.1
Relationships between parameters and level of governability

	Embodiment	Maturity	Convergence	Rate of advance	International diffusion
Low Governability	Intangible information	Not mature	High level of convergence	High rate of advance	High level of international diffusion
Medium Governability	Hybrid (combines intangible information and hardware)	Very mature	Moderate level of convergence	Moderate rate of advance	Moderate level of international diffusion
High Governability	Hardware	Moderately mature	Low level of convergence	Low rate of advance	Low level of international diffusion

the dual-use risks of a technology are fairly well understood, but the number of manufacturers and consumers is limited. Accordingly, when using technological maturity as a parameter of governability, early research and development scores low, advanced development and early commercialization score high, and widespread commercial availability scores medium. In general, technologies that are intangible, inchoate, highly convergent, advancing rapidly, or pervasive have lower levels of governability. The relationships between each of the five parameters and the governability score of an emerging technology are summarized in table 4.1.

In part II of the book, which applies the decision framework to fourteen contemporary case studies of emerging technologies in the biological and chemical fields, each technology is assessed with regard to its risk of misuse and governability using the parameters described in this chapter. The technology-assessment findings for all fourteen case studies, including their aggregate risk of misuse and governability scores, are summarized in chapter 21.

Selection of Governance Measures

Once analysis has identified the subset of emerging dual-use technologies that pose a significant risk of misuse *and* are reasonably susceptible to governance, the next step in the decision framework is to devise a tailored package of governance measures.

As noted in chapters 2 and 3, there are three broad categories of such measures. The first category involves hard-law or legally binding measures, such as national legislation, formal regulations, and statute-based export controls. The second category of technology governance involves soft-law measures that are not legally binding. Examples of such measures are voluntary guidelines promulgated by governments, self-regulatory mechanisms devised by industry associations and nongovernmental organizations, and international standards such as those established by the International Standards Organization (ISO) or the European Committee for Standardization.

Finally, the third category of technology governance involves informal measures, which apply to individuals and groups rather than to countries and institutions. Examples include professional codes of conduct, dual-use education and awareness-building efforts, and the creation of secure channels through which whistle-blowers can report misuse without risking retaliation. Informal measures seek to build a culture of security and responsibility within a relevant professional community, such as research scientists or suppliers of dual-use equipment and services.

Because emerging dual-use technologies have varying levels of risk and governability, they are best managed by tailored packages of hard-law, soft-law, and informal measures. An average governability score of high means that a technology is susceptible to the full range of governance measures, including binding regulations, voluntary guidelines, and professional codes of conduct. An average governability score of medium means that only soft-law and informal measures are feasible. Finally, an average governability score of low means that only informal measures are available. Wherever possible, it is desirable to pursue synergies among the different modes of governance by combining them in mutually reinforcing ways to create a web of prevention.

As the first step in selecting a package of governance measures, it is important to identify any existing laws, regulations, or guidelines that already cover an emerging technology before devising new ones. Such measures may apply either to the technology itself (e.g., the registration of DNA synthesizers) or to the products resulting from its use (e.g., the screening of gene-synthesis orders). It is also important to assess how well the existing laws and regulations are being carried out. Just because a law is on the books does not mean it is being implemented effectively. Important factors include which agency is responsible for regulatory enforcement and whether it suffers from inefficiencies or conflicts of interest.

Empirical research has shown, for example, that the Institutional Biosafety Committees that oversee recombinant-DNA research in the United States vary widely in effectiveness and rigor.[4] Similarly, a U.S. federal advisory committee, the National Science Advisory Board for Biosecurity (NSABB), has a potential conflict of interest because it is administered by the National Institutes of Health, which promotes biomedical research and thus has a strong bias against formal regulations that could reduce scientific productivity, raise the costs of research, and hamper innovation.[5]

Finally, it is important to assess the extent to which existing governance measures can be applied to an emerging technology, as such measures were generally designed with earlier innovations in mind.

Intervention Points and Targets

When thinking about the governance of emerging dual-use technologies, it is useful to visualize the process of using a technology as a stream, with intervention points at various locations along its banks. In the case of genome synthesis, for example, one could intervene upstream by requiring the registration of all new DNA synthesizers, at midstream by licensing commercial suppliers who use DNA synthesizers to produce genetic sequences to order, or downstream by screening DNA synthesis orders to ensure that they do not contain pathogenic sequences. A second type of upstream intervention would be to design DNA synthesizers that are "proliferation resistant" through programming that renders them incapable of synthesizing pathogenic sequences.[6]

Another key variable is the target of a governance measure. Technological research and development takes place in four different settings: private industry, government research laboratories, universities or nonprofit research institutes, and entities outside a formal institutional context, such as do-it-yourself biology (DIYbio). In most of these settings, the R&D process proceeds in secret until the technology is formally disclosed. This disclosure may take various forms: a patent application, the launch of a commercial product, the declassification or licensing of a new technology, the publication of a paper in a scientific journal, or the presentation of findings at a professional conference. A government laboratory, for instance, might declassify a technology and license it for commercial use.

Although most governance measures are relevant to the postdisclosure phase, prediscolsure R&D may also be a suitable target for intervention in some cases. It is therefore possible to imagine a regulatory framework that would require both government and nongovernment laboratories to report any emerging technology that poses dual-use risks before it is publicly released.[7] The U.S. Patent and Trademark Office, for example, reviews patent applications for innovations that may have implications for national security, and the agency has the authority to classify patents in exceptional cases.[8]

Table 4.2 provides a menu of possible governance measures for emerging dual-use technologies, organized according to the mode of governance and the target of the policy intervention, which may be a federal or state government, a company or institution, an individual, a product or piece of hardware, or a piece of intangible knowledge. A tailored package of governance measures for a given technology may incorporate several different targets and intervention points, which ideally should reinforce one another. For example, efforts to govern the synthesis of bioactive

Table 4.2
Menu of possible governance measures

Intervention target	Hard law (civil and criminal statutes, binding regulations)	Soft law (voluntary guidelines, self-governance mechanisms)	Informal (codes of conduct, education and training, etc.)
State	Multilateral treaty Framework convention National export controls End-user certificates Economic sanctions	Voluntary guidelines and best practices Multilateral export-control regimes Oversight mechanisms Advisory Committees (e.g., Recombinant DNA Advisory Committee)	International norms Economic pressure from customers
Institution	Registration, accreditation, or certification Mandatory data declarations Onsite inspections Institutional Review Boards Conditions on funding	Informal oversight mechanisms Industry best practices and voluntary standards Institutional Biosafety Committees	Economic pressure from customers Economic boycotts Outreach and awareness raising
Individual	Security vetting or screening Registration, accreditation, or certification	Voluntary guidelines and best practices	Hippocratic oath, professional codes of conduct Dissent and whistle-blowing channels Education and awareness building
Product	Licensing or registration Select Agent Rules	ISO standards Screening of product orders (e.g., screening of DNA synthesis orders)	Sales awareness training ("know thy customer")
Knowledge	Classification Deemed exports	Prepublication review	Information-sharing Transparency

peptides might target the firms that manufacture and sell peptide synthesizers, the companies that produce peptides to order in significant quantities, and the companies and scientists who work with bioactive peptides in the research, medical, and pharmaceutical communities.

Cost–Benefit Analysis

After identifying a set of potentially suitable governance measures, the decision framework subjects them to a cost–benefit analysis. This step involves weighing the security benefits of each governance measure against the economic and other costs, including whether certain expected benefits of the technology (health, economic, or environmental) must be forgone as a result of the proposed measure. Other factors that should be taken into account when conducting a cost–benefit analysis are the likely effectiveness of the proposed measure at reducing the risk of misuse and the indirect costs of implementation, such as decreased international competitiveness. In general, approaches to governance based on denial, such as export controls and interdiction, are most effective in the early stages of technology development, when few suppliers and users exist. Over time, however, the diffusion and globalization of the technology gradually reduce the utility of this approach.

The cost–benefit analysis of various combinations of governance measures should be performed in an iterative manner, with the goal of identifying a package of hard-law, soft-law, and informal measures that offers the greatest overall benefit at the lowest cost. Technologies for which *only* informal governance measures are being considered need not be subjected to a comprehensive cost–benefit analysis because the costs of such governance measures (e.g., professional codes of conduct, dual-use education and awareness-building efforts, and whistle-blowing channels) are negligible.

While synergies among governance measures are desirable, trade-offs are inevitable. In some cases, only partial steps to lower the risk of misuse may be feasible, because the costs of more stringent measures would outweigh the benefits. When the dual-use risks of a technology are particularly serious and imminent, however, policy makers may be willing to tolerate high governance costs to ensure an adequate level of security.

Another factor influencing the cost–benefit analysis of technology governance measures is the perception of risk. As noted in chapter 2, a large literature in the fields of psychology and behavioral economics describes how nonrational factors, such as cognitive biases, can distort risk judgments. Parochial interests may also influence how risks are portrayed. On the one hand, the developers of a new technology tend to emphasize its benefits, both immediate and hypothetical, while downplaying the risks that it may pose in the future. On the other hand, civil-society groups tend to exaggerate the potential harm from emerging technologies to mobilize public opposition. Biased risk assessments create an uneven playing field in which the burden of scientific proof tends to fall disproportionately on the regulator.

Another real-world constraint on cost–benefit analysis is the fact that any governance mechanism, whether or not it is legally binding, must be implemented in a political environment in which various stakeholders and interest groups will have their say. When devising a package of governance measures for an emerging dual-use technology, policy makers generally take into account the views of major stakeholders, including the willingness of the affected industry to accept a particular regulatory scheme. These competing interests must be balanced through a process of negotiation and compromise. Ideally, the end result is an effective governance strategy that benefits the public good while being acceptable to all the major players. In the real world, of course, suboptimal outcomes are common, and powerful political or economic interests often trump concerns about human health and safety.

In summary, the decision framework for the governance of emerging dual-use technologies consists of the following steps:

(1) Monitor technological developments in academia, government, and private industry with the goal of identifying emerging technologies in the biological and chemical fields that have a potential for misuse.

(2) Assess the risk of misuse of an emerging technology according to four parameters: accessibility, ease of misuse, magnitude of potential harm, and imminence of potential misuse.

(3) If the aggregate risk of misuse is low, there is no urgent need to devise governance measures, but the technology should continue to be monitored in case its potential for misuse increases over time.

(4) If the aggregate risk of misuse is medium or high, go on to assess the governability of the technology according to five parameters: embodiment, maturity, convergence, rate of advance, and international diffusion.

(5) If the aggregate governability of the technology is low, focus on informal governance measures.

(6) If the aggregate governability of the technology is medium, consider soft-law governance measures in addition to informal governance measures.

(7) If the aggregate governability of the technology is high, consider the full spectrum of governance measures: informal, soft law, and hard law.

(8) If the risk of misuse associated with the technology appears to be exceptionally grave and imminent, consider more stringent governance measures than the decision framework would lead one to adopt.

(9) Based on a cost–benefit analysis, assemble a tailored package of governance measures that reduces the risk of misuse at acceptable cost and in a manner that is acceptable to the major stakeholders.

Testing the Decision Framework

To test the utility of the decision framework, part II applies the model to case studies of fourteen dual-use technologies that have recently emerged or are still emerging. The starting point for selecting these cases was a 2006 report by the U.S. National Research Council titled *Globalization, Biosecurity, and the Future of the Life Sciences*. This study, directed by the microbiologists Stanley Lemon and David Relman, looked beyond research with dangerous pathogens to examine a variety of emerging biological and chemical technologies that might be exploited for hostile purposes.[9] The Lemon-Relman report classified these technologies into four categories based on their shared functional characteristics:

(1) Technologies that generate collections of molecules with greater structural and biological diversity than those found in nature (e.g., combinatorial chemistry, directed molecular evolution, bioprospecting);

(2) Technologies that create novel but predetermined molecular or biological diversity (e.g., rational design of small molecules that bind to protein targets, genetic engineering of bacteria and viruses, synthetic biology);

(3) Technologies that facilitate the manipulation of complex biological systems (e.g., systems biology, RNA interference, genomic medicine, modification of homeostatic systems, bioinformatics); and

(4) Technologies for the production, delivery, and packaging of biological products (e.g., production of drugs in transgenic plants, aerosol drug-delivery systems, microencapsulation, microfabrication technologies, nanotechnology, gene therapy).[10]

In addition to a number of emerging technologies identified in the Lemon-Relman report, the fields of chemistry, biochemistry, molecular genetics, biomedicine, and neuroscience provided additional case studies. Viral genome synthesis and parts-based synthetic biology overlap extensively but are also prominent individually, which merited the commissioning of separate case studies for these two fields. Ensuring that the technologies being analyzed were roughly comparable in scope was an important criterion for the selection of cases. Nanotechnology, for example, is a new discipline based on the manipulation of materials at the nanometer scale.[11] Nanotechnology did not make the cut as a case study, however, because it is such a large field, encompassing many applications with different levels of dual-use risk that it does not compare directly to the other emerging technologies.

Another criterion in selecting the cases was to ensure a high level of variance across several parameters to test the utility of the decision framework. Accordingly, the cases were selected to include (1) technologies with different levels of maturity, from early research and development to wide commercial availability; (2) technologies based primarily on hardware, intangible information, and a hybrid of the two; and (3) technologies that are advancing and diffusing at different rates. In addition, some of the chosen technologies are based on cutting-edge science, whereas others involve the application of existing knowledge.

Finally, some of the innovations covered by the case studies are technologies in the strict sense of the word, while others are more science based. Indeed, an important characteristic of the life sciences today is that the traditional distinction between science and technology has become increasingly blurred. According to the traditional paradigm, advances in science—the understanding of how nature works—lead to technological innovations, meaning the application of scientific knowledge to solve practical problems. In fields such as molecular biology, however, the distance between knowledge and application is so short as to make it hard to distinguish between the two. RNA interference, for example, began with the scientific discovery of a natural phenomenon, which was quickly converted into a powerful tool for scientific research. Today RNA interference has a wide variety of medical and industrial applications, some of them with dual-use potential. Similarly, because the field of synthetic biology involves strong elements of both science and technology, it is more accurately described as a "technoscience."

Study Methodology

To facilitate the process of cross-case comparison, the scholars who were invited to prepare case studies for this volume were asked to use a case study template consisting of a standard set of research questions. The initial version of this template was based on a preliminary framework for technology governance that was gradually refined over the course of the study in response to empirical evidence from the case studies. Some of the parameters in the original template proved to have minimal explanatory value and were dropped, and new variables were incorporated in their place. One new variable was the role of tacit knowledge in the ability to exploit a technology for harmful purposes, which became a key element of the ease of misuse. The individual parameter values and the aggregate scores for each technology are provided in chapter 21 of the book.

Because of the numerous changes in variables and terminology over the course of the study, the participating authors were asked to revise their case studies in midcourse to reflect the revised template. Limiting the total number of parameters in the final decision framework yielded a model that is parsimonious yet still sufficiently nuanced to discriminate among technologies that have different levels of risk and governability. Although parsimony necessarily entails some loss of detail, that drawback is more than offset by the increased simplicity and clarity of the model.

In practice, the decision framework is not intended to assess every emerging dual-use technology from start to finish. Whenever a particular technology does not pose a significant risk of misuse, the analytical process will be aborted, and monitoring of the technology will continue until the risk assessment changes. To facilitate the systematic comparison of the case studies, however, the study participants decided to apply the entire decision framework to all fourteen emerging dual-use technologies, including those that currently pose a low risk of misuse. Accordingly, every case study in part II assesses the technology's risk of misuse and governability. Furthermore, to add analytical richness, some of the study participants decided to present a wider range of governance options for their respective technologies than would normally be recommended by the decision framework.

Conclusions

The decision framework presented in this chapter is designed to provide a practical tool for policy makers seeking to decide which emerging dual-use technologies warrant the development of governance measures on a priority basis, and what types and combinations of governance measures would be most effective. Because emerging technologies are "moving targets" that undergo a continual process of evolution and refinement, the selected governance strategies must be flexible enough to permit frequent modification and adaptation at various points in the product development cycle. Moreover, even if the effectiveness of early governance measures declines over time as a technology evolves, the existence of such measures can help to establish and reinforce enduring behavioral norms against deliberate misuse.

Notes

1. Nicholas Sims, "Midpoint between Review Conferences: Next Steps to Strengthen the BWC," *Disarmament Diplomacy*, no. 91 (summer 2009), http://www.acronym.org.uk/dd/dd91/91bwc.htm.

2. Alfred Nordmann, *Converging Technologies: Shaping the Future of European Societies* (Brussels: European Commission, Directorate-General for Research, 2004), 3.

3. Mihail C. Roco, "Possibilities for Global Governance of Converging Technologies," *Journal of Nanoparticle Research* 10, no. 1 (January 2008): 11–29.

4. Margaret S. Race and Edward Hammond, "An Evaluation of the Role and Effectiveness of Institutional Biosafety Committees in Providing Oversight and Security at Biocontainment Laboratories," *Biosecurity and Bioterrorism* 6, no. 1 (March 2008): 19–35.

5. Gregory Koblentz, personal communication to the editor (Tucker), May 27, 2010.

6. Ali Nouri and Christopher F. Chyba, "Proliferation-Resistant Biotechnology: An Approach to Improve Biological Security," *Nature Biotechnology* 27 (2009): 234–236.

7. Leonard S. Spector, James Martin Center for Nonproliferation Studies, personal communication to the editor (Tucker), September 2010.

8. The Invention and Secrecy Act of 1952 (35 U.S.C. 181) authorizes the U.S. Patent and Trademark Office to prevent the release of information contained in a patent on security grounds. The statute specifies that "whenever publication or disclosure by the grant of a patent on an invention in which the Government has a property interest might, in the opinion of the head of the interested Government agency, be detrimental to the national security, the Commissioner upon being so notified shall order that the invention be kept secret and shall withhold the grant of a patent therefor."

9. National Research Council, *Globalization, Biosecurity, and the Future of the Life Sciences* (Washington, DC: National Academies Press, 2006).

10. Ibid., 140–141.

11. Margaret E. Kosal, "The Security Implications of Nanotechnology," *Bulletin of the Atomic Scientists* 66, no. 4 (July–August 2010): 58–69.

II

Contemporary Case Studies

A

Technologies for the Acquisition of Novel Biological or Molecular Diversity

5

Combinatorial Chemistry and High-Throughput Screening

Jonathan B. Tucker

The traditional process of drug discovery was extremely labor-intensive. Teams of medicinal chemists synthesized thousands of different compounds, which were then screened for biological activity to identify promising "lead" molecules for further development. During the 1980s, researchers developed a more efficient approach to drug discovery. Called combinatorial chemistry, or "combi-chem," it involves the systematic mixing and matching of chemical building blocks to generate large collections of structurally related compounds called "libraries." A related technique called high-throughput screening (HTS) is then used to identify specific compounds in the library that have a desired biological activity.

Whereas a traditional organic chemist can synthesize between 100 and 200 different compounds per year by hand, combinatorial chemistry and HTS can generate and screen tens of thousands of structurally related molecules in a matter of weeks. Although combinatorial chemistry and HTS were developed initially as a "brute-force" approach for discovering new lead compounds, today these techniques are used primarily to optimize the structure-function relationship after a lead compound has been identified.

Over the past few decades, combi-chem and HTS have become "an integral component of the drug discovery repertoire."[1] During combinatorial synthesis, highly toxic substances are sometimes created inadvertently and then discarded because of their lack of commercial value. Nevertheless combi-chem and HTS might be misused intentionally to identify and optimize highly toxic compounds as potential chemical warfare (CW) agents. This chapter describes the technology, assesses its potential for misuse, and suggests some possible approaches to governance.

Overview of the Technology

Combi-chem emerged initially from a method for synthesizing peptides, which consist of short chains of amino acid units. In the early 1960s, R. Bruce Merrifield at the Rockefeller University invented a cycle of chemical reactions that added amino acids one by one, in any desired sequence, to growing chains anchored to tiny plastic beads.[2] In the early 1980s, H. Mario Geysen adapted this solid-phase method to create a combinatorial technique called "parallel synthesis," in which a bead-anchored molecular scaffold reacts with various mixtures of amino acids to generate a library of structurally related peptides. Parallel synthesis is usually performed on a microtiter plate: a sheet of molded plastic containing 96 tiny wells in an array of 8 rows by 12 columns. Each well contains a few milliliters of liquid in which the reactions occur. By injecting different combinations of amino acids into each well, it is possible to synthesize 96 peptide variants on a single plate.[3]

Advanced laboratory robotic systems now permit the use of microtiter plates that have 384 wells or more, giving chemists the ability to generate large compound libraries in a single synthesis campaign. The advantage of this approach to combinatorial synthesis is that cleaving the end product from the beads provides high yields without the need for laborious purification steps. Nevertheless, because chemical reactions that are normally performed in solution behave differently in a solid-phase system, reoptimizing the reaction conditions for the solid phase is a time-consuming process.[4]

In the late 1980s, Árpád Furka developed a second combi-chem method called "split-and-pool" synthesis, which can generate much larger compound libraries. In this case, the polymer beads are reacted with chemical building blocks in several different test tubes, creating mixtures of beads with different molecules attached to them. The contents of the test tubes are pooled in a single vessel, randomly distributing the beads; this mixture is then split into several equivalent portions and reacted with another set of chemical building blocks. The process of pooling and splitting is an enormous combinatorial multiplier: the greater the number of reaction cycles, the larger the library of variant molecules produced.[5] Split-and-pool synthesis routinely generates up to a million different molecular structures.

At the end of the process, the synthesized compounds are detached chemically from the beads, and the content of each test tube is screened to determine its average biological activity. The mixture with the highest activity is separated into about a hundred different compounds, which can then be purified and individually screened. The main drawback of the split-and-pool method is the need for the purification step. Since each variant molecule is present in tiny amounts, it can be difficult to sort through an active mixture and determine which compound is responsible for the desired biological activity. For this reason, contemporary medicinal chemists tend to use parallel synthesis to create combinatorial libraries rather than the split-and-pool approach.[6]

Combi-chem is normally employed in conjunction with high-throughput screening (HTS), which can test large compound libraries for a desired biological activity at a rate commensurate with the speed of combinatorial synthesis. Before the advent of HTS, screening assays were conducted in intact experimental animals and were based on therapeutic effect, such as antibacterial or anti-inflammatory action. Today, however, high-throughput screening is performed against an isolated biomolecular target, such as a cell-surface protein, an enzyme, or an ion channel. Ideally, a therapeutic drug should bind with high affinity to a specific receptor site in the body to induce a desired physiological change; if the compound binds to multiple sites, it will most likely have unwanted side effects.

HTS systems are well suited to automation with laboratory robots, making it possible to screen thousands of different compounds in parallel. For example, a cell-surface receptor protein that is a target for drug development can be tagged with a fluorescent molecule that glows in response to binding, enabling researchers to identify drug candidates with a high affinity for the receptor.[7] Because a poorly defined screening target can generate false-positive hits (or worse, false negatives, meaning real matches for the desired biological activity that are not detected), a robust, highly sensitive screening mechanism is essential. When screening a combinatorial library, a medicinal chemist does not want to miss even a modestly potent lead compound that could serve as the starting point for more focused development.

History of the Technology

In 1988 the entrepreneur Alejandro Zaffaroni founded a company called Affymax in Palo Alto, California, which used combi-chem methods to synthesize and screen large peptide libraries to identify potential therapeutic drugs.[8] Because peptides are rapidly broken down by enzymes in the stomach, however, they cannot be administered orally and hence do not make ideal drug candidates. Skeptics doubted that combinatorial synthesis could be used to generate libraries of small-molecule drugs, which can be taken by mouth and are more persistent in the body. Such molecules typically consist of fewer than thirty nonhydrogen atoms, lack toxic or reactive elements, and are stable in the presence of water and oxygen. Other desirable characteristics of small-molecule drugs are the ability to be absorbed through the gastrointestinal tract, solubility in lipids, and a moderate rate of metabolism in the liver, so that the drug can have a useful physiological effect before being broken down.[9] In 1992, Jonathan Ellman and Barry Bunin at the University of California, Berkeley, developed a method for the parallel synthesis of an important class of small-molecule drugs: the benzodiazepines, which are used to treat anxiety.[10]

During the late 1980s and early 1990s, combi-chem and HTS attracted great interest from major pharmaceutical companies, nearly all of which established specialized departments devoted to combinatorial synthesis. Numerous drug-discovery firms were also founded to perform contract work. The golden age of combi-chem, which lasted from 1992 to about 1997, saw rapid improvements in the speed and throughput of the technology. After this initial wave of enthusiasm, however, the growth of combi-chem slowed in the late 1990s because the synthesis and screening of large, quasi-random compound libraries produced low hit rates of compounds with desired biological activity and did not lead to the discovery of any new block-buster drugs, resulting in a sense of disillusionment.[11] It gradually became clear that the first generation of combinatorial libraries was ineffective because of the libraries' excessive complexity, as well as the low purity caused by the presence of unwanted synthetic by-products. Another shortcoming of combinatorial synthesis was that it generated compounds that were too much alike and did not fill enough of the available "molecular space."[12]

In response to this reassessment at the end of the 1990s, many pharmaceutical companies scaled back and reoriented their combi-chem units. Although the initial approach had been to synthesize and screen large, diverse libraries for the discovery of lead compounds, drug companies now began to use combi-chem primarily for optimization, or modifying the molecular structure of a lead compound to enhance its biological activity. According to Dr. William A. Nugent of Vertex Pharmaceuticals in Cambridge, Massachusetts, "Combinatorial chemistry was originally seen as a powerful tool for lead discovery, but it didn't pan out. Instead it's become an important tool for optimization."[13]

Utility of the Technology

During the optimization process, medicinal chemists use combi-chem to synthesize hundreds of structural variants of a lead molecule in an effort to enhance its biological activity and eliminate unwanted side effects.[14] The resulting compound library is then screened with HTS to identify the variant compounds that bind most tightly and selectively to the receptor molecule. When used for optimization, combi-chem and HTS represent only about a third of the activity involved in the process of drug discovery.

One approach to optimization is to employ a computer program to create a virtual library consisting of millions of hypothetical compounds that would result from the reaction of a generic molecular structure called a pharmacophore with a variety of functional groups. To constrain the size of the virtual library, one can use computational filters to select those compounds with the most desirable bulk properties and metabolic and pharmacokinetic profiles. Only this subset of the library is then synthesized for screening purposes. Whereas combinatorial libraries generated by quasi-random synthesis typically contain more than 5,000 variant molecules, libraries based on virtual screening range from 50 to 100.[15] In addition, a method called "click chemistry," developed by the Nobel-laureate chemist K. Barry Sharpless of the Scripps Research Institute in San Diego, uses a few highly efficient reactions to synthesize compound libraries based on pharmacophores.

Each derivative is more than 85 percent pure, instead of a mixture of synthetic by-products.[16]

Combi-chem has also been integrated with rational drug design, which uses computer modeling to generate a more focused library of compounds that are more likely to possess the desired biological activity. First a biochemical target in the body is identified as a potential site of drug action. Researchers then use x-ray crystallography to determine the three-dimensional structure of the molecular complex formed when a natural body chemical (such as a hormone or a neurotransmitter) binds to its receptor. From the configuration of the binding site, pharmacologists can predict the chemical structure of small molecules that are likely to bind tightly and selectively to the receptor.[17] Other legitimate applications of combi-chem include the development of manufacturing processes for commercial drugs by optimizing the sequence of synthetic reactions needed to obtain a pure end product, and the discovery of new catalysts in the polymer and petrochemical industries.[18]

Potential for Misuse

Pharmaceutical and agrochemical companies currently employ combi-chem and HTS to build large compound databases that include information on the chemicals' toxicity to humans, animals, and plants, as well as their physiochemical properties such as stability, volatility, and persistence. Because private companies are interested only in molecules of commercial value, the development of a new drug or pesticide usually ends when it proves to be highly toxic in humans.

Nevertheless a state or terrorist organization seeking to develop novel CW agents might deliberately search such an industry database for toxic compounds suitable for weaponization.[19] Indeed, historical precedent exists for such an effort. Both the G-series and V-series nerve agents were discovered accidentally during industrial pesticide research and developed into military CW agents by the German and British armies, respectively.

Although combi-chem and HTS have some potential for misuse, the precise level of risk is difficult to assess. The technology generally works best for optimizing compounds with a molecular weight of about 700 daltons, yet traditional CW agents such as mustard or sarin are small molecules with a molecular weight below 500 daltons. Combi-chem and HTS may therefore be best suited for the development of mid-spectrum biochemical agents, such as peptide bioregulators, provided that the weaponization and delivery hurdles associated with this class of agents can be overcome. The risk of misuse of this technology is magnified by its theoretical ability to bypass rational design constraints in identifying new CW agents.

Accessibility

Combi-chem and HTS require the use of specialized hardware and software for the automated synthesis of compound libraries and their screening against biomolecular targets. Fewer than ten major vendors of such equipment exist today, nearly all of which are based in the United States, the European Union, and Japan. (Leading U.S. suppliers include Symyx, Tecan, and Caliper.) A turnkey combi-chem and HTS system costs about $1 million, and the commercial market consists almost exclusively of large pharmaceutical companies and start-ups in industrialized countries. According to a former researcher at Symyx, however, company scientists built early prototypes for advanced combi-chem and HTS systems using off-the-shelf components purchased at Home Depot.[20] Given this fact, a small team with the right knowledge, experience, and motivation might be able to assemble a crude combi-chem and HTS system fairly cheaply that could perform reasonably well.

An alternative route would be to gain access to an existing combinatorial database that contains highly toxic molecules. To do so, a proliferant state or terrorist group would have to break into the computer system of a pharmaceutical company, possibly with the aid of an insider such as a disgruntled employee. Nevertheless a good deal of information on drug development in academic and industrial laboratories is available in the public domain. Several open-source databases contain detailed information on the pharmacokinetic properties of newly synthesized compounds, and pharmaceutical companies often publish failed drug-discovery campaigns in the scientific literature while keeping their commercially promising (i.e., nontoxic) lead compounds under wraps.[21]

Ease of Misuse

The greatest obstacle to misuse of combi-chem and HTS is not access to hardware components or database information but the need for individuals with the relevant knowledge and experience. Although skilled technicians can perform some aspects of the technology, a synthetic organic chemist with Ph.D.-level expertise and experience would have to oversee the process, and a pharmacologist with a good understanding of physiology would have to identify a biomolecular target for screening. It has also been estimated that combining rational-design methods with combi-chem and HTS to develop a new drug from scratch would require a multidisciplinary team of forty to sixty people, including biochemists to isolate the target receptor, x-ray crystallographers to determine its molecular structure, and about twenty organic chemists to synthesize lead compounds for optimization.[22] If, however, the sole purpose of the development process is to identify candidate CW agents for military use, there would be no need to worry about harmful side effects, thereby streamlining the process of identification and optimization.[23]

Magnitude and Imminence of Risk

The greatest potential for the misuse of combi-chem and HTS lies with advanced industrial countries that have clandestine CW development programs. Such countries could use the technology to identify lethal or incapacitating chemicals that are militarily more effective or less costly to produce than existing agents. The magnitude of risk is hard to estimate a priori because it depends on whether or not new compounds could be identified that offer significant military advantages over standard CW agents. Theoretically, however, the discovery of enhanced CW agents using combi-chem and HTS is possible. In addition, the imminence of risk is fairly high for state programs but low for nonstate actors.

Awareness of Dual-Use Potential

The dual-use implications of combi-chem and HTS were first discussed in 2002 at an expert workshop convened by the International Union of Pure and Applied Chemistry (IUPAC) to discuss the implications of advances in science and technology for the Chemical Weapons Convention (CWC). Although the experts did not rule out the possibility that combi-chem and HTS might be exploited to discover novel CW agents, they put the threat in perspective by pointing out the technical hurdles involved:

Some new chemicals found by database mining will have toxicity characteristics that could lead to their being considered as chemical weapon agents. . . . [But] unless the compounds are simple and of low molecular weight, considerable effort will be required to devise practical methods to produce sufficient quantities to constitute a threat. Such quantities are likely to be a few tens of kilograms for research and development (or terror applications) and tens or hundreds of tons for military use. Further, unless the new compounds are gases or liquids with suitable volatility characteristics, all the usual problems of dispersing solids so that they could be used effectively as chemical weapons will apply.[24]

A second IUPAC workshop on chemical technologies and the CWC, held five years later in April 2007, concluded that the risk of misuse of combi-chem and HTS for CW purposes was "increasing." The experts noted that among the large number of bioactive chemicals synthesized and screened during pharmaceutical R&D, toxic chemicals will inevitably arise, and some of them could have physiochemical properties that make them candidate CW agents. Here again, however, the experts put the threat in context:

Despite this dramatic increase in knowledge and in the number of chemicals that could have CW utility given their toxicological and chemical profile, the risk to the object and purpose of the CWC posed by these scientific advances may not have increased as much as one might fear. To use a new toxic compound as an effective CW requires a number of developments before it can successfully be used. However, the risks from such novel toxic chemicals should not be ignored.[25]

A new concern raised at the 2007 meeting of the IUPAC panel was the potential use of combi-chem and HTS to develop novel chemical incapacitants for law enforcement and counterterrorism purposes. Such compounds affect the central nervous system to induce persistent sedation, anesthesia, or unconsciousness and can be fatal at high doses. Although scientists have tried to expand the safety margin between the lethal and incapacitating effects of these drugs, it is impossible to exert precise dosage control during tactical operations. As a result, the effects of these agents are fundamentally unpredictable, and their nonlethal nature cannot be assumed.[26]

Characteristics of the Technology Relevant to Governance

Embodiment   Combi-chem and HTS are a hybrid technology involving a combination of specialized hardware and software.

Maturity   The technology is mature and commercially available.

Convergence   The technology is moderately convergent, drawing on fundamental advances in miniaturization, laboratory robotics, and drug-screening technologies.

Rate of advance   Combi-chem and HTS have existed since the late 1980s but have undergone a number of changes. They emerged initially as a spin-off from solid-phase peptide synthesis and were generalized to the broader universe of druglike molecular structures. In the field of drug discovery, combi-chem was originally conceived as a brute-force method for the identification of lead compounds but is now used primarily for optimization, or improving the structure-function characteristics of promising leads. Although the rate of advance was exponential during the 1990s, it has since plateaued.

International diffusion   Combi-chem and HTS equipment is available to countries with a modern pharmaceutical research and development infrastructure (e.g., Australia, Canada, the European Union, Japan, Russia, Singapore, South Africa, and the United States) and, to a lesser extent, countries with a generic drug manufacturing capability (e.g., China, India, Spain, Israel, Hungary, and Brazil). The quality of Chinese pharmaceutical science in particular is improving rapidly, in part because of the large number of Chinese scientists who perform outsourced research and development work for U.S. pharmaceutical firms. Most countries of CW proliferation concern, such as Syria and North Korea, lack a highly developed pharmaceutical infrastructure and are therefore unlikely to possess combi-chem and HTS equipment and know-how. Iran is a possible exception to the rule, however. According to an estimate in 2008 by the U.S. intelligence community, "Tehran maintains dual-use facilities intended to produce CW agent in times of need and conducts research that may have offensive applications."[27]

Susceptibility to Governance

Because combi-chem and HTS are widely used by the pharmaceutical industry and are based on commercially available hardware and software, formal regulation of these technologies would be difficult. The main point of leverage is the relatively small number of suppliers of commercial turnkey systems, which could enable law enforcement to educate the industry about the risk of misuse and establish "trip wire" arrangements for reporting suspicious customers and purchase inquiries.

Relevance of Existing Governance Measures

Combi-chem and HTS have been available since the late 1980s and have diffused widely throughout the advanced industrial countries. At present, no restrictions exist on the use or export of such equipment to countries of CW proliferation concern. This lack of regulation contrasts sharply with the stringent controls imposed by the Australia Group on trade in precursor chemicals for nerve and blister agents and dual-use production equipment, such as chemical reactors made of corrosion-resistant steel alloys such as Hastelloy or lined with glass. Imposing legally binding export controls on combi-chem and HTS technologies would not be cost-effective because any country that is sophisticated enough to perform drug-discovery research probably has the capacity to reverse engineer its own combi-chem and HTS equipment, should it decide to do so.[28] Nevertheless controls on the sale of combi-chem and HTS equipment might conceivably have a deterrent effect by forcing state proliferators or sophisticated terrorist groups to use more labor-intensive methods for identifying and screening candidate CW agents.

Proposed Governance Measures

If a decision is made to regulate combi-chem and HTS technology, the most effective approach would be to control access to the specialized hardware and software. This measure might involve voluntary self-regulation on the part of the major suppliers, perhaps through a set of best practices, much as the gene-synthesis industry has done. Under such a voluntary governance system, governments would urge suppliers of combi-chem and HTS equipment to perform due diligence on new customers before approving the sale of turnkey systems, and to notify customs or law-enforcement authorities if suspicions arise. Because proliferant states might use outsourcing contracts with foreign research laboratories to acquire sensitive technology, equipment suppliers should treat orders from unfamiliar customers and contract labs as potential "red flags" warranting greater scrutiny. To ensure a level playing field for competition, it would be desirable for all the major suppliers of combi-chem and HTS systems to harmonize their policies in this area.

Conclusions

In principle, the capacity to synthesize large libraries of novel compounds and screen them rapidly for biological activity could be exploited to develop novel CW agents. The dual-use potential of combinatorial chemistry and high-throughput screening has been recognized since at least 2002. Although this technology is probably too complex and costly to be exploited by terrorist organizations, it poses a significant risk of misuse by proliferant states with mature CW programs. One possible governance option would be for suppliers to perform due diligence on new customers to prevent sales of combi-chem and HTS equipment and software to countries or substate entities of CW proliferation concern.

Notes

1. Stu Borman, "Combinatorial Chemistry," *Chemical and Engineering News* 80, no. 45 (November 11, 2002): 43–57.

2. R. Bruce Merrifield, "Solid Phase Peptide Synthesis. I. The Synthesis of a Tetrapeptide," *Journal of the American Chemical Society* 85 (1963): 2149.

3. Matthew J. Plunkett and Jonathan A. Ellman, "Combinatorial Chemistry and New Drugs," *Scientific American* 276, no. 4 (April 1997): 68–73.

4. Dawn Verdugo, James Martin Center for Nonproliferation Studies, Monterey Institute of International Studies, Monterey, CA, personal communication to author, August 19, 2009.

5. Mark S. Lesney, "Rationalizing Combi-Chem," *Modern Drug Discovery* 5, no. 2 (February 2002): 26–30.

6. Verdugo, personal communication to author.

7. Joseph Alper, "Drug Discovery on the Assembly Line," *Science* 264 (June 3, 1994): 1399–1401.

8. Robert F. Service, "Combinatorial Chemistry: High-Speed Materials Design," *Science* 277, no. 5325 (July 25, 1997): 474.

9. Simon J. Teague, Andrew M. Davis, Paul D. Leeson, and Tudor Oprea, "The Design of Leadlike Combinatorial Libraries," *Angewandte Chemie International Edition* 38, no. 24 (1999): 3743–3748.

10. Barry A. Bunin and Jonathan A. Ellman, "A General and Expedient Method for the Solid-Phase Synthesis of 1,4-Benzodiazepine Derivatives," *Journal of the American Chemical Society* 114 (1992): 10997–10998.

11. Konrad H. Bleicher, Hans-Joachim Böhm, Klaus Müller, and Alexander I. Alanine, "Hit and Lead Generation: Beyond High-Throughput Screening," *Nature Reviews Drug Discovery* 2 (May 2003): 369–378.

12. Christopher Lipinski and Andrew Hopkins, "Navigating Chemical Space for Biology and Medicine," *Nature* 432 (December 16, 2004): 855–861.

13. Dr. William A. Nugent, Vertex Pharmaceuticals, telephone interview by author, June 23, 2009.

14. Bleicher et al., "Hit and Lead Generation."

15. Thierry Langer and Gerhard Wolber, "Virtual Combinatorial Chemistry and In Silico Screening: Efficient Tools for Lead Structure Discovery?" *Pure and Applied Chemistry* 76, no. 5 (2004): 991–996.

16. Hartmuth C. Kolb and K. Barry Sharpless, "The Growing Impact of Click Chemistry on Drug Discovery," *Drug Discovery Today* 8, no. 24 (December 2003): 1128–1137.

17. Robert F. Service, "Combinatorial Chemistry Hits the Drug Market," *Science* 272 (May 31, 1996): 1266–1268.

18. Dr. Joel M. Hawkins, Pfizer Research and Development Center, Groton, CT, interview by author, June 26, 2009.

19. George W. Parshall, Graham S. Pearson, Thomas D. Inch, and Edwin D. Becker, "Impact of Scientific Developments on the Chemical Weapons Convention (IUPAC Technical Report)," *Pure and Applied Chemistry* 74, no. 12 (2002): 2331.

20. Verdugo, personal communication to author.

21. Ibid.

22. Nugent, telephone interview by author.

23. George W. Parshall, former director of Central Research and Development at the DuPont Company, interview by author, Wilmington, DE, June 8, 2009.

24. Parshall et al., "Impact of Scientific Developments," 2331.

25. Mahdi Balali-Mood, Pieter S. Steyn, Leiv K. Sydnes, and Ralf Trapp, "Impact of Scientific Developments on the Chemical Weapons Convention (IUPAC Technical Report)," *Pure and Applied Chemistry* 80, no. 1 (2008): 184.

26. Alan M. Pearson, Marie Isabelle Chevrier, and Mark Wheelis, eds., *Incapacitating Biochemical Weapons: Promise or Peril?* (Lanham, MD: Lexington Books, 2007).

27. J. Michael McConnell, director of national intelligence, "Annual Threat Assessment of the Intelligence Community for the House Permanent Select Committee on Intelligence," February 7, 2008, 13.

28. Verdugo, personal communication to author.

6

DNA Shuffling and Directed Evolution

Gerald L. Epstein

DNA shuffling is one approach for accelerating the evolutionary process by using molecular-biology techniques to manipulate an organism's genome, the genetic blueprint that determines its inherited characteristics. This manipulation would aim to achieve a practical goal, such as increasing the expression of a protein or improving enzymatic activity. In principle, actors seeking to create novel pathogens or toxins for harmful purposes could misuse this technique, although none are known to have done so. All the tools, materials, and equipment needed to perform DNA shuffling and equivalent techniques are available in modern molecular-biology laboratories and have been used for decades. As a result, the technology is potentially accessible to many thousands of researchers worldwide, although skill and expertise are required to perform the relevant techniques efficiently. No effective options exist to prevent the further spread of these techniques, but a number of governance measures may reduce the risk of malicious use.

Overview of the Technology

The evolution of species, as first described by Charles Darwin in 1859, involves a process in which organisms inherit genes from their parents, resulting in offspring with a range of characteristics. Sexual reproduction produces diversity because each descendant inherits a different combination of traits from its two parents. Additional genetic diversity results from mutations, or mistakes in DNA replication that give offspring traits their parents did not possess. The descendants are then subjected to natural selection, or a test of their ability to survive and reproduce in an environment full of hazards. Organisms that are not well adapted to their environment will either die before reproducing or will not reproduce as prolifically. The more successful organisms will pass on their traits, again with some variability, to their own descendants, increasing the fraction of offspring in the next generation that possess characteristics adapted to that particular environment. In this way, each successive generation is filtered or selected by external factors.

Whereas natural selection has no goal other than adaptation to a changing environment, *directed* evolution takes place under the control of human beings who have specific objectives in mind. Crop breeders, for example, target attributes such as yield, strength, disease resistance, and tolerance to drought or low temperatures. To enhance these traits, they select parent organisms that perform better than their peers, and look for descendants that are better yet. The best-performing members of each generation are then used to sire the next generation. Over the past century, breeders have increased the genetic diversity of crops by raising the mutation rate through irradiation or exposure to certain chemicals.

DNA shuffling and equivalent molecular biological techniques are forms of directed evolution that, unlike breeding, yield genetic variants that are unlikely to arise in nature. When these directed-evolution techniques are performed on microbes, they increase the diversity of each successive generation beyond what natural processes would create, permitting an investigator to screen for rare genetic variants that have a particular desired characteristic. As a result, these techniques can generate bacteria with properties that would be unlikely to evolve naturally, such as the enhanced ability to produce a certain protein. Directed evolution also provides an effective way to engineer enzymes and other proteins because it does not require an understanding of the underlying biological mechanisms. DNA shuffling can be viewed as the biological equivalent of combinatorial chemistry and high-throughput screening.

When used on microorganisms, DNA shuffling provides an accelerated form of directed evolution. Microbes typically reproduce asexually from a single parent by simple division, resulting in two daughter cells with genomes identical to the parent.[1] Normally, errors in the replication of a microbial genome are the only source of diversity in the descendants. DNA shuffling creates far more diversity by starting with several different "parent" genomes and producing "daughter" genomes that randomly incorporate distinct stretches of DNA from the different parents.

Further diversity can be introduced by creating the descendant genomes through a process that deliberately introduces additional mutations, so that not all versions of the DNA sections copied from the same parent are identical. Some of the descendant organisms might not be viable, but the ones that survive will have a wide range of characteristics. A researcher can test these variants manually or with high-throughput assays to see how well they perform with respect to the property being optimized, such as the yield of a given protein. The variants that perform best can then be prepared for another round of shuffling in which *their* genomes are randomly mixed, and the entire process repeated.

Selection schemes that do not rely on high-throughput screening may also be used, such as those in which the culture conditions are designed so that the microbial variants of interest are the only ones that survive and propagate, while all others are inhibited. The strains best suited to thrive in a stressful environment will reproduce most prolifically and will therefore constitute the bulk of the recovered organisms. Several studies have demonstrated the ability of DNA shuffling to improve an organism's resistance to antibiotics, which can serve as a proxy for optimizing its enzymatic activity or protein production.[2]

Other directed-evolution techniques introduce random mutations in DNA through means such as chemicals, radiation, or error-prone replication, and additional diversity can be generated through successive replication stages that randomly combine these mutations. These other approaches and DNA shuffling are all forms of a general process known as *random mutagenesis*. The power of directed evolution using DNA shuffling and other random-mutagenesis techniques derives from the ability to create large numbers of genetic variants, combined with the ability to screen the resulting library with high-throughput methods to identify those variants with the desired property. Repeating the process for multiple generations will further optimize the resulting organisms. As opposed to traditional recombinant DNA methods, which require knowing precisely which molecular changes are needed to achieve a desired improvement, an investigator using directed evolution does not have to know, or even guess, which specific molecular changes will accomplish this objective.

History of the Technology

During the early 1990s, researchers at the Affymax Research Institute in Palo Alto, California, pursued directed-evolution studies by using mutation-inducing chemicals and other means to introduce random errors into the genomes of microorganisms.[3] In 1993, Willem P. C. "Pim" Stemmer invented a combinatorial approach for producing diverse microbial genomes by randomly combining DNA fragments from different sources and screening the resulting offspring for desired properties. This method provided a substantial leap in capability. Stemmer's initial publications demonstrated the ability of DNA shuffling to increase the antibiotic resistance of a bacterium by a factor of 32,000.[4] Other studies created diversity by shuffling analogous genes from different species, such as mouse and human genes coding for the same protein.[5]

In 1997 Stemmer and three colleagues founded a company called Maxygen—an abbreviation of "maximizing genetic diversity"—to commercialize the DNA shuffling technique. At the time of its incorporation, Maxygen had filed twelve patent applications and received one. Different divisions of the company worked on agricultural, chemical, pharmaceutical, and vaccine applications of DNA shuffling.[6] Maxygen researchers investigated the utility of the technique for strain libraries containing up to 10,000 genetic variants. Using high-throughput screening, researchers selected 20 to 40 variants that had the greatest ability to produce a particular enzyme that the original bacterial strain manufactured only at low levels; these variants then provided the basis for the next round of shuffling. Seven rounds of selection produced a strain of E. coli with enzymatic activity 66 times greater than the original strain, showing that even modestly sized strain libraries can evolve effectively.[7]

A paper published in 1998 demonstrated the power of shuffling a pool of related genes from four bacterial species to evolve an enzyme that could degrade an antibiotic, thereby conferring resistance to it.[8] Each of the bacterial genes was shuffled individually, generating a library of genes that differed from each other only by a few point mutations. In each case, a single round of shuffling led to an eightfold increase in enzymatic activity. Yet when all four genes were shuffled together, introducing diversity not only through point mutations but by randomly recombining portions of the four original genes, a single round of shuffling yielded as much as a 540-fold increase in enzymatic activity. In addition, in 2002 a paper in Nature described how two rounds of whole-genome shuffling over the course of a year resulted in a ninefold increase in the ability of a bacterial strain to produce the antibiotic tylosin. The same level of improvement took twenty years to achieve with the conventional technique of successive mutation and screening.[9]

Although the initial applications of DNA shuffling were to optimize protein expression in bacteria, a paper published in 2000 described the use of the technique on viruses. A single round of shuffling six strains of mouse leukemia virus yielded a variant strain that could infect a different target organ in another species: the ovaries of Chinese hamsters.[10] Analysis of the shuffled viral genome revealed changes in the DNA sequence that could not have resulted easily from natural recombination. Another DNA shuffling experiment extended the host range of the human immunodeficiency virus (HIV) to infect the cells of pig-tailed macaques.[11] In this way, DNA shuffling was able to extend the range of tissues and species that a virus could infect, a change that was unlikely to occur naturally. These findings demonstrated the remarkable power of directed evolution, as well as its dual-use potential—that is, its ability to increase an organism's ability to perform either a useful function or a harmful one, depending on what screening or selection criteria were used.

By 2001 the U.S. Defense Advanced Research Projects Agency (DARPA) had invested $20 million in Maxygen for the development of new vaccines. In general, vaccines work by exposing the body to foreign proteins (antigens) similar to those found on the surface of pathogenic microbes, eliciting an immune response that enables the body to recognize and attack organisms that express those proteins the next time they are encountered. Under natural conditions, pathogens never evolve antigens that maximize the immune response, because it is in their survival interest to minimize it. But Maxygen's goal was to use DNA shuffling to modify the viral antigens to increase their immunogenicity and cross-protective range, or their ability to protect against multiple strains of the same pathogen.[12] In laboratory assays, the resulting vaccine candidates exhibited both characteristics.[13] Despite these promising results, however, the efficacy of vaccines produced by DNA shuffling has yet to be demonstrated in clinical trials.

Although much of the methodology of DNA shuffling is under patent protection—Maxygen holds close to a hundred patents[14]—several other approaches to directed evolution have been developed.[15] As with DNA shuffling, these methods produce libraries of diverse variants, which are then subjected to screening or selection techniques to identify organisms with desired properties. Many of these other directed-evolution techniques are under patent protection, but some are not.

Utility of the Technology

Directed evolution—for which DNA shuffling is a major, but not the only, approach—is now the principal means of generating proteins with new or improved properties. It differs sharply from the "rational-design" approach, which uses an in-depth understanding of a protein's three-dimensional structure to improve its functional characteristics.[16] Unlike rational design, directed evolution typically yields functional improvements by multiple, unrelated mechanisms, often in unexpected ways. It is also capable of optimizing multiple aspects of a protein simultaneously, such as potency and level of expression.

Most applications of directed evolution seek to develop better enzymes, antibodies, therapeutic proteins, and other products. Enzymes for industrial use are a particular application, including improving the heat stability of enzymes for use in laundry detergents, or developing means for large-scale production of the amino acid phenylalanine, which led to the commercialization of the nonnutritive sweetener aspartame.[17] Nearly every biotechnology company that engineers proteins uses this approach. Directed evolution is also widely used in academic research, agriculture, and other fields.

Potential for Misuse

In principle, directed evolution could be used to optimize traits that cause harm, whether by toxicity, pathogenesis, resistance to therapeutic or prophylactic countermeasures, environmental hardiness, or other means. Several of the studies described in the previous sections illustrate the ability of DNA shuffling or some other directed-evolution process to increase the antibiotic resistance of a microbe as a proxy for optimizing some other parameter, such as enzymatic activity or protein production. Accordingly, someone with malicious intent might use this method to render a biowarfare agent resistant to medical countermeasures or to enhance the production of a protein toxin. Although it is possible to generate antibiotic-resistant strains with simple selection methods, directed evolution could potentially generate bacteria that are resistant to multiple antibiotics without compromising other militarily desirable attributes.

In one example of DNA shuffling research with obvious dual-use implications, Maxygen's vaccine unit shuffled genes for an immune system protein called interleukin-12 (IL-12) from seven different mammalian species—human, rhesus monkey, cow, pig, goat, dog, and cat—to produce a form of the protein that was 128 times more potent at stimulating the human immune system than ordinary human IL-12.[18] Although this work was performed for the purpose of developing vaccines and other therapeutic applications, it could potentially be misused to suppress the human immune system or turn it against the host. These malicious applications may have been foreshadowed by the identifier that the Maxygen investigators unwittingly chose for the shuffled ("evolved") form of interleukin-12: EvIL.

Accessibility
Some directed-evolution techniques, including DNA shuffling, are patented, but others are not, and the necessary equipment and materials exist in many molecular-biology laboratories. Indeed, directed-evolution techniques are accessible to any laboratory or company that is reasonably proficient in modern molecular-biology techniques—a level of capability possessed by more than 10,000 laboratories around the world.[19]

Ease of Misuse
Because directed-evolution techniques, including DNA shuffling, are rooted in basic molecular-biology skills, state-level biological warfare programs could easily master them, as could nonstate groups or individuals with sufficient expertise, resources, and commitment. Even so, it would be difficult to train members of a terrorist group lacking a background in molecular biology. Although procedures for directed evolution published in the academic and patent literature are sufficient for anyone with a basic set of laboratory skills to create a library of shuffled genes,[20] tacit knowledge acquired through practical experience is essential with respect to the downstream procedures needed to express the genes that have been created and to identify and isolate the variants of interest.[21]

Two factors suggest that tacit knowledge is important for DNA shuffling in particular. First, the technique is not available in the form of a commercially available "kit" that contains all necessary reagents, laboratory vials, and detailed instructions. (One reason that kits do not exist is that many of the methods are patented, and the patent owners prefer to market DNA-shuffling services.) In the absence of a kit, a researcher must know which reagents to use and what procedures to follow.

Studies have also shown that even kits do not necessarily remove the need for tacit knowledge when performing a specific experiment in a particular laboratory context.[22] A second indicator of the importance of tacit knowledge for DNA shuffling is that a set of procedures for the technique is not included in *Current Protocols in Molecular Biology*, a peer-reviewed publication containing a wide variety of methods in biotechnology.[23]

Magnitude and Imminence of Risk

As already described, DNA shuffling and other directed-evolution techniques could permit the development of highly dangerous microorganisms and toxins without requiring an understanding of the underlying biological mechanisms. Nevertheless a major constraint on the immediate application of directed evolution for biological warfare is the need to screen the variant strains to identify those that most strongly express the desired property. If the intent is to develop a pathogen that is more infectious, deadly, or contagious in humans, in vitro assays could not simulate the complicated processes involved in infecting the host, defeating the immune response, causing disease, and transmitting the disease to others. If, however, the developer of the weapon was a ruthless organization that had no qualms about human experimentation, it might identify which of the various strains was most effective by infecting a small number of test subjects, such as prisoners or suicide volunteers, or animals if the extrapolation to humans was well understood.[24]

New technologies may have the prospect for lowering some of these barriers. DARPA has worked on creating an artificial human "immune-system-on-a-chip," intended to accelerate the evaluation of new vaccine candidates and substances to enhance immune responses.[25] Were such systems to be developed and made widely available, actors seeking to develop pathogens as weapons might be able to use them to screen some aspects of a pathogen's interaction with the immune system in vitro, rather than requiring human experimentation. However, these immune chips might not have the throughput to screen a diverse library of possible pathogen candidates, which a directed-evolution approach to pathogen development that did not use human testing might require. Moreover, it is unlikely that such a device could model all the phenomena involved in pathogenesis and transmissibility.

Awareness of Dual-Use Potential

Pim Stemmer, the inventor of DNA shuffling, recognized its dual-use potential almost immediately. "Arguably, it's the most dangerous thing you can do in biology," he observed.[26] DARPA program managers who funded early research on DNA shuffling were aware of its dual-use implications but focused more on its potential to transform research on vaccines and other medical countermeasures.[27] The broader life sciences community, however, did not recognize the dual-use potential of DNA shuffling at the time it was developed. Indeed, according to a National Research Council report, "Only a few in the scientific community had raised concerns about the potential contributions of life sciences research to biological weapons programs and bioterrorism before the anthrax attacks of 2001."[28]

In recent years, DNA shuffling has received far less attention from the policy community than other dual-use biotechnologies. Synthetic biology, for example, became prominent after Eckard Wimmer and his colleagues synthesized poliovirus in 2002.[29] Since then, dozens of surveys, studies, and analyses of synthetic biology have been published, to the point that one observer complained that "synthetic biology represents 5 percent of the risk but is getting 95 percent of the attention."[30] DNA shuffling and other directed-evolution approaches belong to the less-examined 95 percent, and few studies have examined their dual-use implications.

Characteristics of the Technology Relevant to Governance

Embodiment   DNA shuffling and other directed-evolution techniques do not rely on unique or specialized equipment but consist of procedures that use, and build on, standard methods and tools of molecular biology: the ability to cut, transfer, and duplicate DNA; to combine shorter pieces of DNA into longer ones; to alter a DNA sequence; to sort pieces of DNA by size; to insert DNA into a microorganism's genome such that the functions or processes it encodes are carried out; and to screen large numbers of microorganisms rapidly to identify those with properties of interest.

Maturity   Directed-evolution techniques are widely used in research and by biotechnology firms to engineer new proteins. The technology is not available for sale as a "black box" commodity but can be created by practitioners seeking its benefits.

Convergence   DNA shuffling and other directed-evolution techniques are not convergent technologies but rather lie fairly centrally within biotechnology. Some bioinformatics is needed to identify the appropriate genes with which to initiate shuffling or random mutagenesis, and the high-throughput screening of variant organisms may involve the robotic manipulation of large numbers of samples. Even so, the beauty of the approach is that practitioners need not comprehend the underlying biology or conduct sophisticated information processing.

Rate of advance   It is hard to measure the rate of advance of directed-evolution techniques, which have not been defined with sufficient precision to permit monitoring them with quantitative parameters. A possible indicator of how DNA shuffling and other forms of random mutagenesis have developed, if the data were available, would be the approximate number of distinct genetic sequences that can be created.[31] Still, that number is only a partial measure of the power of DNA shuffling because what matters is the *functional* diversity of the sequences, or the number of different ways in which the resulting organisms behave, with respect to the property being sought. Moreover, as the number of genetic variants increases, it may be more difficult to screen for sequences that optimize the characteristic of interest. From a qualitative standpoint, DNA shuffling has progressed in terms of the types of DNA that can be recombined and the methods by which the recombination takes place. More significant than improvements in the technique itself, however, is the development of new applications.

International diffusion   The United States, the European Union, and Japan are the most powerful players in the life sciences and will likely remain so for the next five to ten years.[32] Nevertheless biotechnology is globalizing rapidly because the necessary equipment, materials, and facilities are relatively inexpensive, especially when compared with capital-intensive industries such as semiconductor manufacturing and aerospace. Trained biologists and biotechnologists are increasingly available in developing countries. International academic exchanges and publications are expanding, and research and industrial facilities around the world draw on a supporting infrastructure of companies that supply reagents and other materials.[33]

Because of these trends, biotechnology clusters—geographically proximate and interconnected companies, research institutions, service providers, suppliers, and trade associations—have emerged in Argentina, Brazil, China, Cuba, Hong Kong, India, Malaysia, Singapore, South Africa, and Taiwan.[34] As developing and rapidly industrializing countries grow more adept at using biotechnology, they will be better able to conduct DNA shuffling. Although Maxygen's patents may limit the spread of DNA shuffling among legitimate users, alternative approaches to directed evolution are expected to proliferate widely.

Susceptibility to Governance

The tools, processes, materials, and equipment underlying DNA shuffling and other directed-evolution approaches are all generic to molecular biology, placing the technique within reach of thousands of reasonably equipped laboratories around the world—although a fair amount of skill and expertise is required to apply these approaches successfully. Accordingly, there is no way to regulate them without constraining biotechnology as a whole. Any broad-based controls that restrict the spread and further development of these techniques would be not only impractical but also, given biotechnology's importance for legitimate purposes, counterproductive and arguably immoral. Even if mandatory regulations were imposed, the lack of distinctive equipment, materials, and procedures associated with directed evolution would impede the effectiveness of controls unless its use was self-reported.

Given the many different ways in which directed evolution can be implemented, even defining it in terms specific enough to be regulated would pose a significant challenge.[35]

Devising specific governance measures for directed evolution is also complicated by the difficulty of distinguishing legitimate applications from illicit ones, raising the prospect of penalizing or discouraging legitimate scientific research without a corresponding security benefit. As with other dual-use biotechnologies, the risk of harm from directed evolution is tied not to the technology per se but to the *purpose* for which it is used. Given that even seemingly suspicious activities may have a legitimate purpose, such as developing a gene-therapy vector to evade the human immune system or maximizing the biological activity of a toxin, knowing how directed evolution is being used might not be sufficient to distinguish between legitimate and illicit activities. It is even possible that scientists working in established laboratories could perform directed evolution for malicious purposes without others in the facility having reason to question this activity. These factors do not preclude governance measures, but because their value would be modest, the costs of implementation must not be too high, in terms of both direct investment and forgone opportunities. Thus minimizing the risk of misuse of directed evolution is best addressed with measures that reduce the risk of misuse of biotechnology in general.

Relevance of Existing Governance Measures

The dual-use risks of certain types of experiments are usually discussed in terms of their potential results and not the experimental method being employed. For example, the U.S. National Science Advisory Board for Biosecurity (NSABB) has identified the deliberate alteration of a pathogen's host range as warranting scrutiny from a dual-use perspective. Yet the advisory board did not call specific attention to DNA shuffling, which has been used for that purpose. Indeed, with the exception of synthetic genomics, where specific guidelines have been developed for commercial vendors of synthetic DNA, current approaches to the governance of dual-use biotechnologies are fairly generic. Soft-law and informal governance measures include the review of proposed experiments that raise dual-use concerns, guidance on communicating the results of such experiments, reinforcing norms against malicious use, and efforts to raise the awareness of dual-use concerns among life scientists.

Proposed Governance Measures

Possible governance options for directed evolution, including DNA shuffling, fall into the category of approaches for reducing the risk of misuse of biotechnology in general, rather than of directed evolution specifically. These general options include hard-law measures, such as mandatory reporting.

Reporting requirement   One way to facilitate the difficult task of detecting illicit applications of directed evolution would be to require that certain types of legitimate activity be reported to an appropriate authority. For example, scientists might be required to report all research that could reasonably be expected to enhance a pathogen's virulence and ability to resist countermeasures or to increase the range of species or tissue types it can infect. This reporting requirement would cover all directed-evolution activities that are intended, or could reasonably be expected, to increase pathogenicity.

Nevertheless the concern that DNA sequences from benign microorganisms might be combined or mutated in such a way as to create a pathogenic organism by chance, when pathogenesis is not the trait being selected or screened for, is so far theoretical and difficult to explore analytically or experimentally. A recent National Research Council study concluded that predicting whether or not an organism is pathogenic from its gene sequence alone is not possible now and will not be in the "usefully near future."[36] It would be even more challenging to predict whether a genome constructed by randomly changing or combining DNA from nonpathogens would be pathogenic.

Reports on specific directed-evolution activities might contain not only technical data about the procedures being used but also information about the investigators, possibly including a registration requirement for the personnel involved. Should a reportable activity be discovered that was *not* reported, it would be viewed with considerable suspicion. A reporting scheme would also force malefactors to work secretly or to create an elaborate cover story, exposing them to some risk of detection and exposure. Without a reporting requirement, illegitimate work could be performed openly with little risk of being identified as such.

The benefits of such a reporting scheme would depend on how burdensome the reporting proves to be and whether the criteria are sufficiently clear and objective so that practitioners know when to report. Another critical factor would be the willingness of practitioners to make the necessary reports to the appropriate government entity. If researchers or vendors were concerned that sensitive or proprietary information would be compromised despite guarantees of confidentiality, then they would probably refuse to report relevant activities, and the scheme would fail. Finally, there must be some way to differentiate legitimate from illicit work, or at least to raise questions, on the basis of the information provided. The key to whether any reporting requirement would be acceptable is whether it can provide information that is sufficient to permit raising questions while not proving unacceptably or impracticably intrusive.

Whistle-blowing   It would be useful to create a mechanism by which anyone suspecting that a coworker, supplier, customer, or professional colleague was pursuing illicit activities would be able to bring these concerns to the attention of an institutional or government authority, which could then take appropriate action. The details of such a mechanism would have to be worked out, including who would collect the reports, what steps would be taken to resolve a concern, and how due process and the right of redress for those reported would be preserved.

Restrictions on publication   In some cases, discretion may be advisable when communicating sensitive research results. The genetic variability generated by DNA shuffling and other random-mutagenesis approaches raises the possibility that microbes with properties of interest for hostile purposes, such as a new mechanism of pathogenesis, could emerge in the course of legitimate research. Any experiment *intended* to produce highly pathogenic organisms should receive close scrutiny before being undertaken or communicated, because it would entail some risk of violating the Biological Weapons Convention's ban on developing biological weapons. However, thought should be given to communicating the results of any experiment that *unintentionally* produces microorganisms with increased pathogenicity, even if those results were tangential or irrelevant to the desired objective.

A group of scientific editors and publishers acknowledged in 2003 "On occasion an editor may conclude that the potential harm of publication outweighs the potential societal benefits. Under such circumstances, the paper should be modified, or not be published."[37] Although it would be inappropriate in most cases for governmental authorities to play a formal role in reviewing scientific publications, scientific authors—like editors and publishers—must consider the consequences of their actions. Exercising restraint with respect to the publication of dual-use information, while pejoratively labeled "self-censorship" by some, is more appropriately termed "scientific responsibility." Every day scientists face a choice of what lines of research to pursue, and considering the potential misuse of the results is as legitimate a factor to consider as any.

Conclusions

Directed evolution using DNA shuffling or other techniques of random mutagenesis is a method for generating diverse biological characteristics, even when the underlying biological processes are not known or are poorly understood. As long as a biological outcome can be defined and efficiently screened for, these approaches can help to enhance it. The only limitations are that the desired property must be biologically possible—the limits of which may not be known—and that the few organisms possessing the desired property can be screened efficiently from the great majority that do not. Any technique this powerful has a tremendous potential for benefit but could also be misused to create novel pathogens or toxins that cause harm in ways beyond what is now possible. Accordingly, scientists who perform directed-evolution experiments should consider the consequences of their actions and refrain from pursuing or publishing results that are more likely to facilitate misuse and harm than to improve human welfare.

Acknowledgments

The author would like to thank Drs. Roger Brent, Hal Padgett, Kathleen Vogel, and Christopher Locher for their suggestions, comments, and reviews of this paper or portions of it. Any remaining errors are, of course, solely the responsibility of the author.

Notes

1. Less common processes exist, such as conjugation, through which descendants can acquire genetic material from more than one parent microbe. Antibiotic resistance is transmitted through this mechanism.

2. Willem P. C. Stemmer, "Rapid Evolution of a Protein *in Vitro* by DNA Shuffling," *Nature* 370, no. 6488 (August 4, 1994): 389–391; Ying-Xin Zhang, Kim Perry, Victor A. Vinci, et al., "Genome Shuffling Leads to Rapid Phenotypic Improvement in Bacteria," *Nature* 415, no. 6872 (February 7, 2002): 644–646.

3. Jon Cohen, "'Breeding' Antigens for New Vaccines," *Science* 293, no. 5528 (July 13, 2001): 236–238.

4. Stemmer, "Rapid Evolution of a Protein," 389–391; Willem P. C. Stemmer, "DNA Shuffling by Random Fragmentation and Reassembly: In Vitro Recombination for Molecular Evolution," *Proceedings of the National Academy of Sciences USA* 91, no. 22 (October 1994): 10747–10751.

5. Stemmer, "DNA Shuffling by Random Fragmentation and Reassembly," 10750.

6. Maxygen's first patent for this technology was U.S. patent no. 5,605,793, "Methods for in Vitro Recombination," issued February 25, 1997; "Zaffaroni Announces New Start-Up Company—Maxygen, Inc.," *Maxygen News* (press release), June 2, 1997.

7. Ji-Hu Zhang, Glenn Dawes, and Willem P. C. Stemmer, "Directed Evolution of a Fucosidase from a Galactosidase by DNA Shuffling and Screening," *Proceedings of the National Academy of Sciences USA* 94, no. 9 (April 1997): 4504–4509. The variants of interest were identified because they grew into bacterial colonies that turned blue, with the greatest enzymatic activity resulting in the deepest blue. The bluest colonies were identified visually and harvested.

8. Andreas Crameri, Sun-Ai Raillard, Ericka Bermudez, et al., "DNA Shuffling of a Family of Genes from Diverse Species Accelerates Directed Evolution," *Nature* 391, no. 6664 (January 15, 1998): 288.

9. Ying-Xin Zhang, Kim Perry, Victor A. Vinci, et al., "Genome Shuffling Leads to Rapid Phenotypic Improvement in Bacteria," *Nature* 415 (February 7, 2002): 644–646.

10. Nay-Wei Soong, Laurel Nomura, Katja Pekrun, et al., "Molecular Breeding of Viruses," *Nature Genetics* 25, no. 4 (August 2000): 436–439.

11. Katja Pekrun, Riri Shibata, Tatsuhiko Igarashi, et al., "Evolution of a Human Immuno-deficiency Virus Type 1 Variant with Enhanced Replication in Pig-Tailed Macaque Cells by DNA Shuffling," *Journal of Virology* 76, no. 6 (March 2002): 2924–2935.

12. Cohen, "'Breeding' Antigens for New Vaccines."

13. Christopher P. Locher, Madan Paidhungat, Robert G. Whalen, et al., "DNA Shuffling and Screening Strategies for Improving Vaccine Efficacy," *DNA and Cell Biology* 24, no. 4 (2005): 256–263.

14. Crispin Littlehales, "Profile: Willem 'Pim' Stemmer," *Nature Biotechnology* 27, no. 3 (March 2009): 220.

15. Frances H. Arnold, "Protein Engineering for Unusual Environments," *Current Opinion in Biotechnology* 4, no. 4 (August 1993): 450–455.

16. Ling Yuan, Itzhak Kurek, James English, et al., "Laboratory-Directed Protein Evolution," *Microbiology and Molecular Biology Reviews* 69, no. 3 (September 2005): 374. According to this article, "Our present knowledge of structure-function relationships in proteins is still insufficient to make rational design a robust approach."

17. Claudia Schmidt-Dannert and Frances H. Arnold, "Directed Evolution of Industrial Enzymes," *Trends in Biotechnology* 17, no. 4 (April 1999): 135–136; Roger Brent, former president of the Molecular Sciences Institute, interview by author, August 21, 2009.

18. Steven R. Leong, Jean C. C. Chang, Randal Ong, et al., "Optimized Expression and Specific Activity of IL-12 by Directed Molecular Evolution," *Proceedings of the National Academy of Sciences USA* 100, no. 3 (February 4, 2003): 1163–1168.

19. Estimate by Roger Brent, former editor of *Current Protocols in Molecular Biology* and former president of the Molecular Sciences Institute, e-mail communication to author, August 27, 2009.

20. For example, see Huimin Zhao and Frances H. Arnold, "Optimization of DNA Shuffling for High Fidelity Recombination," *Nucleic Acids Research* 25, no. 6 (1997): 1307–1308.

21. Observations in both sentences from Hal Padgett, e-mail communication to author, August 29, 2009.

22. Michael Lynch, "Protocols, Practices, and the Reproduction of Technique in Molecular Biology," *British Journal of Sociology* 53, no. 2 (June 2002): 203–220; published online December 15, 2003.

23. Frederick M. Ausubel, Roger Brent, Robert E. Kingston, et al., editorial board, *Current Protocols in Molecular Biology* (Wiley Interscience, 2009).

24. Moreover, pathogenesis would depend not only on the characteristics of the pathogen but on factors such as the number of infecting organisms and the mode of entry into the body, introducing additional complexity to the experiments.

25. Brett P. Giroir, DARPA Defense Sciences Office, *The Future of Ideas* (2005), p. 60, http://archive.darpa.mil/DARPATech2005/presentations/dso/giroir-mod.pdf.

26. Littlehales, "Profile: Willem 'Pim' Stemmer," 220.

27. Roger Brent, interview by author, August 21, 2009; Stephen S. Morse, former DARPA program manager, e-mail communication to author, August 27, 2009.

28. National Research Council, *A Survey of Attitudes and Actions on Dual Use Research in the Life Sciences: A Collaborative Effort of the National Research Council and the American Association for the Advancement of Science* (Washington, DC: National Academies Press, 2008), 11.

29. Jeronimo Cello, Aniko V. Paul, and Eckard Wimmer, "Chemical Synthesis of Poliovirus cDNA: Generation of Infectious Virus in the Absence of Natural Template," *Science* 297, no. 5583 (August 9, 2002): 1016–1018.

30. Roger Brent, personal communication to author, 2009.

31. One early paper estimated that DNA shuffling techniques were capable of producing libraries of up to 10^{10} to 10^{15} molecules. See Willem P. C. Stemmer, "Searching Sequence Space," *Nature Biotechnology* 13, no. 6 (June 1995): 549.

32. Committee on Advances in Technology and the Prevention of Their Application to Next Generation Biowarfare Threats, National Research Council, *Globalization, Biosecurity, and the Future of the Life Sciences* (Washington, DC: National Academies Press, 2006), 79.

33. Gerald L. Epstein, *Global Evolution of Dual-Use Biotechnology* (Washington, DC: CSIS Press, 2005), 9.

34. "Select Global Biotechnology and Bioscience Clusters," http://mbbnet.umn.edu/scmap/biotechmap.html. The listed countries are pursuing biotechnology but this source does not specifically discuss DNA shuffling or directed evolution.

35. DNA shuffling, as one specific approach, would be easier to define, and it might be possible after the fact to determine that DNA shuffling had been performed if one had access to the organisms that had been produced and could analyze their genomes to look for the distinctly chimeric nature of a shuffled genome.

36. Committee on Scientific Milestones for the Development of a Gene Sequence–Based Classification System for the Oversight of Select Agents, National Research Council, *Sequence-Based Classification of Select Agents: A Brighter Line* (Washington, DC: National Academies Press, 2010), 2.

37. Journal Editors and Authors Group, "Statement on Scientific Publication and Security," *Science* 299, no. 5610 (February 21, 2003): 1149. Potential restrictions on communicating dual-use research are elaborated further in "Points to Consider in Assessing the Risks and Benefits of Communicating Research Information with Dual Use Potential," included as appendix 5, pp. 53–54, of National Science Advisory Board for Biosecurity (NSABB), *Proposed Framework for the Oversight of Dual Use Life Sciences Research: Strategies for Minimizing the Potential Misuse of Research Information*, June 2007, http://oba.od.nih.gov/biosecurity/pdf/Framework%20for%20transmittal%200807_Sept07.pdf.

B

Technologies for Directed Design

7

Protein Engineering

Catherine Jefferson

Proteins play vital roles in the body as structural components and catalysts in biochemical reactions. A protein molecule consists of a linear chain of amino acid building blocks, of which twenty different varieties exist. Each type of amino acid has a distinct molecular structure and electrical charge, and the complex interactions among these units and the surrounding water molecules cause the linear chain to fold up spontaneously into a unique globular shape that determines the function of the protein.

Protein engineering involves the design and synthesis of tailor-made proteins—modified from nature or created from scratch—for applications in industry, agriculture, and medicine. Because protein engineering could potentially be applied to increase the toxicity of certain proteins for hostile purposes, such as warfare or terrorism, governance measures are needed to prevent the misuse of this technology while preserving its beneficial applications.

Overview of the Technology

There are three approaches to protein engineering. The first, called "rational design," involves modifying the sequence of amino acids in a protein to alter its three-dimensional shape and its functional properties. Techniques such as x-ray crystallography and bioinformatics are used to determine how the linear chain of amino acids folds into a compact, globular structure. Even when the precise folding pattern is known, however, predicting the effect of one or more changes in amino acid sequence on the shape and function of a protein is a challenging task. As a result, rational design usually requires several cycles of modification and testing before it yields an engineered protein with the desired properties.[1]

The second approach to protein engineering, called "directed evolution," was developed in the early to mid-1990s.[2] This technique shuffles parts of the gene coding for a protein to create thousands of mutant proteins with slightly altered structures, which are then subjected to high-throughput screening to identify the proteins that exhibit the desired function or activity. Compared to rational design, directed evolution is a semiautomated, randomized process that requires much less expertise, although specialized screening techniques are needed.[3] Some analysts have raised concerns that directed evolution could be used to generate mutant proteins that are toxic to human cells.[4]

The third approach to protein engineering is the synthesis of artificial proteins. This method expands on the standard set of twenty amino acids by adding unnatural amino acids with novel properties. Because of the difficulty of predicting the folding patterns associated with unnatural amino acids, however, synthesizing artificial proteins that are fully functional is currently beyond the state of the art.[5] Still, researchers are making incremental progress toward this capability, which when realized could have dual-use implications.[6]

The three technical approaches to protein engineering are not mutually exclusive, and some investigators have successfully combined elements of rational design and directed evolution.[7] This chapter focuses primarily on rational design and describes specific areas of research with a potential for misuse.

History of the Technology
Protein engineering first emerged in the early 1980s. It was made possible by advances in X-ray crystallography, which made it possible to determine the precise three-dimensional structure of a protein from the diffraction pattern that results when x-rays pass through a crystallized protein, and by progress in bioinformatics.[8] Since then, the technology has been gradually refined.

Utility of the Technology
Engineered proteins have multiple uses in industry, agriculture, and medicine. One well-known application is the development of heat-resistant proteases (enzymes that break down proteins), which are added to laundry detergent formulations to remove protein-rich stains.[9] Protein engineering has also been used to endow natural proteins with new biological functions. For example, modifying the amino acid sequence of an antibody molecule can change its folding pattern, so that it behaves like an enzyme to catalyze a specific biochemical reaction. Such catalytic antibodies have been created for reactions that have no naturally occurring enzymes.[10]

The pharmaceutical industry has also used protein engineering extensively. One important medical application involves the development of protein "fusion toxins" for therapeutic purposes.[11] Protein toxins are poisons of biological origin; examples include snake, insect, and spider venoms, plant toxins such as ricin, and bacterial toxins such as botulinum toxin and diphtheria toxin. Because of the highly specific manner in which protein toxins interfere with cellular metabolism, they can kill or incapacitate at extremely low doses. Many protein toxins have two functional components: a binding domain that recognizes and binds specifically to a receptor on the surface of the target cell type, and a catalytic domain that enters the cell and exerts a toxic effect, such as blocking protein synthesis.[12]

By means of protein engineering, it is possible to combine the binding and catalytic domains of two different proteins to create a hybrid or "fusion" toxin. For example, combining the catalytic domain of diphtheria toxin or ricin with the binding domain of interleukin-2, an immune-system signaling protein, produces a fusion toxin that can selectively kill cancer cells, which have surface receptors for interleukin-2, whereas healthy cells do not.[13] In one experiment, modifying the binding domain to enhance the affinity for interleukin-2 receptors produced a seventeenfold increase in cell-killing activity.[14] One such fusion toxin has been marketed commercially for the treatment of cutaneous T-cell lymphoma.[15]

Potential for Misuse

Protein engineering could lead to new security challenges, in terms of both harmful products and the creation of new knowledge that could be misused for nefarious purposes.[16]

Increasing the toxicity of known toxins In principle, protein engineering could create protein toxins that have enhanced lethality, target range, and resistance to detection, diagnosis, and treatment. *Bacillus thuringiensis* (Bt), for example, is a soil bacterium that is commonly used as a biological pesticide.[17] It produces a variety of crystalline protein toxins with insecticidal activity (Cry or Cyt toxins) that offer an environmentally friendly means of pest control.[18] Because insects can become resistant to Bt toxins, scientists have used both the rational-design and directed-evolution methods of protein engineering to increase the ability of Bt toxins to kill insect pests.[19] This research is controversial, however, because Bt is closely related to the bacterium *Bacillus anthracis*, which produces protein toxins that cause the lethal human disease anthrax. At least in principle, the same protein-engineering methods that make Bt toxins more effective could also enhance the lethality of anthrax toxin.[20]

Creating harmful fusion toxins In general, the systemic toxicity of fusion toxins produced for medical applications is less than that of either parent toxin because the hybrid molecules are narrowly targeted against malignant or otherwise abnormal cells.[21] It is theoretically possible, however, to increase the lethality of fusion toxins against normal cells for hostile purposes. For example, the extreme toxicity of botulinum toxin might be combined with the stability and persistence of staphylococcal enterotoxin B (SEB) to create a highly lethal fusion toxin that can withstand heat and environmental stresses.[22]

Fusion toxins may have other properties that increase their potential for misuse. For example, when the catalytic domain of tetanus toxin or shiga toxin is combined with the binding domain of anthrax toxin, the resulting fusion toxin can target a broader range of mammalian cell types than either parent toxin.[23] Scientists have also developed a method to produce fusion toxins rapidly and efficiently by engineering mammalian cells to secrete them as properly folded protein molecules.[24] Because a fusion toxin may have different physiological effects than either parent toxin, it could elicit confusing medical signs or symptoms, making detection, diagnosis, and treatment more difficult.[25]

Modifying infectious proteins   Another possible misuse of protein engineering involves the protein-based infectious agents called prions. These proteins are known to cause transmissible spongiform encephalopathies, a family of diseases in animals and humans characterized by the spongy degeneration of the brain, resulting in severe neurological symptoms and death.[26] Although prion diseases resemble genetic disorders, the transmissible particles lack DNA or other genetic material and instead consist exclusively of a modified protein.[27] Prion replication involves the conversion of normal protein molecules into a misfolded conformation.[28] A prion disease that affects cattle, bovine spongiform encephalopathy (BSE), first appeared in the United Kingdom in November 1986 and became popularly known as "mad cow disease." In March 1996 a new variant form of Creutzfeldt-Jakob disease, a prion infection that affects humans, was reported in Britain and linked to the consumption of food contaminated by BSE, demonstrating the transmission of prions from one species to another.[29] Prions have been called "lethal pathogens par excellence—indeed, it is hard to think of other examples of infectious diseases with 100 percent mortality once the earliest clinical signs have developed."[30]

The feasibility of weaponizing prions is doubtful, however, for several reasons. First, prions do not infect reliably. Second, the incubation period of prion diseases is exceptionally long, with a delay of several months to years between infection and the onset of clinical illness, severely reducing the potential utility of prions as biological warfare agents.[31] Finally, the delivery of prions would be problematic because the normal routes of infection are the ingestion of contaminated meat and the intravenous administration of contaminated blood products. Nevertheless recent research suggests that mice can be infected with prions delivered as an aerosol, and attempts by scientists to design artificial prions have raised dual-use concerns.[32] Although the search for curative treatments drives prion research, it is conceivable that protein engineering could be misapplied to develop prions with more rapid harmful effects. In addition, the recent development of technologies for the large-scale synthesis of proteins might be misused to mass-produce infectious prions.[33]

Accessibility

Scientists engaged in rational protein design use specialized computer programs to predict the conformational impact of possible changes in amino acid sequence. These programs are generally proprietary and require licenses, as well as special training with the software, which serves to restrict access to the technology.

Ease of Misuse

The need for both explicit and tacit knowledge to conduct protein engineering by rational design limits the ability of terrorist groups to exploit the technology for harmful purposes.[34] Rational-design approaches to protein engineering are hypothesis driven, based on a large body of knowledge and expertise, and require extensive tacit knowledge obtained through hands-on experience and trial-and-error methods.[35] Typically, the level of expertise required to produce fusion toxins is that of an advanced graduate student in molecular biology.[36]

The directed-evolution approach to protein engineering has a greater potential to result in the inadvertent creation of a protein toxin with greater potency or stability. Although considerable resources are needed to perform directed evolution, the technology is semiautomated and demands much less expertise than rational design—probably the equivalent of an undergraduate degree in biology.[37] Thus the growing popularity of directed-evolution techniques has increased the risk that individuals with relatively little specialized knowledge could create a library of novel protein toxins harmful to humans.

Magnitude and Imminence of Risk

Even if a more lethal engineered or fusion toxin could be developed, the effective weaponization and delivery of the agent would pose additional hurdles, particularly for bioterrorists. In addition, toxins are generally less effective biological weapons than microbial agents. For these reasons, the magnitude and imminence of the risk of misuse associated with protein engineering appear to be moderate.

Awareness of Dual-Use Potential

The risks of misuse of protein engineering were first noted in the 1993 edition of the *World Armaments and Disarmament* yearbook, published by the Stockholm International Peace Research Institute (SIPRI). According to this source, "The ease with which novel engineered bacterial toxins, bacterial-viral toxins and the like can be produced by protein engineering is of military interest, as is the way in which protein engineering enables the changing of the site on a toxin against which antidotes normally are developed."[38]

The danger of engineered prions was mentioned in the 2001 briefing book prepared for the Fifth Review Conference of the Biological Weapons Convention

(BWC) by the Department of Peace Studies at the University of Bradford: "In view of the growing knowledge of the dangers of prion diseases, the increasing capabilities for manipulation of receptors and ligands in the nervous, endocrine and immune systems, and the growing understanding of how proteins may be designed for particular purposes . . . it is recommended that an explanatory sentence should be added at this Review Conference on prions, bioregulators and proteins."[39]

Since these publications, only a few commentators have focused on the security implications of protein engineering.[40] Far more attention has focused on the risks of genome synthesis and synthetic biology.[41] As the preceding discussion suggests, however, the dual-use risks associated with protein engineering warrant further characterization.

In general, the scientific community is aware of the dual-use risks associated with toxin research but has tended to overlook or minimize the risks of protein engineering. The scientists who enhanced the potency of Bt toxin were aware that the same technique might be applied to the anthrax toxin and discussed the issue in an internal committee review, but mechanisms for continual oversight of the dual-use risk were not put in place.[42] In an e-mail interview, a scientist involved in the development of fusion toxins downplayed the risk of misuse. "While the prospect of weaponized toxic proteins is worrisome," he wrote, "I think it is more likely to come in the form of enhanced delivery of readily available, robust toxins like ricin or saporin rather than from engineered proteins. I feel that the risk of my work contributing to weapons technology is minimal."[43]

Characteristics of the Technology Relevant to Governance

Embodiment Protein engineering is based primarily on intangible information and does not require specialized hardware beyond what is available in a standard molecular biology laboratory, although rational design requires access to crystallographic data and specialized modeling software.

Maturity Most applications of protein engineering are in advanced development, although a few engineered enzymes and fusion toxins are commercially available.

Convergence Protein engineering is a convergent technology that draws on advances in several scientific fields, including x-ray crystallography, bioinformatics, recombinant DNA technology, and chemical DNA synthesis. Protein engineering also involves scientists from multiple disciplines, such as chemistry, biochemistry, biology, and engineering.[44]

Rate of advance Although rational-design methods of protein engineering are advancing slowly, use of the directed-evolution approach has grown rapidly in recent years. Because directed-evolution methodology demands less explicit and tacit knowledge, it is potentially more accessible to actors who would exploit the technology for harmful purposes.

International diffusion   The enabling technologies for protein engineering—bioinformatics, DNA synthesis, and genetic engineering—are diffusing world-wide. As a result, the development of engineered proteins is increasingly common in advanced industrialized states. Nevertheless protein-engineering techniques are still beyond the capabilities of most developing countries and terrorist organizations.[45]

Susceptibility to Governance

The governability of protein engineering is limited because the technology is based largely on intangible information, and the directed-evolution method draws on techniques and equipment that are widely available in molecular biology laboratories around the world.

Relevance of Existing Governance Measures

Because toxins are nonliving chemicals produced by living organisms, they are covered by both the 1972 BWC and the 1993 Chemical Weapons Convention (CWC). Although both treaties prohibit the development and production of toxins for hostile purposes, only the CWC has verification mechanisms. Nevertheless routine verification under the CWC covers only two toxins, saxitoxin and ricin, which are listed on schedule 1 in the treaty's Annex on Chemicals. (Of those two, only ricin is a protein toxin.) National export controls, such as those harmonized by the Australia Group, provide another mechanism for managing dual-use risk by seeking to ensure that exports of dangerous toxins, as well as dual-use production equipment, do not fall into the hands of known proliferators.[46]

In the United States, the Select Agent Regulations require all institutions to register with the federal government if they possess, transfer, or use listed microbial agents and toxins of bioterrorism concern. The current list of Select Agents includes several protein toxins used in fusion-toxin research, such as botulinum toxin, ricin, shiga toxin, and staphylococcal enterotoxin B, as well as the prion responsible for BSE because of its potential use as an agent of agroterrorism.[47] Even so, specific governance strategies for protein engineering per se do not yet exist.

Proposed Governance Measures

The hardware, people, and processes involved in dual-use research and development are often conflated into the term "technology." Unpacking this concept makes clear that distinct but complementary measures are needed to govern the various aspects. Hardware (equipment and materials) can be governed through export controls, screening of DNA synthesis orders, and the licensing of dual-use equipment, whereas people can be governed through systems of vetting, education, awareness raising, and codes of conduct. It is also useful to distinguish among different intervention points, which may involve the upstream inputs for protein-engineering research or the downstream products.

Governing upstream inputs   To reduce the risk that key materials for protein-engineering research will be diverted to hostile purposes, governance measures should be in place to regulate the transfer and use of dangerous toxins or prions. Recipients of these items should be vetted and registered to confirm their scientific bona fides, as is the case today with items regulated by the U.S. Select Agents and Toxins List. Governments should also adopt legislation requiring academic institutions or private companies to be licensed if they conduct research with dual-use equipment and materials. Finally, the authorized users of such equipment should be held responsible for restricting access to legitimate scientists.[48] All these measures would focus primarily on controlling access to toxins and prions and would therefore regulate protein engineering only indirectly. Moreover, although the proposed measures would help to mitigate the risk of misuse of protein engineering, other problems remain: ensuring that the legitimate users of the technology do not permit its diversion for hostile purposes and managing potentially sensitive research results.

Governing downstream outputs   Oversight of protein-engineering research would seek to identify high-risk outputs at a stage when governance measures are still feasible and effective. The oversight mechanism would review and oversee all research involving the engineering of protein toxins or the development of fusion toxins and would have "the legal authority to block funding of specific projects, or to constrain the publication of sensitive scientific results, whenever the dangers to society clearly outweigh the benefits."[49] Such an oversight board could be modeled on the U.S. Department of Homeland Security's Compliance Review Group, which reviews the department's biodefense projects to ensure their compliance with the BWC and monitors projects as they evolve.[50] In addition to top-down oversight mechanisms, scientists could consult online portals that enable them to discuss the dual-use implications of planned research projects with biosecurity experts.[51]

Governing dual-use knowledge   Governing the dual-use potential of knowledge is a more difficult challenge than governing the transfer and use of hardware. One approach is to introduce a system for vetting scientific personnel who work with toxins. For example, the United Kingdom recently introduced the Academic Technology Approval Scheme (ATAS), which requires postgraduate students from outside the European Economic Area who are interested in security-sensitive fields to obtain clearance before they can apply for a visa to study in Britain.[52] Academic institutions and private companies in all countries that work with toxins should adopt similar vetting schemes.

The scientific community, aided by civil society, also has an important role to play in managing dual-use risk. Professional codes of conduct and other forms of self-regulation are important to raise awareness of dual-use issues and to support laboratory biosafety and biosecurity measures.[53] Universities should also encourage scientific responsibility through the inclusion of ethical and security issues in the life sciences curriculum.[54]

Strengthening international norms   At the root of all governance measures lies the fundamental ethical norm that toxic chemicals and disease agents should not be used as weapons. This norm codifies an ancient, cross-cultural taboo against poison weapons and is the cornerstone of the biological disarmament regime.[55] Although the BWC has no formal verification measures, it urges all member states to enact national legislation to prohibit and prevent activities that violate the convention. A number of states have done so, but the full implementation of such measures and their harmonization among member states remain major challenges. One response to these shortcomings is a proposal to negotiate an international convention that would criminalize the development and use of biological and chemical weapons.[56] This treaty would give national courts jurisdiction over individuals present in the territory of a member state, regardless of their nationality or official position, who order, direct, or knowingly render substantial assistance to the development and use of biological and chemical weapons anywhere in the world. The concept of individual criminal responsibility for the hostile use of biology would support the initiatives considered here.

Conclusions

Protein engineering offers many potential benefits, particularly in the field of advanced therapeutics. Nevertheless it is not outside the bounds of possibility that rogue states or technologically sophisticated terrorist groups could misuse the technology to create enhanced biological weapons. Potential consequences of the misuse of protein engineering include the creation of modified toxins with increased toxicity, potency, stability, and effects that defy diagnosis and treatment. To balance the risks of such misuse against the potential costs of regulation to the scientific enterprise, governance measures should focus on three dimensions: hardware, products, and people. Combining formal oversight mechanisms with informal self-governance by the scientific community would create a web of measures to prevent the misuse of protein engineering.

Notes

1. Jonathan B. Tucker and Craig Hooper, "Protein Engineering: Security Implications," *EMBO Reports* 7, no. S1 (July 2006): 14–17.

2. Willem P. C. Stemmer, "Rapid Evolution of a Protein *in Vitro* by DNA Shuffling," *Nature* 370 (August 1994): 389–391; Cara A. Tracewell and Frances H. Arnold, "Directed Enzyme Evolution: Climbing Fitness Peaks One Amino Acid at a Time," *Current Opinion in Chemical Biology* 13 (2009): 3–9.

3. Dr. Neil Crickmore, Department of Biochemistry, University of Sussex, interview by author, June 16, 2009.

4. Ibid.

5. Steven A. Benner and Michael Sismour, "Synthetic Biology," *Nature Reviews: Genetics* 6 (July 2005): 533–543.

6. Charles B. Millard, "Medical Defense against Protein Toxin Weapons: Review and Perspective," in *Biological Weapons Defense: Infectious Diseases and Counterterrorism*, ed. Luther E. Lindler, Frank J. Lebeda, and George W. Korch (Totowa, NJ: Humana Press, 2005), 255–283.

7. Tucker and Hooper, "Protein Engineering," 15.

8. Kevin M. Ulmer, "Protein Engineering," *Science* 219 (February 1983): 666–671.

9. David A. Estell, "Engineering Enzymes for Improved Performance in Industrial Applications," *Journal of Biotechnology* 28 (1993): 25–30; C. von der Osten, S. Branner, S. Hastrup, et al., "Protein Engineering of Subtilisins to Improve Stability in Detergent Formulations," *Journal of Biotechnology* 28 (March 1993): 55–68.

10. James A. Branningan and Anthony J. Wilkinson, "Protein Engineering 20 Years On," *Nature Reviews Molecular Cell Biology* 3 (December 2002): 964–970.

11. Tucker and Hooper, "Protein Engineering," 14.

12. Millard, "Medical Defense against Protein Toxin Weapons," 273.

13. D. P. Williams, K. Parker, P. Bacha, et al., "Diphtheria Toxin Receptor Binding Domain Substitution with Interleukin-2: Genetic Construction and Properties of a Diphtheria Toxin-Related Interleukin-2 Fusion Protein," *Protein Engineering* 1 (1987): 493–498; Arthur E. Frankel, Chris Burbage, Tao Fu, et al., "Characterization of a Ricin Fusion Toxin Targeted to the Interleukin-2 Receptor," *Protein Engineering* 9 (1996): 913–919.

14. Tetsuyuki Kiyokawa, Diane P. Williams, Catherine E. Snider, et al., "Protein Engineering of Diphtheria-Toxin-Related Interleukin-2 Fusion Toxins to Increase Cytotoxic Potency for High-Affinity IL-2-Receptor-Bearing Target Cells," *Protein Engineering* 4 (1991): 463–468.

15. Francine M. Foss, "Interleukin-2 Fusion Toxin: Targeted Therapy for Cutaneous T Cell Lymphoma," *Annals of the New York Academy of Science* 941 (2006): 166–176.

16. Caitríona McLeish, "Reflecting on the Problem of Dual-Use," in *A Web of Prevention: Biological Weapons, Life Sciences, and the Governance of Research*, ed. Brian Rappert and Caitríona McLeish (London: Earthscan, 2007), 189–203.

17. Crickmore, interview by author.

18. Donald H. Dean, "Biochemical Genetics of the Bacterial Insect-Control Agent *Bacillus thuringiensis*: Basic Principles and Prospects for Genetic Engineering," *Biotechnology and Genetic Engineering Reviews* 2 (October 1984): 341–363.

19. E. Schnepf, N. Crickmore, J. Van Rie, et al., "*Bacillus thuringiensis* and Its Pesticidal Crystal Proteins," *Microbiology and Molecular Biology Reviews* 62 (September 1998): 775–806; Chandi C. Mandal, Srimonta Gayen, Asitava Basu, et al., "Prediction-Based Protein Engineering of Domain I of Cry2A Entomocidal Toxin of *Bacillus thuringiensis* for the Enhancement of Toxicity against Lepidopteran Insects," *PEDS* 20 (November 2007): 599–606; Xinyan S. Liu and Donald H. Dean, "Redesigning *Bacillus thuringiensis* Cry1Aa Toxin into a Mosquito Toxin," *PEDS* 19 (January 2006): 107–111; Hiroshi Ishikawa, Yasushi Hoshino, Yutaka Motoki, et al., "A System for the Directed Evolution of the Insecticidal Protein from *Bacillus thuringiensis*," *Molecular Biotechnology* 36 (June 2007): 90–101.

20. Crickmore, interview by author.

21. Dr. Benjamin E. Rich, researcher, Department of Dermatology, Brigham and Women's Hospital, Boston, MA, e-mail correspondence with author, July 29, 2009.

22. Tucker and Hooper, "Protein Engineering," 15–16.

23. Naveen Arora, Lura C. Williamson, Stephen H. Leppla, et al., "Cytotoxic Effects of a Chimeric Protein Consisting of Tetanus Toxin Light Chain and Anthrax Toxin Lethal Factor in Non-neuronal Cells," *Journal of Biological Chemistry* 269 (October 1994): 26165–26171; Naveen Arora and Stephen H. Leppla, "Fusions of Anthrax Toxin Lethal Factor with Shiga Toxin and Diphtheria Toxin Enzymatic Domains Are Toxic to Mammalian Cells," *Infection and Immunity* 62 (November 1994): 4955–4961.

24. S. Shulga-Morskoy and Benjamin E. Rich, "Bioactive IL7-Diphtheria Fusion Toxin Secreted by Mammalian Cells," *PEDS* 18 (March 2005): 25–31.

25. Janet R. Gilsdorf and Raymond A. Zilinskas, "New Considerations in Infectious Disease Outbreaks: The Threat of Genetically Modified Microbes," *Clinical Infectious Diseases* 40 (April 2005): 1160–1165.

26. Charles Weissman, "The State of the Prion," *Nature Reviews: Microbiology* 2 (2004): 861–871.

27. See Stanley B. Prusiner, "Prions," *Proceedings of the National Academy of Sciences* 95 (November 1998): 13363–13383, for more on the prion concept and the now largely discredited alternative viral hypothesis.

28. Ibid.

29. Patrick van Zwanenberg and Erik Millstone, *BSE: Risk, Science, and Governance* (Oxford: Oxford University Press, 2005).

30. John Collinge and Anthony R. Clarke, "A General Model of Prion Strains and Their Pathogenicity," *Science* 318 (November 2007): 935.

31. Congressional Research Service, "Agroterrorism: Threats and Preparedness," *CRS Report to Congress*, August 13, 2004.

32. Johannes Haybaeck, Mathias Heikenwalder, Britta Klevenz, et al., "Aerosols Transmit Prions to Immunocompetent and Immunodeficient Mice," *PLOS Pathogens* 7 (January 2011): 1–14; Giuseppe Legname, Ilia V. Baskakov, Hoang-Oanh B. Nguyen, et al., "Synthetic Mam-

malian Prions," *Science* 305 (July 2004): 673–676; Lev Z. Oscherovich, Brian S. Cox, Mick F. Tuite, et al., "Dissection and Design of Yeast Prions," *PLOS Biology* 2 (April 2004): 442–451.

33. Gabriela P. Saborio, Bruno Permanne, and Claudio Soto, "Sensitive Detection of Pathological Prion Protein Cyclic Amplification of Protein Misfolding," *Nature* 411 (June 2001): 810–813; Claudio Soto, Gabriela P. Saborio, and Laurence Anderes, "Cyclic Amplification of Protein Misfolding: Application to Prion-Related Disorders and Beyond," *Trends in Neurosciences* 25 (August 2002): 390–394; Nathan R. Deleault, Brent T. Harris, Judy R. Rees, et al., "Formation of Native Prions from Minimal Components *in Vitro*," *Proceedings of the National Academy of Sciences* 104 (June 2007): 9741–9746.

34. Caitríona McLeish and Paul Nightingale, "Biosecurity, Bioterrorism, and the Governance of Science: The Increasing Convergence of Science and Security Policy," *Research Policy* 36 (December 2007): 1635–1654; see esp. p. 1645.

35. Crickmore, interview with author.

36. Rich, e-mail correspondence with author.

37. Ibid.

38. Tomas Bartfai, S. J. Lundin, and Bo Rybeck, "Benefits and Threats of Developments in Biotechnology and Genetic Engineering," in *SIPRI Yearbook 1993: World Armaments and Disarmament*, ed. Stockholm International Peace Research Institute (SIPRI) (Oxford, UK: Oxford University Press, 1993), 297.

39. Malcolm R. Dando and Simon M. Whitby, "Article 1: Scope," in *Key Points for the Fifth Review Conference*, ed. Graham S. Pearson, Malcolm R. Dando, and Nicholas A. Sims (University of Bradford, Department of Peace Studies, November 2001), http://www.brad.ac.uk/acad/sbtwc.

40. Millard, "Medical Defense against Protein Toxin Weapons," 255–283; Tucker and Hooper, "Protein Engineering," 14–17.

41. Dr. Neil Davison, senior policy adviser to the Royal Society, London, e-mail correspondence with author, June 19, 2009.

42. Crickmore, interview with author.

43. Rich, e-mail correspondence with author.

44. Stefan Lutz and Uwe Théo Bornscheuer, eds., *Protein Engineering Handbook* (Weinheim: Wiley-VCH, 2008), xxvii.

45. Rich, e-mail correspondence with author.

46. Robert J. Mathews, "Chemical and Biological Weapons Export Controls and the 'Web of Prevention': A Practitioner's Perspective," in *A Web of Prevention: Biological Weapons, Life Sciences, and the Governance of Research*, ed. Brian Rappert and Caitríona McLeish (London: Earthscan, 2007), 163–169.

47. U.S. Department of Health and Human Services and U.S. Department of Agriculture, "Select Agents and Toxins," http://www.selectagents.gov.

48. Ronald M. Atlas and Malcolm Dando, "The Dual-Use Dilemma for the Life Sciences: Perspectives, Conundrums, and Global Solutions," *Biosecurity and Bioterrorism* 4 (2006): 281.

49. Tucker and Hooper, "Protein Engineering," 17.

50. Matthew Meselson, "Your Inbox, Mr. President," *Nature* 457 (January 2009): 259–260.

51. Barry Bergman, "Goldman School Portal Takes the Worry Out of 'Experiments of Concern,'" *University of California Berkeley News*, April 2, 2009.

52. The scheme is operated by the Foreign and Commonwealth Office. See House of Commons, Foreign Affairs Committee, Fourth Report of Session 2008–9, *Global Security: Non-Proliferation*, June 14, 2009, Evidence, 261–263.

53. Brian Rappert, "Responsibility in the Life Sciences: Assessing the Role of Professional Codes," *Biosecurity and Bioterrorism* 2 (2004): 164–174; James Revill and Malcolm Dando, "A Hippocratic Oath for Life Scientists," *EMBO Reports* 7 (2006): 55–60.

54. Brian Rappert, "Education for the Life Sciences: Choices and Challenges," in *A Web of Prevention: Biological Weapons, Life Sciences, and the Governance of Research*, ed. Brian Rappert and Caitríona McLeish (London: Earthscan, 2007), 51–65.

55. Catherine Jefferson, "The Chemical and Biological Weapons Taboo: Nature, Norms, and International Law," D.Phil. diss., University of Sussex, 2009.

56. Matthew Meselson and Julian Perry Robinson, "A Draft Convention to Prohibit Biological and Chemical Weapons under International Criminal Law," *Fletcher Forum of World Affairs* 28 (winter 2004): 57–71.

8

Synthesis of Viral Genomes

Filippa Lentzos and Pamela Silver

The emerging field of synthetic biology seeks to create a rational framework for manipulating the DNA of living organisms through the application of engineering principles.[1] This chapter focuses on a key enabling technology for synthetic biology: the ability to synthesize strands of DNA from off-the-shelf chemicals and assemble them into genes and microbial genomes. When combined with improved capabilities for the design and assembly of genetic circuits that perform specific tasks, synthetic genomics has the potential for revolutionary advances. At the same time, it could permit the reconstruction of dangerous viruses from scratch, as well as genetic modifications designed to enhance the virulence and military utility of pathogens.

The potential misuse of gene synthesis to re-create deadly viruses for biological warfare or terrorism would require the integration of three processes: the automated synthesis of DNA segments, the assembly of those segments into a viral genome, and the production and weaponization of the synthetic virus. Each of these steps differs with respect to the maturity of the technologies involved, the ease with which it could be performed by nonexperts, and the associated threat. Even with access to synthetic DNA, assembling the DNA segments into a synthetic virus and converting the virus into a deliverable weapon would pose significant technical hurdles. This chapter reviews the security concerns related to DNA synthesis technology and suggests some measures to limit the risk of misuse.

Overview of the Technology

DNA molecules consist of four fundamental building blocks: the nucleotide bases adenine (A), thymine (T), guanine (G), and cytosine (C), which can be linked together in any sequence to form a linear chain that encodes genetic information. A DNA molecule may consist of a single strand of nucleotide bases along a sugar backbone or two mirror-image strands that pair up to form a double helix, with adenine (A) always complementary to thymine (T) and guanine (G) complementary to cytosine (C). A second type of nucleic acid called RNA differs from DNA in the structure of its sugar backbone and the fact that one of the four nucleotide bases is uracil (U), which replaces thymine as the complementary base for adenine. An infectious virus consists of a long strand of single-stranded or double-stranded DNA or RNA, encased in a protein shell.

There are at least three ways to acquire a synthetic viral genome. The first and most straightforward approach is to order the entire viral genome from a commercial gene-synthesis company by entering the DNA sequence on the company's Web site. (A leading commercial supplier, Blue Heron Biotechnology in Bothell, Washington, has synthesized DNA molecules up to 52,000 base pairs long.) The genomic sequence would be synthesized in a specialized facility using proprietary technology that is not available for purchase, packaged in a living bacterial cell, and shipped back to the customer. The second option would be to order oligonucleotides (single-stranded DNA molecules less than 100 nucleotides in length) from one or more providers and then stitch them together in the correct order to create an entire viral genome. The advantage of this approach is that one can obtain more accurate DNA sequences, avoid purchasing expensive equipment, and outsource the necessary technical expertise. The third option would be to synthesize oligonucleotides with a standard desktop DNA synthesizer and then assemble the short fragments into a genome. This approach would require acquiring a DNA synthesizer (purchased or custom built) and a relatively small set of chemicals.

Although the chemical synthesis of oligonucleotides up to 120 base pairs is now routine, accurately synthesizing DNA sequences greater than 180 base pairs remains somewhat of an art. For this reason, the de novo synthesis of most viruses is still more difficult than stealing a sample from a laboratory or isolating the agent from nature.[2] It is just a matter of time, however, before technological advances further reduce costs and the frequency of errors, making genome synthesis readily affordable and accessible.[3]

History of the Technology

The field of synthetic genomics dates back to 1979, when the first gene was synthesized by chemical means.[4] The Indian American chemist Har Gobind Khorana and seventeen coworkers at the Massachusetts Institute of Technology took several years to produce a small gene made up of 207 DNA nucleotide base pairs. In the early 1980s, two technological developments facilitated the synthesis of DNA constructs: the invention of the automated DNA synthesizer and the polymerase chain reaction (PCR), which can copy any DNA sequence many millionfold. By the end of the 1980s, a DNA sequence of 2,100 base pairs had been synthesized chemically.[5]

In 2002 the first functional virus was synthesized from scratch: poliovirus, whose genome is a single-stranded RNA molecule about 7,500 nucleotide base pairs long.[6] Over a period of several months, Eckard Wimmer and his coworkers at the State University of New York at Stony Brook assembled the poliovirus genome from customized oligonucleotides, which they had ordered from a commercial supplier. When placed in a cell-free extract, the viral genome then directed the synthesis of infectious virus particles. The following year, Hamilton Smith and his colleagues at the J. Craig Venter Institute in Maryland published a description of the synthesis of a bacteriophage, a virus that infects bacteria, called φX174. Although this virus contains only 5,386 DNA base pairs (fewer than poliovirus), the new technique greatly improved the speed of DNA synthesis. Compared with the more than a year that it took the Wimmer group to synthesize poliovirus, Smith and his colleagues made a precise, fully functional copy of the φX174 bacteriophage in only two weeks.[7]

Since then, the pace of progress has been truly remarkable. In 2004 researchers synthesized DNA sequences 14,600 and 32,000 nucleotides long.[8] In 2005 scientists at the U.S. Centers for Disease Control and Prevention used sequence data derived from the frozen or paraffin-fixed cells of victims to reconstruct the genome of the "Spanish" strain of influenza virus, which was responsible for the flu pandemic of 1918–19 that killed tens of millions of people worldwide. The rationale for resurrecting this extinct virus was to gain insights into why it was so virulent. In late 2006, scientists resurrected a "viral fossil," a human retrovirus that had been incorporated into the human genome around five million years ago.[9]

In 2008 a bat virus related to the causative agent of human SARS was re-created in the laboratory.[10] That same year, the J. Craig Venter Institute synthesized an abridged version of the genome of the bacterium *Mycoplasma genitalium*, consisting of 583,000 DNA base pairs.[11] In May 2010, scientists at the Venter Institute announced the synthesis of the entire genome of the bacterium *Mycoplasma mycoides*, consisting of more than one million DNA base pairs, and its successful replication in bacterial cells.[12] The total synthesis of a bacterial genome from chemical building blocks was a major milestone in the use of DNA synthesis techniques to create more complex and functional products.

Utility of the Technology
Synthesizing a genome from scratch is a significant methodological shift from recombinant DNA technology, which involves cutting and splicing preexisting genetic material. Because chemical synthesis can create any conceivable DNA sequence, the technique allows for greater efficiency and versatility in existing areas of research while opening new paths of inquiry and innovation that were previously constrained.

Potential for Misuse

Only a few viral pathogens have any real military utility. Traditional effectiveness criteria for antipersonnel agents are infectivity (the ability to infect humans reliably and cause disease), virulence (the severity of the resulting illness), persistence (the length of time the pathogen remains infectious after being released into the environment), and stability when dispersed as an aerosol cloud. Early U.S. developers of biological weapons preferred veterinary diseases such as anthrax, tularemia, and Venezuelan equine encephalitis, which are not contagious in humans, because they made a biological attack more controllable. The Soviet Union, in contrast, weaponized contagious diseases such as pneumonic plague and smallpox for strategic attacks against distant targets in the belief that the resulting epidemic would not boomerang against the Soviet population. The choice of pathogen also depends on the intended use, such as whether the aim is to kill or incapacitate, contaminate an area for long periods, or trigger a major epidemic.

Of the pathogenic viruses that can be found in nature, some are easier to isolate than others. Filoviruses such as Marburg and Ebola have animal reservoirs that are unknown, poorly understood, or accessible only during active outbreaks. As a result, isolating these viruses from a natural source would require skill, good timing, and the ability to transport the virus safely from the site of an outbreak. Because it is not easy to isolate natural strains with the desired characteristics, most pathogens developed in the past as biological weapons were either deliberately bred or genetically modified.

The increased accessibility and affordability of DNA synthesis techniques could eventually make it easier for would-be bioterrorists to acquire dangerous viral pathogens, particularly those that are restricted to a few high-security labs (such as the smallpox virus), difficult to isolate from nature (such as the Ebola and Marburg viruses), or extinct (such as the Spanish influenza virus). In theory, DNA synthesis techniques might also permit the creation of bioengineered agents more deadly and communicable than those that exist in nature, but this scenario appears unlikely. As Tucker and Zilinskas note:

To create such an artificial pathogen, a capable synthetic biologist would need to assemble complexes of genes that, working in union, enable a microbe to infect a human host and cause illness and death. Designing the organism to be contagious, or capable of spreading from person to person, would be even more difficult. A synthetic pathogen would also have to be equipped with mechanisms to block the immunological defenses of the host, characteristics that natural pathogens have acquired over eons of evolution. Given these daunting technical obstacles, the threat of a synthetic 'super-pathogen' appears exaggerated, at least for the foreseeable future.[13]

Accordingly, the most immediate risk of misuse associated with DNA synthesis technology is the re-creation of known viral pathogens, rather than the creation of entirely new ones. (Since bacterial genomes are generally far larger than viral genomes, synthesizing them is more difficult and time-consuming.) Although the primary threat of misuse comes from state-level biological warfare programs, two possible scenarios involving individuals provide cause for concern. The first scenario involves a "lone operator," such as a highly trained molecular biologist who is motivated to do harm by ideology or personal grievance. The second scenario involves a "biohacker" who does not necessarily have malicious intent but seeks to create bioengineered organisms out of curiosity or to demonstrate technical prowess, a common motivation of many designers of computer viruses. As synthetic biology training becomes increasingly available to students at the college and even high-school levels, a "hacker culture" may emerge, increasing the risk of reckless or malevolent experimentation.[14]

Accessibility

The synthetic DNA strands needed for the synthesis of many viruses can be ordered from commercial suppliers over the Internet. In addition, benchtop DNA synthesizers are becoming more efficient and are increasingly available for purchase at an affordable cost.

Ease of Misuse

In assessing the potential misuse of DNA synthesis, it is important to examine the role of tacit knowledge in synthesizing a pathogen at the laboratory bench. The construction of a pathogenic virus by assembling pieces of synthetic DNA requires extensive training in basic molecular-biology techniques, such as ligation and cloning, including hands-on experience that is not "reducible to recipes, equipment, and infrastructure."[15] This requirement for tacit knowledge is what the U.S. National Science Advisory Board for Biosecurity (NSABB) meant when it noted: "The technology for synthesizing DNA is readily accessible, straightforward and a fundamental tool used in current biological research. In contrast, the science of constructing and expressing viruses in the laboratory is more complex and somewhat of an art. It is the laboratory procedures downstream from the actual synthesis of DNA that are the limiting steps in recovering viruses from genetic material."[16]

Indeed, a would-be bioterrorist seeking to re-create a known viral pathogen from scratch would face a number of challenging technical hurdles. First, the synthesis of an infectious viral genome requires an accurate genetic sequence. Although DNA or RNA sequences are available for many pathogenic viruses, the quality of the sequence data varies. Genomes published in publicly available databases often contain errors, some of which may be completely disabling, while others would reduce the virulence of a synthetic virus. In addition, some of the published sequences are derived not from natural viral strains but from cultures that have spent many generations in the lab and have lost virulence through attenuating mutations.

A second difficulty with synthesizing a highly pathogenic virus is ensuring its infectivity. For some viruses, such as poliovirus, the genetic material is directly infectious, so that introducing it into a susceptible cell results in the production of complete virus particles. For other viruses, such as causative agents of influenza and smallpox, the viral genome itself is not infectious and requires additional components (such as enzymes involved in replicating the genetic material), whose function must be replaced.

A third technical hurdle relates to the characteristics of the viral genome. Viruses with large genomes are harder to synthesize than those with small genomes. In addition, RNA viruses with one positive strand are easier to construct than RNA viruses with one negative strand, which in turn are easier to synthesize than double-stranded DNA viruses. Thus poliovirus is relatively easy to synthesize because it has a small genome made of positive-stranded RNA, whereas the smallpox (variola) virus is hard to synthesize because it has a very large genome made up of double-stranded DNA. Synthesizing the Marburg and Ebola viruses would be moderately difficult: although their genomes are relatively small, they are not directly infectious, and inducing them to produce virus particles would be challenging.[17] Despite these hurdles, the risk of misuse of DNA synthesis is expected to increase over time. One analyst has predicted that ten years from now, it may be easier to synthesize almost any pathogenic virus than to obtain it by other means.[18]

Magnitude and Imminence of Risk

Although the de novo synthesis of viral pathogens is relatively difficult today, rapid improvements in the cost, speed, and accuracy of DNA synthesis suggest that the risk of misuse of this technology will increase over time—although by how much remains a matter of debate. For the next five years, the greatest risk will involve the synthesis of a small number of highly pathogenic viruses that are currently extinct or otherwise difficult to obtain. Access to stocks of the smallpox virus and the Spanish influenza virus is tightly controlled; samples of the smallpox virus are stored at two authorized repositories in the United States and Russia, and samples of Spanish influenza exist only in a few laboratories. Synthesizing the smallpox virus would be difficult because its genome is one of the largest of any virus and is not directly infectious. Although the genome of the 1918 influenza virus is relatively small and has been reconstructed and published, constructing the virus from scratch would be moderately difficult because the genome is not directly infectious.[19]

Awareness of Dual-Use Potential

In response to media and public interest in synthetic genomics, European countries and the United States have assessed the technology's dual-use risks. In August 2006, after journalists reported how easy it was to order pathogenic DNA sequences over the Internet, the British government convened an interdepartmental meeting to consider the feasibility and potential risks of de novo virus synthesis. This meeting concluded that "although there is a theoretical risk, the likelihood of misuse of this kind at the moment, and in the foreseeable future, is very low," and "additional regulation would be inappropriate at the present time." The British government acknowledged, however, that DNA synthesis techniques "will advance such that pathogenic organisms could be constructed or (more likely) modified more easily," and that the issue should be kept under review.[20] Also in 2006, the Dutch government asked the Commission on Genetic Modification (COGEM) to assess whether existing risk-management and security measures under the regulatory framework for genetically modified organisms (GMOs) were sufficient to cover developments in synthetic biology. COGEM concluded that the existing measures were adequate but that it would continue to monitor advances in the field.[21]

In the United States, however, gene-synthesis technology is viewed with greater concern. In a 2006 report, the NSABB highlighted the potential misuse of DNA synthesis to re-create Select Agent viruses in the laboratory.[22] The board urged the development of obligatory standards and practices for screening DNA synthesis orders and interpreting the results, and for retaining records on gene-length orders.

Characteristics of the Technology Relevant to Governance

Embodiment   Much of the technology involved in de novo DNA synthesis is intangible, based on specialized knowledge acquired through experimentation. The sequencing of microbial genomes, for example, is a key element of the process. At the same time, one item of specialized hardware, an automated DNA synthesizer, greatly reduces time and cost requirements.

Maturity   DNA synthesis is a mature technology that is commercially available, although synthesizing accurate DNA sequences greater than 180 base pairs long and stitching them together into genome-length sequences still entail significant technical hurdles. Nevertheless the successful synthesis of a bacterial genome consisting of more than one million DNA base pairs suggests that it will soon become possible to synthesize almost any microbial genome for which an accurate genetic sequence is available.

Convergence   Genome synthesis is convergent because it draws on several other technologies, including bioinformatics, chemical DNA synthesis, and standard molecular biology techniques.

Rate of advance   Over the past two decades, the length and accuracy of the DNA sequences that can be made by chemical synthesis, and the speed of production, have advanced at an exponential rate, while the costs of synthesis have dropped at a comparable pace. According to one estimate, the cost per DNA base pair has already fallen fiftyfold and is continuing to halve every thirty-two months.[23]

International diffusion   Although a few decades ago DNA synthesis techniques were accessible only to a handful of top research groups working in state-of-the-art facilities, these methods have become more widely available as they are refined and simplified. A 2007 survey estimated that at least twenty-four firms in the United States and an additional twenty-one firms worldwide can manufacture gene-length stretches of DNA, and the industry continues to grow and expand internationally.[24]

Susceptibility to Governance

Much can be done at the national or regional level to manage the risk of misuse of DNA synthesis. That only about fifty companies worldwide currently possess the advanced know-how and technical infrastructure needed to produce gene-length DNA molecules offers a window of opportunity to introduce governance measures.

Relevance of Existing Governance Measures

In Europe, concerns about genome synthesis have focused on biosafety, the nature and integrity of life, and equity and intellectual property rather than on security and deliberate misuse.[25] Typical of this approach is the European Commission's assessment that the most pressing need is "to examine whether existing safety regulations for the management of engineered microorganisms provide adequate protection against inadvertent release of 'synthetic' pathogens. In particular, who is responsible for ascertaining and quantifying risks, and for implementing any clean-up measures that might be undertaken?"[26] Two European countries, the United Kingdom and the Netherlands, stand out as having considered the biosecurity aspects of synthetic genomics in some detail, and both have concluded that the current regulatory frameworks are adequate to address the risk of misuse.

The United States has been far more aggressive in addressing the security dimensions of gene synthesis. In November 2009, the Department of Health and Human Services published a draft "Screening Framework Guidance for Synthetic Double-Stranded DNA Providers" in the *Federal Register*, and the finalized guidelines were published a year later, in October 2010.[27] These guidelines call for subjecting all requests for synthetic double-stranded DNA to a process of customer and sequence screening. Upon receiving a DNA synthesis order, the supplier should review the information provided by the customer to verify its accuracy and check for red flags suggestive of illicit activity. If the information provided raises concerns, the supplier should ask the customer for additional information. Screening the requested DNA sequence to identify any sequences derived from or encoding Select Agents is also recommended. If the customer or the sequence raises concerns, providers are urged to clarify the intended end use. In cases where follow-up screening does not resolve the concern, providers are encouraged to seek advice from designated government officials. The U.S. guidance document also recommends that providers retain electronic copies of customer orders for at least eight years, the duration of the statute of limitations for prosecution. Although adherence to the screening framework is considered voluntary, the guidance reminds providers of their legal obligations under existing export control regulations.

Recognizing the security concerns associated with synthetic DNA, a number of gene-synthesis companies have already begun screening customers and orders on their own initiative.[28] The International Association of Synthetic Biology (IASB), a consortium of mainly German companies, launched its "Code of Conduct for Best Practices in Gene Synthesis" on November 3, 2009.[29] Like the U.S. government guidance document, the IASB Code of Conduct recommends an initial screen to confirm the customer's bona fides, followed by an automated screen of the sequence order using a computer program to search for similarities between gene sequences.[30] Any hits from the automated screen are then assessed by human experts. If the hits are judged to be real and not false positives, follow-up screening is done to verify the legitimacy of the customer before the order is filled.[31]

Shortly before IASB implemented its Code of Conduct, two companies that had initially been involved in the process dropped out and established their own group, the International Gene Synthesis Consortium (IGSC). This body includes five of the world's leading gene-synthesis companies and claims to represent 80 percent of the industry.[32] Because of its large market share, the IGSC asserts it has the experience to develop workable screening measures and has put forward a "Harmonized Screening Protocol" to rival the IASB Code of Conduct.[33] As a result, gene-synthesis companies have been left to decide whether to adopt one of the three competing standards, to devise their own by mixing and matching various elements, or to ignore the process altogether.

Proposed Governance Measures

Previous surveys on the effectiveness of voluntary self-governance in the biotechnology industry have highlighted inconsistencies in the way the regimes are implemented.[34] For example, biotechnology companies vary greatly in how they use Institutional Biosafety Committees (IBCs) to review recombinant-DNA research, including the structure of the committees, the frequency of meetings, the quality of minutes produced, and whether or not the committees approve individual projects or groups of projects. The IBCs also differ in how they interpret their purpose and responsibilities.[35]

Similarly, most providers of synthetic DNA are sensitive to security concerns and would probably agree to implement some sort of screening practices if they are not doing so already, but what the minimum standards should be is unclear. Who decides if the DNA sequence database used for screening purposes is adequate? Is it sufficient to retain order records in the form of broad statistics, or must the details of each individual order be kept? Is five years long enough to retain records, rather than eight? One way to settle such questions is to establish a set of minimum screening standards through statutory legislation rather than voluntary guidance.

In devising a governance framework for the de novo synthesis of viral genomes, it is useful to think of regulation as a process that operates through three different mechanisms to influence both formal and informal behavior.[36] The "coercive" mode regulates companies by explicitly imposing certain practices through statutory legislation, the "normative" mode regulates companies by standardizing particular practices, and the "mimetic" mode regulates companies by designating particular practices as necessary for success. Compliance with the three forms of regulation confers organizational legitimacy on companies and helps to ensure their survival. The most effective regulatory frameworks include all three modes, so that companies are directed coercively, normatively, and mimetically to behave in a similar fashion.

Much of the discussion on the regulation of gene synthesis has focused on ensuring that the burgeoning gene-synthesis industry does not bear unnecessary burdens. Yet regulatory law can benefit suppliers if it increases public confidence in the technology. This advantage is particularly relevant in the biotechnology field because private biotech companies ultimately depend on social support for the creation of new markets. Moreover, a regulatory regime that leads companies to act in a responsible manner (and to be seen as doing so) may actually be more beneficial to the participants than a less restrictive regime that generates constant controversy and hostile campaigning.

Michael Porter has argued that strict environmental regulations, rather than shifting external costs onto companies and burdening them relative to competitors in countries with less stringent regulations, can result in a "win-win" situation in which the companies' private costs are reduced along with the external costs they impose on the environment.[37] Porter concludes, "Strict environmental regulations do not inevitably hinder competitive advantage against foreign rivals, indeed, they often enhance it." Thus the synthetic DNA industry could potentially benefit from a regulatory regime that carefully balances stringency with legitimacy, although this solution would require companies to accept a certain regulatory burden.

Arguing for statutory legislation is not meant to imply that voluntary measures have no merit. Self-governance may provide incentives for companies to behave in a responsible way. The reward for adopting screening practices, for example, is inclusion in the club of companies that are seen as reputable and doing the right thing, sending a positive signal to customers and investors. In this way, successful companies can help to regulate others by designating screening practices as necessary for economic success. If, however, the screening guidelines are not generally adhered to, then a reliance on self-governance may discourage other companies from implementing them, especially when costs are involved. In this type of situation, the force of law can be particularly persuasive. Indeed, the gene-synthesis industry has recognized the problem of noncompliance with voluntary guidelines. A workshop report from the IASB notes, "Ultimately, the definition of standards and the enforcement of compliance with these is a government task."[38]

Statutory legislation is also important for managing rogue companies. Commenting on the IASB's early efforts to develop a code of conduct, an editorial in *Nature* argued that they were "laudable first steps," but synthetic DNA providers "still need independent oversight" in the form of statutory legislation. "There have been, and will probably continue to be, companies that are not interested in cooperating with any industry group, and that are happy to operate in the unregulated grey area. The ultimate hope is that customers will put economic pressure on those non-compliers to fall in line, or else lose all but the most disreputable business. But that is just a hope. As the recent meltdowns on Wall Street have indicated, industry self-policing can sometimes fail dramatically. When bad business practices can have grave effects for the public, regulators should be firm and proactive."[39]

Another approach is professionalization, which lies between self-governance and statutory measures. In most jurisdictions, professional practitioners such as physicians and lawyers are licensed and belong to an association established by law, which sets the standards of practice for its members to align them with the public good. The officers of the association are elected by the members and are expected to be advocates for the profession. This combination of a legislated mandate and collegial self-governance provides accountability for the profession as a whole and for its individual practitioners. Weir and Selgelid argue that the professionalization of synthetic biology would establish educational standards for its members and define normative standards of practice, with the aim of ensuring competence and preventing misconduct. By combining self-governance and legally authorized governance, this approach avoids the polarization between the two approaches that has characterized much of the debate about the regulation of synthetic biology.[40]

Conclusions

DNA synthesis is a powerful enabling technology that has many beneficial applications but also entails a significant risk of misuse. An optimal strategy to limit this risk would entail applying the three modes of governance (coercive, normative, and mimetic) to DNA synthesis so that (1) national governments regulate companies by imposing a baseline of minimum security measures that all providers of synthetic DNA must adopt; (2) the DNA synthesis community reinforces the statutory legislation through a professional code of conduct that regulates gene-synthesis companies across borders and encourages universal adherence despite differing national assessments of the risk of misuse; and (3) role-model companies, such as commercial suppliers that have adopted the IASB or IGSC protocols, regulate other companies by designating screening practices as necessary for economic success, much as ISO accreditation and other nonstatutory regimes have become accepted as requirements to operate in other fields.

Notes

1. Royal Academy of Engineering, *Synthetic Biology: Scope, Applications, and Implications* (London: Royal Academy of Engineering, 2009), 13.

2. Gerald Epstein, "The Challenges of Developing Synthetic Pathogens," *Bulletin of the Atomic Scientists* Web site, May 19, 2008, http://www.thebulletin.org/web-edition/features/the-challenges-of-developing-synthetic-pathogens.

3. National Academies of Sciences, *Globalization, Biosecurity, and the Future of the Life Sciences* (Washington: National Academies of Sciences, 2006), 143–145.

4. Har Gobind Khorana, "Total Synthesis of a Gene," *Science* 203, no. 4381 (February 16, 1979): 614–625.

5. Wlodek Mandecki, Mark A. Hayden, Mary Ann Shallcross, et al., "A Totally Synthetic Plasmid for General Cloning, Gene Expression, and Mutagenesis in *Escherichia coli*," *Gene* 94, no. 1 (September 28, 1990): 103–107.

6. Jeronimo Cello, Aniko V. Paul, and Eckard Wimmer, "Chemical Synthesis of Poliovirus cDNA: Generation of Infectious Virus in the Absence of Natural Template," *Science* 297, no. 5583 (August 9, 2002): 1016–1018.

7. Hamilton O. Smith, Clyde A. Hutchinson III, Cynthia Pfannkoch, et al., "Generating a Synthetic Genome by Whole Genome Assembly: φX174 Bacteriophage from Synthetic Oligonucleotides," *Proceedings of the National Academy of Sciences* 100, no. 26 (November 3, 2003): 15440–15445.

8. Jingdong Tian, Hui Gong, Nijing Sheng, et al., "Accurate Multiplex Gene Synthesis from Programmable DNA Microchips," *Nature* 432 (December 23/30, 2004): 1050–1054; Sarah J. Kodumai, K. G. Patel, R. Reid, et al., "Total Synthesis of Long DNA Sequences: Synthesis of a Contagious 32-kb Polyketide Synthase Gene Cluster," *Proceedings of the National Academy of Sciences* 101, no. 44 (September 17, 2004): 15573–15578.

9. Martin Enserink, "Viral Fossil Brought Back to Life," *Science Now*, November 1, 2006, http://news.sciencemag.org/sciencenow/2006/11/01-04.html.

10. Nyssa Skilton, "Man-Made SARS Virus Spreads Fear," *Canberra Times*, December 24, 2008.

11. Daniel G. Gibson, Gwynedd A. Benders, Cynthia Andrews-Pfannkoch, et al., "Complete Chemical Synthesis, Assembly, and Cloning of *Mycoplasma genitalium* Genome," *Science* 319, no. 5867 (February 29, 2008): 1215–1220.

12. Daniel G. Gibson, John I. Glass, Carole Lartique, et al., "Creation of a Bacterial Cell Controlled by a Chemically Synthesized Genome," *Science* 329 (July 2, 2010): 52–56. See

also Elizabeth Pennisi, "Synthetic Genome Brings New Life to Bacterium," *Science* 328 (May 21, 2010): 958–959.

13. Jonathan B. Tucker and Raymond A. Zilinskas, "The Promise and Perils of Synthetic Biology," *New Atlantis* 25 (spring 2006): 38.

14. Ibid., 40–42.

15. Kathleen Vogel, "Bioweapons Proliferation: Where Science Studies and Public Policy Collide," *Social Studies of Science* 36, no. 5 (2006): 676.

16. National Science Advisory Board for Biosecurity (NSABB), *Addressing Biosecurity Concerns Related to the Synthesis of Select Agents* (Bethesda, MD: National Institutes of Health, 2006), 4.

17. Ibid.

18. Epstein, "Challenges of Developing Synthetic Pathogens."

19. Michele S. Garfinkel, Drew Endy, Gerald L. Epstein, and Robert M. Friedman, *Synthetic Genomics: Options for Governance* (Rockville, MD: J. Craig Venter Institute, October 2007), 13–14.

20. United Kingdom, Department for Business, Enterprise, and Regulatory Reform, *The Potential for Misuse of DNA Sequences (Oligonucleotides) and the Implications for Regulation* (2006), http://www.dius.gov.uk/partner_organisations/office_for_science/science_in _government/key_issues/DNA_sequences.

21. Commission on Genetic Modification (COGEM), *Biological Machines? Anticipating Developments in Synthetic Biology*, CGM/080925-01 (September 2008). See also Rinie van Est, Huib de Vriend, and Bart Walhout, *Constructing Life: The World of Synthetic Biology* (The Hague: Rathenau Institute, 2007). The Rathenau Institute, the national technology assessment organization in the Netherlands, encourages social debate and development of political opinion on technological and scientific advances.

22. National Science Advisory Board for Biosecurity (NSABB), *Addressing Biosecurity Concerns Related to the Synthesis of Select Agents* (Bethesda, MD: National Science Advisory Board for Biosecurity, 2006), http://oba.od.nih.gov/biosecurity/biosecurity _documents.html.

23. Garfinkel, Endy, Epstein, and Friedman, *Synthetic Genomics*, 10.

24. Ibid., 2.

25. Daniel Feakes, "Synthetic Biology and Security: A European Perspective," *WMD Insights* (December 2008–January 2009); Filippa Lentzos, "Synthetic Biology in the Social Context: The UK Debate to Date," *BioSocieties* 4, nos. 3–4 (2009): 303–315; Markus Schmidt, "Public Will Fear Biological Accidents, Not Just Attacks," *Nature* 441 (June 29, 2006): 1048.

26. European Commission, *Synthetic Biology: Applying Engineering to Biology*, Report of a NEST High-Level Expert Group (Brussels: Directorate General Research, 2005), 18.

27. U.S. Department of Health and Human Services, "Screening Framework Guidance for Synthetic Double-Stranded DNA Providers," *Federal Register* 74, no. 227 (November 27, 2009): 62319–62327.

28. Industry Association Synthetic Biology (IASB), report on the workshop "Technical Solutions for Biosecurity in Synthetic Biology," held in Munich, Germany, April 3, 2008, p. 8, http://www.synbiosafe.eu/uploads///pdf/iasb_report_biosecurity_syntheticbiology.pdf.

29. Corie Lok, "Gene-Makers Put Forward Security Standards," *Nature Online*, November 4, 2009, http://www.nature.com/news/2009/041109/full/news.2009.1065.html. See the IASB Web site for the code itself, http://www.ia-sb.eu/go/synthetic-biology.

30. One such screening program is the U.S. National Center for Biotechnology Information's Basic Local Alignment Search Tool (BLAST).

31. IASB, report on the workshop "Technical Solutions for Biosecurity in Synthetic Biology," 4.

32. Erika Check Hayden, "Gene-Makers Form Security Coalition," *Nature Online*, November 18, 2009, http://www.nature.com/news/2009/091118/full/news.2009.1095.html. For the Harmonized Screening Protocol, see the International Gene Synthesis Consortium (IGSC) Web site, http://www.genesynthesisconsortium.org.

33. Yudhijit Bhattacharjee, "Gene Synthesis Companies Pledge to Foil Bioterrorists," *Science*Insider, November 19, 2009, http://news.sciencemag.org/scienceinsider/2009/11/gene-synthesis.html.

34. Filippa Lentzos, "Managing Biorisks: Considering Codes of Conduct," *Nonproliferation Review* 13, no. 2 (2006): 211–226.

35. Ibid., 220.

36. Filippa Lentzos, "Countering Misuse of Life Sciences through Regulatory Multiplicity," *Science and Public Policy* 35, no. 1 (February 2008): 55–64; Filippa Corneliussen, "The Legal Trichotomy: Biotech Company Perspectives on Institutional Elements Constraining Research Activities," *Zeitschrift für Rechtssoziologie* 22, no. 2 (2001): 1–18.

37. Michael E. Porter and Claas van der Linde, "Green and Competitive: Ending the Stalemate," *Harvard Business Review*, September–October 1995, 120–134. See also H. Landis Gabel and Bernard Sinclair-Desgagné, "The Firm, Its Procedures, and Win-Win Environmental Regulations," *INSEAD Working Paper*, N 99/05/EPS (1999).

38. IASB, report on the workshop "Technical Solutions for Biosecurity in Synthetic Biology," 14.

39. "Pathways to Security" [editorial], *Nature* 455 (September 25, 2008): 432.

40. Lorna Weir and Michael J. Selgelid, "Professionalization as a Governance Strategy for Synthetic Biology," *Systems and Synthetic Biology* 3 (2009): 91–97.

9

Synthetic Biology with Standard Parts

Alexander Kelle

Nuclear physics was the leading science of the twentieth century, but biology is poised to dominate the twenty-first, with synthetic biology perhaps its most ambitious manifestation. This emerging discipline involves "the synthesis of complex, biologically based (or inspired) systems which display functions that do not exist in nature."[1] If synthetic biology delivers on the promises of its visionaries, it will turn biology into a mechanistic science, triggering a paradigm shift comparable to the way the invention of the periodic table transformed chemistry. Although synthetic biology promises beneficial applications in several fields, this new technoscience (meaning a field where the boundary between science and technology is blurred) carries with it a risk of misuse. Accordingly, the governance measures developed for synthetic genomics should be broadened to cover all aspects of synthetic biology.

Overview of the Technology

Not surprisingly for a field still in its formative stages, synthetic biology has several different definitions.[2] This chapter focuses on a particular vision of synthetic biology involving "the design and construction of new biological parts, devices, and systems, and the re-design of existing, natural biological systems for useful purposes."[3] The goal of this emerging discipline is to build operational genetic circuits out of a set of standard biological parts, namely, pieces of synthetic DNA whose functions have been well characterized so as to minimize unintended interactions.

Parts-based synthetic biology aims to be a transformative technology. Researchers at the forefront of the effort to develop standard parts conceive of it as providing a toolbox that will enable users to design and build a multitude of biological systems, much as transistors, capacitors, and other electronic components can be assembled into an enormous variety of functional devices. Standard biological parts that have undergone quality controls will be used to build genetic modules, which can then be assembled into novel biological systems. A report by the Royal Academy of Engineering describes the design cycle for synthetic biology.[4] The cycle starts with the specification of biological parts, followed by a design step that involves detailed computer modeling. During the implementation stage, a strand of synthetic DNA corresponding to the genetic circuit is assembled and inserted into bacterial or yeast cells. In the final validation stage, the intended functionality of the circuit is verified.

A key goal of parts-based synthetic biology is to create building blocks that have a standard interface and can be assembled like Lego blocks, without having to understand their internal structure. In other words, the parts themselves will be "black-boxed," meaning that their DNA sequences can be safely ignored by the user, much as one does not have to understand how a microprocessor works to use word-processing software on a personal computer. The creation of standard genetic parts that can be manipulated at a higher level of abstraction will ultimately make synthetic biology methods more accessible to nonbiologists and amateurs.[5]

History of the Technology

The effort to convert biology into a predictive science by incorporating elements of the engineering design cycle was initially called "open-source biology" or "intentional biology." Rob Carlson, then a research fellow at the Molecular Sciences Institute of the University of California, Berkeley, developed a vision of open-source biological manufacturing driving future industry. In 2001 he wrote: "When we can successfully predict the behavior of designed biological systems, then an intentional biology will exist. With an explicit engineering component, intentional biology is the opposite of the current, very nearly random applications of biology as technology."[6]

The Massachusetts Institute of Technology was an early hub of synthetic biology. During the 1990s, electrical engineering professor Thomas Knight set up a biology laboratory within the MIT Laboratory for Computer Science and started to develop standard genetic parts called "BioBricks" with funding from the U.S. Defense Advanced Research Projects Agency (DARPA) and the Office of Naval Research.[7] Since then the BioBricks have been incorporated into a Registry of Standard Biological Parts, an open-source database that is available to all bona fide researchers in the field.[8]

The main vehicle for expanding the number of publicly available biological parts is the International Genetically Engineered Machine (iGEM) competition, which is held annually at MIT by the BioBricks Foundation and involves teams of undergraduate university students from around the world.[9] The goals of iGEM are "to enable the systematic engineering of biology, to promote the open and transparent development of tools for engineering biology, and to help construct a society that can productively apply biological technology."[10] This annual competition also serves as an indicator of the international diffusion of parts-based synthetic biology.[11] Starting in 2003 with a small group of teams from the United States, iGEM has since become a global event. In 2005, when foreign students began to participate, 13 undergraduate student teams from 4 countries (Canada, Switzerland, the United Kingdom, and the United States) participated.[12] Four years later, in 2009, there were 112 student teams from nearly 20 countries around the globe.[13] In 2010, 118 teams from 26 countries participated.[14]

The iGEM teams not only use existing biological parts but design and build new ones, which may then be incorporated into the registry. The 2006 competition introduced more than 700 new parts, the 2007 competition about 800 parts, the 2008 competition nearly 1,400 parts, the 2009 competition 1,348 parts, and the 2010 competition 1,863 parts.[15] Over the next several years, the number of standard biological parts is expected to continue its exponential growth. In February 2010, scientists at Stanford University and the University of California, Berkeley, received a $1.4 million grant from the National Science Foundation to establish a laboratory called BIOFAB, which will design, fabricate, and test some 3,000 standard genetic parts, including DNA control elements.[16]

Utility of the Technology

Optimistic assessments of synthetic biology, such as a 2005 report to the European Commission by the expert group on New and Emerging Science and Technology (NEST), predict that it will "drive industry, research, education and employment in the life sciences in a way that might rival the computer industry's development during the 1970s to the 1990s."[17] The NEST report identified six areas that could benefit from synthetic biology techniques: biomedicine, synthesis of biopharmaceuticals, sustainable chemical industry, environment and energy, production of smart materials and biomaterials, and measures to counter bioterrorism.[18]

The most frequently cited example of a high-value application of synthetic biology is the work by Jay Keasling and his colleagues at the University of California, Berkeley, to insert an engineered metabolic pathway into yeast to produce artemisinic acid, the immediate precursor of artemisinin, a key antimalaria drug.[19] The goal of this project is to reduce the production cost for artemisinin and thereby increase its availability to people in developing countries.[20] Synthetic biologists have also made progress using bioengineered microbes to produce biofuels.[21] Although the main focus of synthetic biology is on the design and engineering of artificial microorganisms that perform useful tasks, some biologists emphasize the field's contribution to "achieving a better understanding of life processes."[22] Such improved understanding can better the human condition by leading to the development of new diagnostic tools, therapeutic drugs, and other beneficial applications.

Potential for Misuse

The newfound ability to understand, modify, and ultimately create new life-forms could bring with it a substantial potential for misuse. Once the standardization of genetic parts and modules has progressed to the point where they function reliably and can be inserted into a simplified bacterial genome to achieve specific applications, the technology will cross a threshold of dual-use potential. Precisely when this transition occurs will depend on the speed at which the field progresses.

In principle, malicious actors could exploit parts-based synthetic biology to increase the efficiency, stability, and usability of classical biological warfare agents or to create new ones. To date, however, the primary concern with synthetic biology has been the synthesis of known pathogenic viruses rather than the use of standard parts to create novel pathogens. Because parts-based synthetic biology is still a cutting-edge technology, bioterrorists would have to overcome formidable technical challenges to exploit it for harmful purposes.[23] Nevertheless the increasing speed, accuracy, and accessibility of gene-synthesis technology, and its explicit deskilling agenda, are likely to lower these barriers over time.[24]

Accessibility

Access to the Registry of Standard Biological Parts is currently limited to recognized research laboratories, regardless of their geographic location, although this rule may change in the future. The registry is open source, and no one has attempted to patent the genetic components or restrict access on the basis of nationality or other criteria. Some laboratory supply companies have also begun to sell synthetic-biology reagent kits and how-to manuals to interested biologists, both professional and amateur.[25] Examples include the "BioBrick Assembly Kit" and the "BioBrick Assembly Manual," distributed jointly by Ginko BioWorks and New England Biolabs.[26]

Ease of Misuse

Misuse of parts-based synthetic biology is difficult because of the current absence of weaponizable parts and devices. Because synthetic biology requires considerable explicit and tacit knowledge, the greatest risk of misuse probably resides in state-level offensive biological warfare programs. As hands-on experience with synthetic biology continues to spread internationally, however, the nature of the risk will change. In addition to the annual iGEM competition, which has helped to popularize synthetic biology, leading synthetic biologists have pursued a number of deliberate deskilling efforts, including the dissemination of do-it-yourself synthetic-biology kits and how-to protocols. As a result of these efforts, the field of synthetic biology will gradually become easier for nonspecialists, potentially enabling nonstate actors to use genetic parts and modules for nefarious purposes.

Magnitude and Imminence of Risk

Given the early stage of development of parts-based synthetic biology, the risk that a terrorist or criminal group could order standard biological parts and construct an artificial pathogen for harmful purposes is extremely low. As a result, the threat is not imminent. Should the technology proliferate widely, however, the magnitude of the potential consequences of deliberate misuse could be considerable.

Awareness of Dual-Use Potential

In 2006 the National Research Council, the policy analysis arm of the U.S. National Academies, tasked an expert committee chaired by the microbiologists Stanley Lemon and David Relman to analyze the security implications of the revolution in the life sciences.[27] The Lemon-Relman committee concluded that the de novo synthesis of existing pathogens poses a greater near-term threat than the creation of artificial organisms through the parts-based approach.[28] Similarly, the National Science Advisory Board on Biosecurity (NSABB) and its Synthetic Biology Working Group have focused primarily on the use of DNA synthesis technology to re-create Select Agents, defined as biological agents and toxins that pose a severe threat to public, animal, or plant health, or to animal or plant products.[29] In sum, the primary focus of risk assessment to date has been on the synthesis of Select Agent viruses rather than the construction of functional genetic modules from standard biological parts.

Characteristics of the Technology Relevant to Governance

Embodiment   Parts-based synthetic biology relies both on automated DNA synthesizers (hardware) and on powerful bioinformatics tools (software) that allow the modeling of desired biological functions and the design of genetic circuits. Even so, the dichotomy between hardware and intangible information is not as clear as it is for other emerging dual-use technologies. After all, the basic building blocks of the parts, devices, and systems created by synthetic biology are pieces of DNA, which is sometimes described as a computer code for programming living matter.

Maturity   Parts-based synthetic biology is still at an early stage of research and development, although it is emerging from the proof-of-principle phase.

Convergence   Parts-based synthetic biology is a convergent technology that seeks to apply engineering concepts to living systems and also draws on two key enabling technologies, chemical DNA synthesis and bioinformatics, for the design and redesign of genetic circuits.

Rate of advance   At first glance, the field's rate of progress appears dramatic, with an exponential increase in the number of standardized parts available in the

MIT Registry of Standard Biological Parts. A few caveats are in order, however. Practitioners have faced several difficulties in "taming the complexity of living systems," such as unintended interactions among genes and a tendency of components to mutate spontaneously.[30] According to the registry's director, Randy Rettberg, "About 1,500 registry parts have been confirmed as working by someone other than the person who deposited them."[31] More skeptical observers claim, however, that only 5 to 10 percent of the standard parts actually function as predicted.[32] According to this estimate, at most 500 of the roughly 5,000 parts in the registry may actually be usable, sharply reducing the number of possible devices that could be built from them.

Another progress-limiting factor is the issue of standardization, or defining the specific technical parameters of the biological parts in the registry. This issue has been the focus of a series of workshops sponsored by the BioBricks Foundation, including discussions from both the technical and the legal standpoints.[33] To date, however, the field of synthetic biology has not been able to develop an unambiguous set of standards for biological parts. Adam Arkin, a professor of bioengineering at the University of California, Berkeley, notes that one reason for the lack of progress is that "unlike many other engineering disciplines, biology does not yet possess a theory of what the minimal information about a biological part should be."[34] As a result, according to Stanford University professor Drew Endy and his colleagues, "the design and construction of engineered biological systems remains an ad hoc process for which costs, times to completion, and probabilities of success are difficult to estimate accurately."[35] Until practitioners of synthetic biology can agree on a uniform set of standards for biological parts, it will not be possible to elevate activities such as the biosynthesis of artemisinin from traditional metabolic engineering into the realm of parts-based synthetic biology.[36]

International diffusion   In parallel with the rapid growth of the iGEM movement, the international Synthetic Biology Conferences held since 2004 have attracted a growing number of participants from a variety of countries. That the second and third conferences were held in Zurich and Hong Kong, respectively, is an indicator of the rapid global diffusion of the discipline. A more recent development is the so-called do-it-yourself (DIY) movement. One of the more prominent groups in this field, DIYbio, describes itself as an "organization that aims to help make biology a worthwhile pursuit for citizen scientists, amateur biologists, and DIY biological engineers who value openness and safety."[37] Noteworthy from a dual-use perspective, however, is the absence of "security" in the group's list of objectives. Several DIYbio groups have emerged, most of them in the United States, but a few in Europe and Asia.[38]

Susceptibility to Governance

Certain areas of synthetic biology are susceptible to governance measures. Most proposals have focused on the bottleneck of gene-length DNA synthesis, an enabling technology that is currently the most mature part of the supply chain for the synthesis of biological parts. Although some proposals emphasize government involvement more strongly than others, all try to minimize the negative impact of top-down regulation on scientific progress and technological development. The notion of making the Registry of Standard Biological Parts an object of governance has yet to receive much attention. Instead policy discussion has focused on whether to keep the registry open source or to allow large biotechnology companies to claim intellectual property rights over certain biological parts to exploit them for commercial purposes.[39] If the open-source movement prevails, it could unintentionally increase the risk of misuse by making synthetic-biology capabilities more widely available.

Relevance of Existing Governance Measures

From the formative days of synthetic biology, the question of governance mechanisms and their relative utility has been a feature of the unfolding discourse about the discipline. At least in part, the early engagement of the scientific community with the potential risks of synthetic biology can be traced back to earlier debates about the ethical, legal, and social implications (ELSI) of genetic engineering. One of the first proposals to address the dual-use implications of synthetic biology was made by George Church of Harvard Medical School in 2004. He proposed a system for licensing DNA synthesizers and reagents and for screening DNA sequence orders sent to commercial suppliers to determine the extent of homology with Select Agent genes.[40]

Although Church's proposals foresaw some form of government involvement or oversight in both areas, subsequent policy analysts have emphasized self-governance by the scientific community. The Second International Conference on Synthetic Biology, held in May 2006 at the University of California, Berkeley, devoted a full day to the societal issues surrounding synthetic biology, including "biosecurity and risk."[41] The discussion during this session was informed by a white paper prepared by the Berkeley SynBio Policy Group titled "Community-Based Options for Improving Safety and Security in Synthetic Biology."[42] Although the paper was intended to serve as the basis of a code of conduct for the synthetic-biology community, more than three dozen civil-society groups publicly criticized the paper because they had been excluded from its preparation and believed that the proposed measures were too weak. Synthetic biology practitioners attending the meeting also declined to endorse the proposed code of conduct.[43] Instead the final conference declaration merely called for "ongoing and future discussions with all stakeholders for the purpose of developing and analyzing governance options . . . such that the development and application of biological technology remains overwhelmingly constructive."[44]

A subsequent analysis of governance options for synthetic biology, conducted jointly by MIT, the J. Craig Venter Institute, and the Center for Strategic and International Studies (CSIS), followed the trend of focusing narrowly on DNA synthesis technology, with a primary emphasis on commercial suppliers of gene-length DNA sequences. This report, released in 2007, concluded that a combination of screening of DNA synthesis orders and the certification of such orders by a laboratory biosafety officer would provide the greatest benefit in preventing the misuse of synthetic DNA.[45] The ideas in the report were subsequently pursued by two industry associations in the field of gene synthesis, as well as by the U.S. government. It should be noted, though, that all three processes resulted in slightly different sets of voluntary screening guidelines.

Given the lack of legally binding governance measures for synthetic biology, it is desirable to review existing domestic legislation and regulations in the field of genetic engineering to assess their applicability to biological parts and modules.[46] One of the limiting factors for such an expansion is that the U.S. approach to regulation of genetic engineering focuses on the product rather than the process, resulting in downstream risk assessments that are poorly suited for assessing dual-use biosecurity risks in general, and risks stemming from parts-based synthetic biology in particular. Because governance measures should seek to prevent the misuse of biological parts before it occurs, intervention is required further upstream.

Relevant to this process are the efforts under way to standardize biological parts.[47] The data sheets produced for standard biological parts and devices should include information about their dual-use potential. In cases where the risk of misuse is judged to be high, additional governance measures may be warranted. More generally, the Select Agent Regulations may have to be expanded with a view to covering not only pathogenic DNA sequences that have a natural equivalent or a parent organism but also sequences whose origin is completely synthetic.

Proposed Governance Measures

A complete or even partial ban on parts-based synthetic biology is unlikely to find much political support. Unless an accident or deliberate attack involving an artificial microorganism results in serious harm or economic losses, the beneficial applications of synthetic biology will generally be perceived as outweighing the risks. Nevertheless exclusive reliance on self-governance by industry associations and professional societies is unlikely to provide sufficient reassurance to the public that synthetic biology will not someday be misused for nefarious purposes.

Governance measures focusing primarily on commercial providers of synthetic DNA are necessary but not sufficient. Instead what is required is a balanced mix of top-down and bottom-up governance measures. I have termed this approach the "5P" strategy because it includes five policy intervention points: the principal investigator, the project, the premises, the provider, and the purchaser. Various governance measures can be applied at each of these intervention points, including awareness raising, education and training, guidelines, codes of conduct, regulations, national laws, and international agreements.[48]

For bottom-up measures to be effective, it will first be necessary to raise the awareness of dual-use issues among practitioners of synthetic biology. In 2007 interviews with twenty leading European synthetic biologists revealed that their awareness of key elements of the biosecurity discourse (such as the Lemon-Relman report) was extremely low.[49] Although several conferences on synthetic biology have provided opportunities to discuss dual-use issues, they have not added up to a systematic effort to raise the awareness of practitioners. It is also essential to educate future generations of synthetic biologists about the dual-use implications of the field they are about to enter. A survey of 142 courses at 57 universities in 29 countries found that only a small number of courses in the life sciences include a biosecurity component, and in every case this component is optional and not part of the core curriculum.[50]

In contrast to academic practitioners of synthetic biology, industrial scientists who work for DNA synthesis companies are more aware of the dual-use potential of their work, as evidenced by the biosecurity measures that the gene-synthesis industry has implemented on a voluntary basis. According to an industry official, oversight and regulation offer two advantages: they "reassure the public that biosafety and biosecurity concerns are addressed," and "provide legal security to the industry by defining clear compliance rules."[51]

Similar governance measures will have to be formulated for the wider circle of users of standard biological parts. Such rules could take the form of end-user certificates that researchers would have to submit before being granted access to the registry. There might also be a formal requirement for the registration of laboratories or companies involved in assembling biological parts into devices or more complex biological systems. Although such a step would be motivated primarily by biosafety concerns, it could also help to prevent deliberate misuse.

At the international level, countries participating in the Australia Group have agreed to harmonize their national export controls on genetic material from listed pathogens. It would be desirable to expand this measure to cover standard biological parts that could be used for harmful purposes. Although the Biological Weapons Convention (BWC) bans the acquisition of genetically modified as well as natural pathogens and toxins for hostile purposes, the treaty does not restrict research but prohibits the development, production, acquisition, and stockpiling of biological weapons. Creating a comprehensive governance mechanism for parts-based synthetic biology would help to fill this gap in the biological weapons prohibition. To achieve the international harmonization of national governance mechanisms, states should consider negotiating a framework convention on preventing the misuse of synthetic biology that could be updated fairly easily to keep pace with rapid technological advances.

Conclusions

Synthetic biology is one of the most dynamic new subfields of biotechnology, with the potential to engineer biological parts, devices, and systems for socially useful purposes. By applying the tools of engineering and information technology to biology, it may be possible to realize a wide range of useful applications. Some of these potential benefits could be transformative, such as the use of engineered bacteria to produce fine chemicals, low-cost drugs, and biofuels. At the same time, synthetic biology entails significant risks of deliberate or accidental harm and can be described as a prototypical dual-use technoscience.

Early attempts to formulate governance mechanisms for synthetic biology have focused almost entirely on a key enabling technology, the commercial synthesis of gene-length DNA sequences. In contrast, a governance system for parts-based synthetic biology is still in its infancy. Developing such a system will require awareness-raising efforts that reach all practitioners of synthetic biology, the integration of dual-use education into life science and bioengineering curricula, and the formulation of codes of conduct that go well beyond the screening of DNA synthesis orders. Domestic governance measures could include the mandatory registration of all individuals who are granted access to standard biological parts, and international governance might involve the negotiation of a framework convention to prevent the misuse of synthetic biology.

Notes

1. New and Emerging Science and Technology (NEST), *Synthetic Biology: Applying Engineering to Biology*, report of a NEST High-Level Expert Group (Brussels: European Commission, 2005), 5.

2. Markus Schmidt, "Do I Understand What I Can Create?" in *Synthetic Biology: The Technoscience and Its Societal Consequences*, ed. Markus Schmidt, Alexander Kelle, Agomoni Ganguli-Mitra, and Huib de Vriend (Dordrecht, Netherlands: Springer, 2009), 81–100. Other typologies have been proposed: Maureen O'Malley, Alexander Powell, Jonathan F. Davies, and Jane Calvert, "Knowledge-Making Distinctions in Synthetic Biology," *BioEssays* 30, no. 1 (2007): 57–65; and Anna Deplazes, "Piecing Together a Puzzle: An Exposition of Synthetic Biology," *EMBO Reports* 10, no. 5 (2009): 428–432.

3. See http://syntheticbiology.org/Who_we_are.html.

4. Royal Academy of Engineering, *Synthetic Biology: Scope, Applications, and Implications* (London, 2009), 18–21.

5. Jeanne Whalen, "In Attics and Closets, 'Biohackers' Discover Their Inner Frankenstein," *Wall Street Journal*, May 12, 2009, http://online.wsj.com/article/SB124207326903607931 .html.

6. Rob Carlson, "Open Source Biology and Its Impact on Industry," *IEEE Spectrum*, May 2001, 15–17.

7. Thomas F. Knight, *DARPA BioComp Plasmid Distribution 1.00 of Standard BioBrick Components*, 2002, http://dspace.mit.edu/handle/1721.1/21167.

8. Registry of Standard Biological Parts, http://partsregistry.org/wiki/index.php/Main_Page.

9. BioBricks Foundation Web site, http://biobricks.org.

10. iGEM Web site, http://parts.mit.edu/wiki/index.php/About_iGEM.

11. iGEM competition, http://2009.igem.org/About.

12. The only non-U.S. teams were from ETH Zurich and the University of Toronto. For a complete list, see http://parts.mit.edu/wiki/index.php/Igem_2005.

13. Nations represented in 2009 were Belgium, Canada, the People's Republic of China, Colombia, Denmark, France, Germany, India, Italy, Japan, Korea, Mexico, the Netherlands, Poland, Spain, Slovenia, Sweden, Taiwan, the United Kingdom, and the United States. See the list of teams at http://ung.igem.org/Team_List?year=2009.

14. Valda Vinson, "Inventive Constructions Using BioBricks," *Science* 330, no. 6011 (December 17, 2010): 1629.

15. For a description of previous iGEM competitions, see http://2010.igem.org/ Previous_iGEM_Competitions.

16. Robert Sanders, "NSF Grant to Launch World's First Open-Source Genetic Parts Production Facility," *Genetic Engineering and Biotechnology News*, January 20, 2010, http://www .genengnews.com/industry-updates/nsf-grant-to-launch-world-s-first-open-source-genetic-parts-production-facility/73430839.

17. New and Emerging Science and Technology (NEST), *Synthetic Biology: Applying Engineering to Biology* (Brussels: European Commission, 2005), 13.

18. Ibid., 13–17.

19. Jay Keasling et al., "Production of the Antimalarial Drug Precursor Artemisinic Acid in Engineered Yeast," *Nature* 440 (April 13, 2006): 940–943.

20. Michelle C. Y. Chang and Jay D. Keasling, "Production of Isopronoid Pharmaceuticals by Engineered Microbes," *Nature Chemical Biology* 2, no. 12 (2006): 674–681.

21. Shotsa Azumi, Taizo Hanai, and James C. Liao, "Non-fermentative Pathways for Synthesis of Branched-Chain Higher Alcohols as Biofuels," *Nature* 451 (January 3, 2008): 86–90; Jay D. Keasling and Howard Chou, "Metabolic Engineering Delivers Next-Generation Biofuels," *Nature Biotechnology* 26, no. 3 (2008): 298–299.

22. Toward a European Strategy for Synthetic Biology (TESSY), *Synthetic Biology in Europe*, http://www.tessy-europe.eu/index.html.

23. Gerald L. Epstein, "The Challenges of Developing Synthetic Pathogens," *Bulletin of the Atomic Scientists* Web site, May 19, 2008, http://www.thebulletin.org/web-edition/features/ the-challenges-of-developing-synthetic-pathogens.

24. Michele S. Garfinkel, Drew Endy, Gerald L. Epstein, and Robert M. Friedman, *Synthetic Genomics: Options for Governance*, October 2007, 12, http://www.jcvi.org/cms/fileadmin/site/research/projects/synthetic-genomics-report/synthetic-genomics-report.pdf.

25. See Aaron Rowe, "Cloning Kits: More Fun Than a Chemistry Set?" *Wired*, May 2008, http://www.wired.com/wiredscience/2008/05/cloning-kits-mo.

26. Ginkgo BioWorks, "BioBrick Assembly Kit," http://ginkgobioworks.com/biobrickassemblykit.html.

27. National Research Council, *Globalization, Biosecurity, and the Future of the Life Sciences* (Washington, DC: National Academies Press, 2006).

28. Ibid., 109.

29. National Science Advisory Board for Biosecurity (NSABB), *Addressing Biosecurity Concerns Related to the Synthesis of Select Agents* (Washington, DC: National Institutes of Health, 2006), http://oba.od.nih.gov/biosecurity/pdf/Final_NSABB_Report_on_Synthetic_Genomics.pdf.

30. Roberta Kwok, "Five Hard Truths for Synthetic Biology," *Nature* 463 (January 21, 2010): 288.

31. Ibid.

32. Victor de Lorenzo, "Not Really New," interview in *Lab Times*, March 2009, 21–23.

33. BioBricks Foundation Web site, http://biobricks.org.

34. Adam Arkin, "Setting the Standard in Synthetic Biology," *Nature Biotechnology* 26, no. 7 (July 2008): 772.

35. Barry Canton, Anna Labno, and Drew Endy, "Refinement and Standardization of Synthetic Biological Parts and Devices," *Nature Biotechnology* 26, no. 7 (July 2008): 787.

36. Quoted in Kwok, "Five Hard Truths for Synthetic Biology," 289.

37. DIYbio Web site, http://diybio.org/about.

38. See map at http://diybio.org.

39. Kenneth A. Oye and Rachel Wellhausen, "The Intellectual Commons and Property in Synthetic Biology," in *Synthetic Biology: The Technoscience and Its Societal Consequences*, ed. Markus Schmidt, Alexander Kelle, Agomoni Ganguli-Mitra, and Huib de Vriend (Dordrecht, Netherlands: Springer, 2009), 121–140.

40. George Church, "Synthetic Biohazard Nonproliferation Proposal," http://arep.med.harvard.edu/SBP/Church_Biohazard04c.htm.

41. Synthetic Biology 2.0 conference, http://syntheticbiology.org/SB2.0/Biosecurity_and_Biosafety.html.

42. Stephen M. Maurer, Keith V. Lucas, and Starr Terrell, *From Understanding to Action: Community-Based Options for Improving Safety and Security in Synthetic Biology* (Berkeley: Goldman School of Public Policy, University of California), Draft 1.1, April 15, 2006, http://gspp.berkeley.edu/iths/UC%20White%20Paper.pdf.

43. ETC Group, "Global Coalition Sounds the Alarm on Synthetic Biology," news release, May 19, 2006, http://www.etcgroup.org/en/node/8.

44. Revised public draft of the SB2.0 declaration, http://openwetware.org/wiki/Synthetic_Biology/SB2Declaration.

45. Garfinkel, Endy, Epstein, Friedman, *Synthetic Genomics: Options for Governance.*

46. Michael Rodemeyer, *New Life, Old Bottles: Regulating First-Generation Products of Synthetic Biology* (Washington, DC: Woodrow Wilson International Center for Scholars, March 2009), http://www.synbioproject.org/library/publications/archive/synbio2.

47. Canton, Labno, and Endy, "Refinement and Standardization."

48. Alexander Kelle, "Ensuring the Security of Synthetic Biology: Towards a 5P Governance Strategy," *Systems and Synthetic Biology* 3, nos. 1–4 (December 2009): 85–90.

49. Alexander Kelle, *Synthetic Biology and Biosecurity Awareness in Europe* (Vienna, Austria: Organization for International Dialogue and Conflict Management, 2007), http://www .synbiosafe.eu/uploads///pdf/Synbiosafe-Biosecurity_awareness_in_Europe_Kelle.pdf.

50. Giulio Manchini and James Revill, *Fostering the Biosecurity Norm: Biosecurity Education for the Next Generation of Life Scientists*, University of Bradford (UK), Report No. 1, November 2008, http://www.brad.ac.uk/acad/sbtwc/dube/publications/index.html.

51. SYNBIOSAFE, *Compilation of All SYNBIOSAFE e-Conference Contributions* (Vienna, Austria: Organization for International Dialogue and Conflict Management, 2008), http:// www.synbiosafe.eu/uploads/pdf/Synbiosafe_e-conference_all_contributions.pdf.

C

Technologies for the Manipulation of Biological Systems

10

Development of Psychoactive Drugs

Malcolm R. Dando

Many neuroscientists believe that research over the next few decades will yield an integrated, mechanistic understanding of the human brain and behavior. Such an understanding could lead to effective treatments for people suffering from schizophrenia, depression, and other major mental illnesses, but it could also create new possibilities for misuse. Given the well-documented efforts by military scientists and intelligence agencies during the Cold War to employ psychoactive drugs as truth serums and incapacitating agents, it would be naive to assume that such activities have ended. Although top-down government regulation to prevent the hostile exploitation of neuroscience is unlikely anytime soon, improved governance may be possible through bottom-up initiatives.

Overview of the Technology

The standard textbook *Psychopharmacology: Drugs, the Brain, and Behavior* provides some useful definitions: "*Neuropharmacology* is concerned with drug-induced changes in the functioning of cells in the nervous system, while *psychopharmacology* emphasizes drug-induced changes in mood, thinking, and behavior. . . . In combination, the goal of *neuropsychopharmacology* is to identify chemical substances that act upon the nervous system to alter behavior that is disturbed due to injury, disease, or environmental factors."[1]

Neuropsychopharmacology can also be defined as the convergence of three disciplines: medicinal chemistry (the synthesis of new drugs), pharmacology (the study of the action of drugs on living systems), and neuroscience (the study of how genetics and the environment affect neurotransmitter/receptor systems in the brain, and how these systems in turn affect behavior). For example, the antidepressant drug fluoxetine (Prozac) selectively inhibits the reuptake into nerve endings of the neurotransmitter chemical serotonin, increasing the availability of this messenger substance in the brain of depressed people and thereby improving mood.

Despite considerable progress over the past few decades, the treatment of many mental illnesses remains rudimentary. Writing in 2001, at the dawn of the new century, the psychiatrist Nancy Andreasen divided the major mental illnesses into four categories: dementias, schizophrenia, mood disorders, and anxiety disorders. For each category, she assigned a grade of A through D for syndrome definition, understanding of the causes of the illness, and treatment. Only the mood disorders received an A, and even in that case, a detailed understanding of what goes wrong in the brain is still lacking.[2]

Contemporary neuroscience seeks to unravel the functioning of the brain in health and disease by working at several levels of analysis. During the 1990s, the use of molecular biology techniques opened the way to an improved understanding of neurotransmitters and their receptors. At the same time, advanced imaging technologies such as functional magnetic resonance imaging (fMRI) have made it possible to observe the activity of the intact human brain.[3] Over the next few decades, Andreasen believes, these different levels of analysis will converge. When that happens, she writes, "We will understand how the cells in our brains go bad when their molecules go bad, and we will understand how this is expressed at the level of systems such as attention and memory so that human beings develop diseases such as schizophrenia and depression."[4]

History of the Technology
The scientific study of the nervous system did not begin until the early twentieth century, when it became clear that the brain and spinal cord are made up of functional cells called neurons and supporting cells called glia. In the 1920s, the German physiologist Otto Loewi showed that information is transferred between neurons by chemical messenger substances called neurotransmitters, which travel across a gap called the synapse between the transmitting and the receiving cell. Loewi also identified the first such neurotransmitter substance, acetylcholine.

In 1936, the same year that Loewi shared the Nobel Prize for his discovery, the German industrial chemist Gerhard Schrader was developing new insecticides when he accidentally discovered a carbon-phosphorus compound that was extraordinarily toxic to the nervous system. It is now known that this class of chemical blocks a key enzyme in the synapse that breaks down acetylcholine after it has transmitted a nerve impulse. The excess acetylcholine leads to an overstimulation of the nervous system, resulting in convulsions, flaccid paralysis, and death. Schrader's compound later became the basis for a new generation of highly lethal chemical weapons, which Germany produced but fortunately did not use during World War II: the nerve agents tabun, sarin, and soman.

After World War II, a series of serendipitous discoveries led to the first therapeutic drugs for people suffering from schizophrenia, improving the prospects of those who had previously been treated only by incarceration. This breakthrough provided a strong incentive for additional basic and applied research in neuropsychopharmacology. At the time, a wide gulf existed between psychiatrists exploring the behavioral aspects of mental illness and neuroscientists studying the chemistry and physiology of the nervous system. During the second half of the twentieth century, however, research in neuropsychopharmacology shed light on the mode of action of many psychoactive drugs.

In one key advance, Arvid Carlsson of the University of Gothenburg in Sweden discovered that Parkinson's disease is associated with the degeneration of specific neurons in the brain that release the neurotransmitter dopamine. The resulting low levels of dopamine disrupt the brain's ability to control movement. This finding led to the therapeutic use of the dopamine precursor L-DOPA to compensate for the loss of the dopamine-producing neurons. The success of this therapy increased confidence that a mechanistic understanding of the brain was possible. Nevertheless, in a clear indication that the new knowledge had military as well as therapeutic applications, several countries developed new types of chemical weapons that targeted the brain. During the 1960s, for example, the United States produced and stockpiled the potent hallucinogen BZ as an incapacitating agent, although it was never used in combat because of its unpredictable effects.[5]

Beginning in the 1990s, scientists obtained a new understanding of neurotransmitter receptors, which are large proteins embedded in the outer membrane of neurons. The binding of a neurotransmitter changes the shape of the receptor, which in turn induces functional changes in the receiving cell. Because each receptor protein is specified by the neuronal DNA, researchers were able to use molecular-genetic techniques to identify the various classes of neurotransmitter receptors and the subtypes within each class. As a result of these studies, it is now known that the brain has far more neurotransmitters than were previously thought to exist. Many of these substances are not small molecules like acetylcholine and dopamine but rather neuropeptides, consisting of short chains of amino acids. During the late 1990s, for example, the study of a serious sleep disorder called narcolepsy led to the discovery of two new peptide neurotransmitters: the hypocretins (also called orexins), which are produced by cells in the hypothalamus.[6] Hypocretin-containing neurons provide excitatory input to a brain region called the locus coeruleus, which is crucial for maintaining wakefulness, and people with narcolepsy have low or nonexistent brain levels of hypocretin.[7]

Several other important advances in the understanding of brain chemistry have occurred in recent years. For example, scientists have found that small-molecule neurotransmitters and neuropeptides can be colocated in individual neurons, toppling the long-standing dogma that each neuron produces only one type of neurotransmitter. Certain brain chemicals called neuromodulators also affect neuronal activity over relatively long time intervals, complementing rapid synaptic transmission. Further surprises and paradigm shifts are to be expected as scientists gradually elucidate the vast complexity of the human brain.

Utility of the Technology

Neuropsychopharmacology is an applied science whose goal is to develop drugs to treat mental illnesses with a minimum of side effects. To that end, the pharmaceutical industry seeks to identify compounds that act on specific receptor subtypes in the brain, serving either as "agonists" to stimulate the receptors or as "antagonists" to block them. For example, a class of neurotransmitter receptors known as G protein–coupled receptors (GPCRs) are important targets for drug development.[8]

Designing drug molecules that bind selectively to particular subtypes of neurotransmitter receptors has been difficult because the binding sites tend to be very similar across the various receptor subtypes.[9] To get around this problem, scientists have synthesized drugs that bind to other locations on the surface of the receptor protein called "allosteric" sites, which induce conformational changes that modulate the action of the neurotransmitter. Because the allosteric sites are more variable than the neurotransmitter binding site, they provide greater opportunity for tailored drug development. For example, the development of allosteric modulators for certain subtypes of acetylcholine receptors could lead to new treatments for Alzheimer's disease and other dementias.[10]

The study of neuropeptide systems in the brain is another active area of research. Oxytocin, for example, is a neuropeptide that for decades has been associated with social bonding behavior and reproduction. In 2005 researchers demonstrated experimentally that aerosolized oxytocin, administered through the nose, has a pronounced effect in increasing trust.[11] Another study using fMRI found that a whiff of oxytocin reduces the ability of threatening images to activate a brain region called the amygdala, which generates fear in response to danger. This finding suggests that oxytocin regulates social anxiety.[12] The structure of oxytocin and a related neuropeptide called vasopressin are highly conserved across a variety of mammals, and there is growing evidence that the genes coding for these peptides and their receptors influence social behavior, both within and between species. For example, research has shown that mice whose oxytocin gene has been "knocked out" are unable to recognize other mice, even after repeated social encounters, although their sense of smell and general learning ability are unaffected. Injecting low doses of oxytocin into the amygdala, however, restores the capability for social recognition.[13] On the basis of such experiments, some scientists contend that "the molecular basis of social behavior is not beyond the realm of our understanding."[14]

Indeed, the intimate relationship between brain and behavior is suggested by a bizarre phenomenon from the annals of parasitology. A protozoan parasite called

Toxoplasma gondii reproduces in the gut of cats, which excrete the parasite's eggs in their feces. Rats then eat the cat feces and become infected with the protozoa. Finally, a cat eats an infected rat, and the life cycle of the parasite begins anew. Normally rats are afraid of cats and avoid the smell of cat urine. Research has shown, however, that the *Toxoplasma* parasite forms cysts in the amygdala of the rat brain, altering the animal's behavior, so that instead of avoiding cat urine, the rat finds it attractive. As a result, parasite-infected rats are more likely to encounter cats and be eaten, to the parasite's advantage. The researchers concluded that "the loss of fear is remarkably specific."[15] Whether new psychoactive drugs can be developed that manipulate human cognition and behavior with such a high degree of precision remains to be seen.

Potential for Misuse

During the Cold War, the major powers devoted considerable effort to developing "improved" chemical warfare agents, both lethal and incapacitating.[16] Given that history, the recent advances in neurochemistry could lead to a renewed search for ways to assault the brain through pharmacological intervention. Indeed, the discovery of several classes of neurotransmitters and receptors during the 1990s was not lost on researchers interested in developing new chemical agents designed to incapacitate rather than kill. A study in 2000 by researchers at Pennsylvania State University listed several classes of drugs as potential "calmative" agents, along with the receptor types and subtypes affected by them.[17] Along similar lines, a report on cognitive neuroscience by the U.S. National Academies observed in 2008:

> If agonists of a particular system enhance cognition, it is mechanistically plausible that antagonists might disrupt cognition; conversely, if antagonists of a particular neurotransmitter enhance, its agonists might disrupt. Examples of the former might include dopamine agonists, which enhance attention, and dopamine antagonists, which disrupt it; examples of the latter might include the suspected cognitive enhancing effects of cannabinoid antagonists and the disrupting effects of agonists like THC.[18]

A key indicator of dual-use risk is the extent to which the development of new drugs for cognitive manipulation has been assimilated into military forces, operations, and doctrines in various parts of the world. In fact, the signs are increasingly ominous. The most immediate danger comes from the efforts by states to exploit the so-called law-enforcement exemption in Article II.9(d) of the Chemical Weapons Convention (CWC), which states that toxic chemicals may be used for "law enforcement including domestic riot control purposes." This exemption enables countries to conduct judicial executions by lethal injection and allows police to use tear gas and pepper spray to suppress riots. Although the scope of the permitted use of toxic chemicals under this clause is a matter of debate, the interpretation under which most CWC member states are operating is that the exemption allows the development of potential incapacitating chemicals for law-enforcement purposes, in addition to standard riot-control agents such as CS tear gas.[19]

In October 2002, for example, Russian security forces employed a derivative of the opiate anesthetic fentanyl to break the siege of a Moscow theater by Chechen rebels, killing all the hostage takers and 130 of the roughly 800 hostages. Although the incapacitating and lethal doses of fentanyl are so close that the term "nonlethal agent" is a misnomer, the reluctance of other CWC member states to challenge the legality of Russia's actions in the Moscow theater incident is an ominous harbinger of the future.[20]

Indeed, it is not difficult to find Western military officers who advocate the development of chemical incapacitating agents.[21] A study for the European Defense Agency in 2007 suggested that "calmative" drugs could be used to clear facilities, structures, and areas, indicating that a normative threshold may have already been crossed.[22] At present, a major obstacle to the wider use of incapacitating agents for law enforcement is the fact that police have not been adequately trained to deal with the consequences.[23]

Accessibility

A program to develop novel incapacitating agents would probably require the technical and financial resources of a state rather than a nonstate actor. However, existing psychoactive agents are available and employed therapeutically and experimentally, and purchasing such drugs is within the means of an individual, group, or state. If the state use of psychoactive drugs for law-enforcement and counterterrorism operations were to become widespread, it would substantially increase the risk that such agents could fall into the hands of terrorists and other nonstate actors.

Ease of Misuse

Exploiting advances in neuroscience to develop new types of chemical agents that affect the brain would require a great deal of explicit and tacit knowledge on the part of highly trained scientists. The most immediate risk of misuse involves state-level efforts to develop novel chemical incapacitating agents for counterterrorism operations. In contrast, the risk that nonstate actors could develop such agents is extremely low.

Magnitude and Imminence of Risk

The field of neuropsychopharmacology is already sufficiently advanced that some of its products pose a risk of misuse. It is also possible that an unexpected discovery could lead to the development of a highly accessible means of incapacitation. As the British chemical warfare expert Julian Perry Robinson has observed, "If a new molecule is discovered that can exert novel disabling effects on the human body at low dosage, attempts to weaponize it may well ensue."[24] Such a development would have serious adverse consequences for international security and the future of the chemical disarmament regime.

Awareness of Dual-Use Potential

The misuse of advances in the understanding of brain chemistry to develop new types of chemical incapacitating agents would undermine the international norm against the hostile use of the life sciences. In particular, certain neuropeptides that exist naturally in the brain offer a potential means of manipulating consciousness, cognition, and emotion. Although this dual-use risk is not widely recognized by the scientific community, some members of the national security establishment have grasped it with great clarity. In 1991, for example, the U.S. contribution to a background paper on the implications of advances in science and technology for the Biological Weapons Convention (BWC) stated in a section on neuropeptides, "Even a small imbalance in these natural substances could have serious consequences, including fear, fatigue, depression or incapacitation. These substances are extremely difficult to detect but could cause serious consequences or even death if used improperly."[25]

Characteristics of the Technology Relevant to Governance

Embodiment Although the field of neuropsychopharmacology consists primarily of intangible knowledge, researchers use a variety of sophisticated tools, such as functional brain imaging.

Maturity Many psychoactive drugs are commercially available, while others are still in research and development. Despite considerable progress over the past few decades, the efficacy of many drugs for the treatment of mental illness is controversial, and serious side effects are common.

Convergence Neuropsychopharmacology is a convergent technology because advances are occurring simultaneously from the bottom up (molecular genetics) and from the top down (the visualization of brain function).

Rate of advance Indicative of the rapid pace of progress in understanding the functional chemistry of the brain has been a dramatic increase in the number of known neurotransmitter receptor systems and ion channels, of which some fifty different classes have been identified to date. In 1990 a standard listing of receptors and ion channels filled 30 pages, but by 1999 it filled 106 pages.[26]

International diffusion Mental illness is a major problem for all countries, leading to a strong interest on the part of the pharmaceutical industry in developing more effective psychoactive drugs. Research and development in the field is increasingly international in scope.

Susceptibility to Governance

Because research in neuropsychopharmacology produces knowledge that is widely available in the scientific literature, the technology is not readily susceptible to hard-law governance. Nevertheless soft-law and informal measures such as education and awareness raising may help to prevent misuse.

Relevance of Existing Governance Measures

Bottom-up governance measures, such as peer review, professional codes of conduct, oversight mechanisms, and prepublication review of sensitive research findings, require a fully aware and engaged scientific community to operate them. Yet the large majority of life scientists have little understanding of the potential for misuse of the materials, knowledge, and technologies they are developing.[27] Recent surveys of biosecurity education in Europe[28] and Japan[29] strongly suggest that the problem of dual-use research is rarely discussed in university courses.

Proposed Governance Measures

Several experts have recommended limiting the law-enforcement exemption in Article II.9(d) of the CWC to prevent the development of a new generation of chemical incapacitating agents. Nevertheless, given that the first two CWC review conferences in 2003 and 2008 declined even to address the issue, top-down governance measures will not be achieved quickly or easily. Although such efforts should be pursued, much can be done from the bottom up.

In particular, it is essential to educate researchers in the field of neuropsychopharmacology about dual-use risks by incorporating these issues into the university curriculum, perhaps in the expanding number of bioethics courses. Greater awareness of the problem will help motivate scientists to develop workable codes of conduct, oversight mechanisms, and other control measures.[30] In the United States, a federal advisory committee, the National Science Advisory Board for Biosecurity (NSABB), has developed a strategic plan for raising awareness of dual-use issues among life scientists.[31] Combined governmental and nongovernmental action on biosecurity education could contribute to improving many areas of governance.

Conclusions

There is every reason to believe that in the coming decades, neuroscience will provide an increasingly mechanistic and integrated understanding of the human brain and behavior. It is also likely that the new understanding of brain chemistry will be misused for hostile purposes unless strong preventive measures are taken. Because no "silver bullet" exists for effective governance, an integrated web of preventive policies should be developed, including the effective use of education and training to inculcate researchers with an ethos of personal responsibility.[32]

Notes

1. Jerrold S. Meyer and Linda F. Quenzer, *Psychopharmacology: Drugs, the Brain, and Behavior* (Sunderland, MA: Sinauer Associates, 2005), 4.

2. Nancy C. Andreasen, *Brave New Brain: Conquering Mental Illness in the Era of the Genome* (Oxford: Oxford University Press, 2001), 173.

3. Brain-imaging technologies include computerized tomography (CT), positron emission tomography (PET), magnetic resonance imaging (MRI), and functional MRI.

4. Andreasen, *Brave New Brain*, 173.

5. Malcolm R. Dando, *A New Form of Warfare: The Rise of Non-Lethal Weapons* (London: Brassey's, 1996).

6. Alexander Kelle, Kathryn Nixdorff, and Malcolm R. Dando, "Behaviour under Control: The Malign Misuse of Neuroscience," chap. 5 in *Controlling Biochemical Weapons: Adapting Multilateral Arms Control for the 21st Century* (Basingstoke: Palgrave, 2006).

7. Craig W. Berridge, "Noradrenaline Modulation of Arousal," *Brain Research Reviews* 38 (2007): 1–17.

8. Laren T. May, Katie Leach, Patrick M. Sexton, et al., "Allosteric Modulation of G Protein–Coupled Receptors," *Annual Review of Pharmacology and Toxicology* 47 (2007): 1–51.

9. Christopher J. Longmead, Jennette Watson and Charlie Revill, "Muscarinic Acetylcholine Receptors as CNS Drug Targets," *Pharmacology and Therapeutics* 117 (2008): 232–243.

10. These receptor subtypes are known as muscarinic acetylcholine M1 and M4 receptors. P. Jeffrey Conn, Carrie K. Jones, and Craig W. Lindsley, "Subtype-Selective Allosteric Modulators of Muscarinic Receptors for the Treatment of CNS Disorders," *Trends in the Pharmacological Sciences* 30, no. 3 (2008): 148–155.

11. Michael Kosfeld et al., "Oxytocin Increases Trust in Humans," *Nature* 435 (2005): 673–676.

12. Peter Kirsch, Christine Esslinger, Qiang Chen, et al., "Oxytocin Modulates Neural Circuitry for Social Cognition and Fear in Humans," *Journal of Neuroscience* 25, no. 49 (December 7, 2005): 11489–11493.

13. J. T. Winslow and T. R. Insel, "The Social Deficits of the Oxytocin Knockout Mouse," *Neuropeptides* 36, nos. 2–3 (2002): 221–229.

14. Zoe R. Donaldson and Larry J. Young, "Oxytocin, Vasopressin, and the Neurogenetics of Sociality," *Science* 322 (2008): 900–904.

15. Ajai Vyas, Seon-Keyong Kim, and Nicholas Giacomini, "Behavioral Changes Induced by *Toxoplasma* Infection of Rodents Are Highly Specific to Aversion of Cat Odors," *Proceedings of the National Academy of Sciences* 104 (2007): 6442–6447.

16. Malcolm R. Dando and Martin Furmanski, "Midspectrum Incapacitant Programs," in *Deadly Cultures: Biological Weapons since 1945*, ed. Mark L. Wheelis, Lajos Rózsa, and Malcolm R. Dando (Cambridge, MA: Harvard University Press, 2006), 236–251.

17. Joan M. Lakoski, W. Bosseau Murray, and John M. Kenny, *The Advantages and Limitations of Calmatives for Use as a Non-Lethal Technique* (College Park: Pennsylvania State University, College of Medicine and Applied Research Laboratory, October 3, 2000). See also Neil Davison, *"Non-Lethal" Weapons* (Basingstoke: Palgrave, 2009).

18. National Research Council, Committee on Military and Intelligence Methodology for Emergent Neurophysiological and Cognitive/Neural Science Research in the Next Two Decades, *Emerging Cognitive Neuroscience and Related Technologies* (Washington, DC: National Academies Press, 2008).

19. David P. Fidler, "Incapacitating Chemical and Biochemical Weapons and Law Enforcement under the Chemical Weapons Convention," 171–194, and Adolf von Wagner, "Toxic Chemicals for Law Enforcement Including Domestic Riot Control Purposes under the Chemical Weapons Convention," 195–207, in *Incapacitating Biochemical Weapons: Promise or Peril?* ed. Alan M. Pearson, Marie Isabelle Chevrier, and Mark Wheelis (New York: Rowman and Littlefield, 2007).

20. Julian Perry Robinson, "Difficulties Facing the Chemical Weapons Convention," *International Affairs* 84, no. 2 (2008): 223–239.

21. George N. T. Whitbred, Commander, U.S. Navy, *Offensive Use of Chemical Technologies by U.S. Special Operational Forces in the Global War on Terror: The Nonlethal Option*, Maxwell Paper No. 37 (Montgomery, AL: Air War College, Maxwell Air Force Base, 2006).

22. Michael J. A. Crowley, *Dangerous Ambiguities: Regulation of Riot Control Agents and Incapacitants under the Chemical Weapons Convention* (Bradford: University of Bradford, 2009), available at http://www.dual-usebioethics.net.

23. Ross Kirby, "Paradise Lost: The Psycho Agents," *CBW Conventions Bulletin*, no. 71 (2006): 1–5.

24. Robinson, "Difficulties Facing the Chemical Weapons Convention," 223–239.

25. United Nations, *Background Document on New Scientific and Technological Developments Relevant to the Convention on the Prohibition of the Development, Production, and Stockpiling of Bacteriological (Biological) and Toxin Weapons and on Their Destruction*, BWC/CONF.III/4 (Geneva: United Nations, August 26, 1991).

26. Malcolm R. Dando, "Specificity of Receptors," chap. 6 in *The New Biological Weapons: Threat Proliferation and Control* (Boulder: Lynne Rienner, 2001).

27. Simon Whitby and Malcolm R. Dando, *Effective Implementation of the BTWC: The Key Role of Awareness Raising and Education* (Bradford: University of Bradford, 2010), available at http://www.brad.ac.uk/acad/sbtwc.

28. Giulio Mancini and James Revill, *Fostering the Biosecurity Norm: Biosecurity Education for the Next Generation of Life Scientists* (Como, Italy: Landau Network–Centro Volta; and Bradford: University of Bradford, 2008), available at http://www.dual-usebioethics.net.

29. Masamichi Minehata and Naryoshi Shinomiya, *Biosecurity Education: Dual-Use Education in Life-Science Degree Courses at Universities in Japan* (Tokorozawa, Japan: National Defense Medical College; Bradford: University of Bradford, 2009), available at http://www.dual-usebioethics.net.

30. The Federation of American Scientists (FAS) has posted several useful case studies of dual-use research at http://www.fas.org/biosecurity/education/dualuse/index.html.

31. National Science Advisory Board for Biosecurity, *Strategic Plan for Outreach and Education on Dual Use Issues* (Washington, DC: NSABB, 2008).

32. Malcolm R. Dando, "Bioethicists Enter the Dual-Use Debate," *Bulletin of the Atomic Scientists*, web edition, April 20, 2009.

11

Synthesis of Peptide Bioregulators

Ralf Trapp

Bioregulators are naturally occurring chemicals that help to ensure the proper functioning of vital physiological systems in living organisms, such as respiration, blood pressure, heart rate, body temperature, consciousness, mood, and the immune response.[1] Recent advances in drug delivery have made bioregulators and the chemical analogs derived from them more attractive as potential medicines. Indeed, the growing understanding of these compounds and their functions in the body is likely to bring about profound changes in medicine by increasing the ability to intervene *selectively* in fundamental biological processes. At the same time, excessive doses of bioregulators can cause severe physiological imbalances, including "heart rhythm disturbances, organ failure, paralysis, coma and death," giving them a potential for misuse.[2]

This case study assesses the implications of recent scientific and technological developments involving bioregulators and evaluates governance measures to prevent their application for hostile purposes. Because many bioregulators are peptides, the chapter also examines technologies for synthesizing peptides in industrial quantities.

Overview of the Technology

Bioregulators modulate the functions of the nervous, endocrine, and immune systems. In terms of chemical structure, they are extremely diverse, ranging from relatively simple molecules in the case of certain hormones or neurotransmitters, to complex macromolecules such as proteins, polypeptides, or nucleic acids. Their physiological action is not limited to any single mechanism, regulatory system, or organ. Malcolm Dando points out that "host homeostatic systems are controlled by midspectrum bioregulators such as hormones and neurotransmitters and the defense system is controlled by the cytokines of the immune system."[3]

For example, neurotransmitters such as acetylcholine, norepinephrine, serotonin, or GABA transfer information through the synaptic cleft between neurons, as well as from motor neurons to muscle cells. These messenger chemicals bind to, and interact with, specific receptors on the postsynaptic membrane. Several different neurotransmitters can be released from a single nerve terminal, including neuropeptides. These compounds can act as neurotransmitters in their own right or function as cotransmitters by activating specific pre- or postsynaptic receptors to alter the responsiveness of the neuronal membrane to the action of other neurotransmitters.

Further, bioregulators play different physiological roles in various tissues. Indeed, several bioactive peptides in the nervous system were first discovered in the intestine. Additional complexity arises because the nervous, endocrine, and immune systems interact. Modulating the concentration of a bioregulator or interfering with its receptors in one system can affect the function of the other systems.[4] Recent research has also provided insights into the role of bioregulator receptors in generating diverse physiological responses and has suggested how they might be manipulated. Because bioregulators maintain equilibrium in body systems, it should be possible, at least in principle, to design molecular analogs that affect body temperature, sleep, and even consciousness in a selective manner.

Many bioregulators are peptides, or short chains of amino acids. Examples include angiotensin (which raises blood pressure), vasopressin (which regulates the body's water balance), Substance P (which transmits pain signals from peripheral receptors to the brain), and bradykinin (which triggers inflammatory responses).[5] Minor chemical modifications of peptide bioregulators can create analogs with markedly different physiological properties. The duration of action can also be extended by means of structural modifications that slow its rate of degradation in the body.

The understanding of bioregulators has grown at an astonishing pace in recent years, stimulated by new investigation techniques and the interaction of hitherto separate scientific disciplines. Today the functional chemistry of the brain is one of the fastest-growing areas of research in the life sciences.[6] A significant part of this progress has been driven by the rapid increase in knowledge about neuropeptides and their receptor and subreceptor systems.

Advances in the understanding and use of peptide bioregulators have gone hand in hand with developments in peptide synthesis. Today companies produce peptides to order in quantities ranging from milligrams for laboratory use to hundreds of kilograms for industrial applications. The choice of production method depends largely on the size of the peptide, its amino acid sequence, and the presence of modifications or protective groups. Overall, the chemical synthesis of peptides remains the most common method for industrial-scale production.[7] Peptides can be produced in solution (liquid phase) or on the surface of tiny plastic beads (solid phase). One approach involves synthesizing peptide fragments eight to fourteen amino acids long on a solid-phase resin, removing and purifying the fragments, and coupling them together to form longer chains.[8] In the future, microwave synthesis, in which single-frequency microwaves are used to speed up the coupling reactions and achieve better purity and higher yields, may become a method of choice.[9] It may also be possible to produce large quantities of peptides in recombinant micro-organisms or in transgenic plants and animals.[10]

History of the Technology

Before 1975, manufacturing bioactive peptides in large quantities was not possible.[11] Since then, however, new methods have transformed the situation. In 1993 the liquid-phase and solid-phase methods were combined for the first time to produce complex peptides on a metric-ton scale, enabling peptide synthesis to evolve from a niche market into a mainstream business. As of 2010, more than 40 peptides were being marketed worldwide, and hundreds more were in some stage of preclinical or phased clinical development. For example, Fuzeon (enfuvirtide), an anti-HIV drug consisting of a chain of 36 amino acids, is currently synthesized in quantities exceeding 3,500 kilograms per year.[12]

In 2011 about 120 companies were involved in the commercial synthesis of peptides, a good many of them in the multiple-kilogram to multi-ton range.[13] In some ways, peptide synthesis can be compared to DNA synthesis, because specialized companies provide contract manufacturing services to customer specifications, and customers can place peptide orders over the Internet. At the same time, peptide synthesizers can be purchased for in-house automated benchtop and larger-scale synthesis. At the beginning of 2011, eighteen companies advertised their peptide synthesizers on a prominent industry Web site.[14] The most significant trend in the evolution of the customer base has been an increase in pharmaceutical industry clients caused by progress in novel peptide formulations and innovative delivery systems.

Utility of the Technology

Scientists are studying bioregulators and their synthetic analogs to obtain a deeper understanding of the physiology of organisms in health and disease. In addition to the quest for new knowledge, important economic and social pressures are driving research and development on bioregulators and the regulatory circuits in which they play a central role. Such research is expected to lead to safer and more targeted medicines, including treatments for diseases that affect the nervous, endocrine, and immune systems.

Peptide bioregulators and their synthetic derivatives are attractive drug candidates for treating a variety of ailments, including asthma, arthritis, cancer, diabetes, growth impairment, cardiovascular disease, inflammation, pain, epilepsy, gastrointestinal diseases, and obesity.[15] Bioregulator research is contributing significantly to a better understanding of aging. Research has also led to promises for new drugs that enhance human performance, for example with regard to learning and memory. Advantages of bioregulators include their high activity and specificity, low toxicity, lack of persistent accumulation in organs, and low immunogenicity compared to monoclonal antibodies. Bioregulators also have potential applications in agriculture, including the regulation of growth and development in livestock, staple crops, and fruit, as well as the development of compounds with insecticidal or fungicidal properties for crop protection and pest control. For all these reasons, the diffusion of knowledge about bioregulators and their production technologies is bound to continue.

Compared to small-molecule drugs, bioactive peptides have a number of drawbacks: they are less stable in bodily fluids, expensive to manufacture, and rapidly degraded by enzymes in the body, requiring continuous administration that greatly increases the cost of drug therapy. Peptide bioregulators also tend to be fairly large molecules with an electrical charge and an affinity for water, properties that hamper their transport across cell membranes. For these reasons, the therapeutic use of peptide bioregulators depends on the ability to manufacture these substances at acceptable cost and in the required purity, store them as needed, and deliver them to the right targets in the human body. Chemical modification of peptides, for example, can dramatically increase their persistence in the bloodstream.

Potential for Misuse

Bioregulators could potentially be developed into biochemical weapons that incapacitate, alter moods, trigger psychological imbalances, cause a variety of other types of physiological reactions, or kill.[16] Studies have found that bioactive peptides can induce profound physiological effects within minutes of exposure.[17] Other advantages of bioregulators as weapons include their ability to elicit a variety of physiological effects and lack of susceptibility to existing medical countermeasures. According to a paper prepared by the U.S. government, "While naturally-occurring threat agents, such as anthrax, and 'conventionally' genetically engineered pathogenic organisms are the near-term threats we must confront today, the emerging threat spectrum will become much wider and will include biologically active agents such as bioregulators."[18]

Until recently, the hostile use of bioregulators was considered unlikely for a number of reasons, including the fact that they are expensive to produce, rapidly inactivated by high or low temperatures, and difficult to deliver. According to a paper prepared by the British government, "Delivery of sufficient quantities to the appropriate target cells or tissues is a significant challenge to the development of therapeutic peptides, with delivery across the blood/brain barrier, for example, remaining a significant problem. Difficulties in delivering bioactive molecules would also affect the utility of such compounds as [biological warfare] agents."[19] Nevertheless new technologies have improved the ability to deliver bioregulators effectively, changing the assessment of their dual-use risk.[20]

For example, although delivering peptide bioregulators in the form of an inhalable aerosol is challenging because peptides are sensitive to acidic conditions and rapidly degraded by enzymes in the lungs, there are ways to overcome these hurdles. One way to facilitate aerosol delivery is through microencapsulation, in which solid particles or liquid droplets are coated with a substance that protects them from evaporation, contamination, oxidation, and other forms of chemical degradation.[21] Another approach is to create porous particles that can carry peptides into the deep regions of the lungs.[22] An international workshop found that "the spray-drying equipment needed to create such particles is relatively cheap and widely available— yet the optimization of a well-engineered particle requires considerable time and skill."[23]

The potential misuse of bioregulators has historical precedents. Several countries have sought to develop such compounds as incapacitating agents.[24] During the Cold War, difficulties with the manufacture, stability, and dissemination of peptide bioregulators meant that smaller psychoactive molecules were favored, such as BZ and certain benzilates and glycolates. Nevertheless the Soviet Union launched a top-secret project code-named Bonfire to develop bioregulators as biochemical weapons.[25] Another research program under the Soviet Ministry of Health, code-named Flute, sought to develop both lethal and nonlethal psychotropic and neurotropic agents for use in KGB operations, which included research on bioactive peptides.[26]

More recently, several countries have shown a renewed interest in the use of incapacitants for law-enforcement purposes, including China, the Czech Republic, France, Russia, the United Kingdom, and the United States.[27] In 1997 the U.S. Department of Defense established the Joint Non-Lethal Weapons Directorate to coordinate the development of a variety of weapons technologies and systems. This effort includes certain chemical agents, which some argue can be employed for law-enforcement purposes under a provision of the 1993 Chemical Weapons Convention (CWC).[28] In 2000, for example, the Applied Research Laboratory of Pennsylvania State University assessed several compounds that might serve as the basis for developing "calmative" agents. The list included two peptide bioregulators: corticotrophin-releasing factor (CRF) and cholecystokinin (CKK).[29]

Slavko Bokan and his colleagues have prepared a list of agents that might be suited for military or terrorist use, including Substance P, endorphins, endothelins, sarafotoxins, bradykinin, vasopressin, angiotensins, enkephalins, somatostatin, bombesin, neurotensin, oxytocin, thyoliberins, and histamine-releasing factors.[30]

The emergence of bioregulator-based weapons would add a new and frightening dimension to modern warfare, not only threatening the lives of troops but potentially altering their perception, provoking severe bodily malfunctions, and altering their emotional state and behavior. Bioregulators might also be employed to enhance the lethality of conventional weapons. The acceptance by some states of this new type of warfare would tend to legitimate it for others and would undermine the norm against the use of toxic chemicals in warfare.[31]

In addition to the possibility that peptide bioregulators might be developed for warfare purposes, one can imagine several other questionable applications, such as their use by occupying forces to repress the local population under the guise of law enforcement and riot control. It might be tempting to employ bioregulators that affect perception, cognition, mood, or trust to render angry civilians more docile. Such use would, of course, raise a host of legal and ethical issues. There is also the possibility that military, intelligence, or police forces could use bioregulators in domestic situations to control crowds, render prisoners more compliant and trusting, or induce acute depression, pain, or panic attacks as an instrument for influencing behavior and enforcing compliance. Although any nonconsensual use of biochemical agents would be illegal, the perception that the use of certain chemicals for law-enforcement purposes is "legitimate" could weaken such legal protections, not just under despotic regimes but in democratic societies as well.

Accessibility

State actors would have no difficulty setting up programs to manufacture peptide bioregulators should they decide to do so. For nonstate actors, the challenges are considerably greater. In theory, a terrorist or criminal organization could purchase a peptide synthesizer or order customized peptides from a commercial supplier, although both of these options might require the use of front companies or access to proliferation networks. If, however, state law-enforcement agencies were to stockpile weapons for the dispersal of incapacitating agents based on peptide bioregulators, such stockpiles might be vulnerable to theft or diversion by nonstate actors.

Ease of Misuse

At least at the state level, access to information on bioregulators is not a limiting factor. Basic knowledge about these compounds and their properties has been published in the scientific literature, presented at conferences, and distributed through other channels. For proprietary reasons, fewer data are available on the early (preclinical) phases of drug development, and information on compounds that pharmaceutical companies have screened but not selected for clinical testing is generally confidential.

For nonstate actors, reliable information on the design of biochemical weapons is hard to come by because it is classified. In particular, formulating peptide bioregulators to enhance their utility as a weapon and devising an effective delivery system are hurdles that terrorists would have great difficulty overcoming. To use a bioregulator-based agent for purposes of interrogation or abduction, terrorists would need to know how to administer an agent to achieve the desired effect without killing the victim.

Magnitude and Imminence of Risk

The possibility exists that states could adopt peptide bioregulators (or their structural analogs) as biochemical weapons designed to incapacitate rather than kill, for purportedly legitimate purposes such as law enforcement. Moreover, given the changing nature of armed conflict in the twenty-first century, with counterinsurgency and urban-warfare scenarios becoming ever more prominent, some countries are interested in developing incapacitating agents for such operations. Exploratory research programs are known to exist, and there have been a few isolated cases of actual use. At the same time, the likelihood of developing a truly nonlethal weapon based on bioregulators is remote, for several reasons. Not only would the margin of safety between the incapacitating and lethal doses have to be very large, but getting the agent across the blood-brain barrier remains a formidable challenge. Finally, the ability to control the dose under field conditions would have to be far superior to what is technically feasible today.[32]

At present, the risk that terrorist or criminal organizations could exploit bioregulators for hostile purposes appears low, because of the limited availability of these agents, the unpredictability of their effects on a target group, and the difficulties related to agent dissemination. It is also unlikely that terrorists would invest the effort and resources needed to develop a bioregulator-based weapon unless it offered clear advantages over weapons that they already possess. Nevertheless, if bioregulator-based drugs become widely available for therapeutic purposes or if they are employed as "nonlethal weapons" by law-enforcement agencies, terrorists or criminals might be able to acquire and use them to facilitate hostage taking, incapacitate security guards, render hostages docile, extract information during interrogations, or control a group of people in a confined space such as an aircraft.

Overall, the dual-use potential of bioregulators is likely to grow in the coming years as the functional understanding of these natural body chemicals increases, along with advances in related areas such as bioinformatics, systems biology, receptor research, and neuroscience. Progress in these fields tends to be incremental, but what really counts is the cross-fertilization and synergies among them. Although the effect of these interactions is hard to predict, one cannot rule out unexpected discoveries that would transform the dual-use potential of bioregulators.

Awareness of Dual-Use Potential

The incident in 2002 at the Dubrovka Theater in Moscow, in which Russian security forces used a potent anesthetic agent (a fentanyl derivative) against a group of Chechen rebels holding some 800 theatergoers, causing the collateral death of 129 hostages, has raised concern about the future role of incapacitating agents. At the international diplomatic level, awareness of the potential misuse of bioregulators has increased in recent years, and several states that are parties to the Biological Weapons Convention (BWC) referred to the issue in papers prepared for the Sixth Review Conference in 2006.

Bioregulators have not yet been discussed directly in the context of the Chemical Weapons Convention (CWC), but several member states and the director general of the Organization for the Prohibition of Chemical Weapons (OPCW) have proposed holding informal discussions about the use of incapacitants for law-enforcement purposes under Article II.9(d) of the treaty.[33,34] Outside the OPCW context, the International Committee of the Red Cross (ICRC) has hosted an expert meeting on technical and legal questions pertaining to the potential acquisition and use of incapacitants for law-enforcement purposes.[35]

Although a few nongovernmental organizations have called attention to the potential misuse of bioregulators, the scientific, medical, academic, and industry communities appear to be largely unaware of the issue. This lack of awareness applies more broadly to the potential for misuse inherent in certain advances in the life sciences. Despite modest attempts to introduce educational materials on dual-use issues into university curricula and the availability of a few instructional modules on the Internet, this topic has yet to be integrated into mainstream science education.

Characteristics of the Technology Relevant to Governance

Embodiment Technologies related to the production and application of peptide bioregulators are hybrid technologies. Although applications of these compounds are based largely on intangible knowledge, the large-scale production of peptides requires automated synthesizers and other sophisticated hardware.

Maturity Drugs based on bioregulators are in advanced development, and peptide manufacturing is a mature technology. Automated peptide synthesizers are commercially available, and companies in many parts of the world provide peptide synthesis services on an industrial scale.

Convergence The study of bioregulators draws on advances in several fields, including systems biology, receptor research, brain research, computational biology, and synthetic biology. Advances in delivery technologies, such as microencapsulation and aerosolization, may also play an important role in the practical applications of synthetic bioregulators.

Rate of advance   Significant progress has been made in the large-scale synthesis and purification of peptides, including peptide bioregulators. Most therapeutic peptides in use today are from 10 to 50 amino acids long, but improvements in peptide chemistry have pushed the maximum length that can be produced in quantity to as many as 80 to 100 amino acids.

International diffusion   Although the biotechnology industry was once concentrated in Western countries, today Brazil, China, Cuba, India, Singapore, and South Korea all have high-quality biotechnology firms, and large Western corporations have outsourced part of their research and development to manage risks and cut costs. Innovative research is also taking place in universities, government laboratories, and small firms established as spin-offs of university research.[36] These observations hold true with respect to research on peptide bioregulators and the manufacture of these compounds and their synthetic derivatives.

Susceptibility to Governance

The bottom-up governance measures developed by the gene-synthesis industry are relevant to peptide synthesis because of the similarities between the scientific, technological, and industrial aspects of the two fields.[37] If one employs a similar approach to that proposed for synthetic genomics, the various intervention points could target (1) firms that manufacture peptides to order on a multiple-kilogram and larger scale, (2) firms that sell peptide synthesizers, and (3) scientists who work with bioregulators in the research, medical, and pharmaceutical communities. Other possible targets for governance measures include manufacturers of delivery systems (e.g., aerosol generators) and organizations that specialize in particle engineering for biomedical purposes.

Relevance of Existing Governance Measures

Both the BWC and the CWC ban the development, production, and retention of bioregulators for hostile purposes. Nevertheless, although the CWC is designed to prevent the emergence of new forms of chemical warfare, the debate about how the law-enforcement exemption in Article II.9(d) should be interpreted with respect to incapacitating agents has yet to be resolved. Accepting the use of these agents for law enforcement would open the door to the introduction of a new category of biochemical weapons. The associated risks would include providing a cover for illicit intent, diminishing national control over weaponized chemicals, requiring the use of personal protective equipment during combat operations, and potentially expanding the scope of law enforcement to include counterinsurgency, counterterrorism, and special-forces operations. It would be only a small step from there to accepting biochemical weapons back into national force structures for a range of military applications, with the potential to move down a slippery slope toward the remilitarization of chemistry.[38] Given these risks, the states that are parties to the CWC should strive to reach consensus on acceptable ways to clarify and limit the scope of Article II.9(d).

Proposed Governance Measures

Overall, managing the risk of misuse of bioregulators will require a multistakeholder, multidisciplinary, and multidimensional approach.[39] Given the dynamics and complexities of the developments in this field of science and technology, it is unlikely that traditional means of governmental regulation alone, such as specific legislation or export and technology transfer controls, would be effective. Possible new governance measures for peptide bioregulators could include establishing legal norms through formal regulations (licensing, guidelines, and export controls), self-regulation by the user community, and the involvement of civil society in monitoring compliance with the applicable norms.

Governance should be built from the bottom up, based on an increased awareness of dual-use risks on the part of the scientists and companies pursuing the application of bioregulators and a willingness to manage these risks through appropriate self-regulation, oversight, reporting and problem-resolution mechanisms, and other measures to promote responsible conduct. Because research on bioregulators offers so many potential beneficial applications for medicine, agriculture, and biotechnology, exchanges of basic scientific information must remain unhindered. At the same time, it is essential that governments agree to limit the use of bioregulators in law enforcement, based on existing international norms, and enforce this agreement within their jurisdictions. It may also be helpful to restate categorically that the prohibition of the development, production, acquisition, transfer, and use of toxic chemicals for warfare purposes extends to incapacitating agents, including bioregulators.

Conclusions

Bioregulators are an important element of the revolution in the life sciences and are expected to yield beneficial applications in medicine, agriculture, and other fields. At the same time, bioregulators have a potential for misuse as biochemical weapons, both on the battlefield (directly or to enhance the effects of other weapons) and as a means to manipulate and coerce human behavior. For governance to be effective, it will be essential to reach a broad international agreement on which applications of these chemicals are acceptable and which constitute a violation of existing norms, including those established by the CWC. In addition, more efforts are needed to improve governance structures and mechanisms in research and industry and to ensure an effective dialogue between government, the research and industry communities, and other nongovernmental stakeholders.

Notes

1. Kathryn Nixdorff and Malcolm R. Dando, "Developments in Science and Technology: Relevance for the BWC," *Biological Weapons Reader* (Geneva: BioWeapons Prevention Project, 2009), 39.

2. The Netherlands, "Scientific and Technological Developments Relevant to the Biological Weapons Convention," paper submitted to the Sixth BWC Review Conference, http://www.unog.ch/80256EDD006B8954/(httpAssets)/018F68EC1656192FC12571FE004982A6/$file/BWC-6RC-S&T-NETHERLANDS.pdf.

3. Malcolm Dando, "Advances in Neuroscience and the Biological and Toxin Weapons Convention," *Biotechnology Research International* 2011, article no. 973851 (2011): 1–9.

4. British Royal Society, "Report of the RS-IAP-ICSU International Workshop on Science and Technology Developments Relevant to the Biological and Toxin Weapons Convention," RS policy document 38(06) (London: Royal Society, November 2006).

5. Jonathan B. Tucker, "The Body's Own Bioweapons," *Bulletin of the Atomic Scientists* 64, no. 1 (March–April 2008): 16–22.

6. Steven Rose, "Prospects and Perils of the New Brain Science: A Twenty Year Time Scale," Royal Society Neuropolicy Lab (London: Royal Society, 2009).

7. Lars Andersson, Lennart Bloomberg, Martin Flegel, et al., "Large-Scale Synthesis of Peptides," *Biopolymers (Peptide Science)* 55 (2000): 227–250.

8. Susan Aldridge, "Peptide Boom Puts Pressure on Synthesis—Drugs Already on the Market and in Clinical Studies Drive Novel Method Development," *Genetic Engineering and Biotechnology News* 28, no. 13 (July 2008), http://www.genengnews.com/gen-articles/peptide-boom-puts-pressure-on-synthesis/2534.

9. "Peptide Manufacturers See Increased Growth—Pharma's Interest in Peptide Drugs Drives This Market," *Genetic Engineering and Biotechnology News* 25, no. 13 (July 2005), http://www.genengnews.com/gen-articles/peptide-manufacturers-see-increased-growth/1001.

10. United Kingdom, "Scientific and Technological Developments Relevant to the Biological Weapons Convention," paper submitted to the 6th BWC Review Conference, para. 54, http://www.unog.ch/80256EDD006B8954/(httpAssets)/5B93AF9D015AD633C12571FE0049ADAF/$file/BWC-6RC-S&T-UK.pdf.

11. Canada, *Novel Toxins and Bioregulators: The Emerging Scientific and Technological Issues Relating to Verification and the Biological and Toxin Weapons Convention* (Ottawa, Canada: Department of External Affairs and International Trade, 1991).

12. Brian L. Bray, "Large-Scale Manufacture of Peptide Therapeutics by Chemical Synthesis," *Nature Reviews Drug Discovery* 2 (July 2003): 587–593; Thomas Bruckdorfer, Oleg Marder, and Fernando Albericio, "From Production of Peptides in Milligram Amounts for Research to Multi-ton Quantities for Drugs of the Future," *Current Pharmaceutical Biotechnology* 5 (2004): 29–43.

13. Peptide Resource Page, http://www.peptideresource.com.

14. Ibid.

15. Anil Seghal, "Peptides 2006: New Applications in Discovery, Manufacturing, and Therapeutics," *Drug and Market Development Report*, no. 9214 (June 2006): 3.

16. Elliot Kagan, "Bioregulators as Instruments of Terror," *Clinics in Laboratory Medicine* 21, no. 3 (2001): 607–618.

17. Elliot Kagan, "Bioregulators as Prototypic Nontraditional Threat Agents," *Clinics in Laboratory Medicine* 26, no. 2 (2006): 421–444.

18. United States of America, "Scientific and Technological Developments Relevant to the Biological Weapons Convention," paper submitted to the Sixth BTWC Review Conference, Geneva, Switzerland, November 20–December 8, 2006, http://www.opbw.org.

19. United Kingdom, "Scientific and Technological Developments Relevant to the Biological Weapons Convention," paper submitted to the Sixth BTWC Review Conference, Geneva, Switzerland, November 20–December 8, 2006, http://www.opbw.org.

20. National Research Council, *Globalization, Biosecurity, and the Future of the Life Sciences* (Washington, DC: National Academies Press, 2006), 180.

21. UN Secretariat, "Background Information Document on New Scientific and Technological Developments Relevant to the Convention," BWC/CONF.VI/INF.4, September 28, 2006, http://www.opbw.org/rev_cons/6rc/docs/inf/BWC_CONF.VI_INF.4_EN.pdf.

22. Jennifer Fiegel, "Advances in Aerosol Drug Delivery," presentation at the IUPAC workshop "Impact of Scientific Developments on the CWC," Zagreb, Croatia, April 22–25, 2007.

23. Mahdi Balali-Mood, Pieter S. Steyn, Leiv K. Sydnes, and Ralf Trapp, "Impact of Scientific Developments on the Chemical Weapons Convention (IUPAC Technical Report)," *Pure and Applied Chemistry* 80, no. 1 (2008): 175–200.

24. Neil Davison, "'Off the Rocker' and 'On the Floor': The Continued Development of Biochemical Incapacitating Agents," *Bradford Science and Technology Report No. 8*, University of Bradford (UK), August 2007; Martin Furmanski, "Historical Military Interest in Low-Lethality Biochemical Agents," in *Incapacitating Biochemical Weapons*, ed. Alan M. Pearson, Marie Isabelle Chevrier, and Mark Wheelis (Lanham, MD: Lexington Books, 2007), 35–66; Alan Pearson, "Late and Post–Cold War Research and Development of Incapacitating Biochemical Weapons," in Pearson, Chevrier, and Wheelis, *Incapacitating Biochemical Weapons*, 67–101.

25. Ken Alibek with Stephen Handelman, *Biohazard* (New York: Random House, 1999), 154–155.

26. Ibid., 171–172.

27. Michael Crowley, *Dangerous Ambiguities: Regulation of Riot Control Agents and Incapacitants under the Chemical Weapons Convention*, Bradford Non-lethal Weapons Research Project (Bradford: University of Bradford, 2009).

28. Pearson, "Late and Post–Cold War Research and Development," 75.

29. Joan M. Lakoski, W. Bosseau Murray, and John M. Kenny, *The Advantages and Limitations of Calmatives for Use as a "Non-lethal" Technique* (State College: Pennsylvania State University, 2000).

30. Slavko Bokan, John G. Breen, and Zvonko Orehovec, "An Evaluation of Bioregulators as Terrorism and Warfare Agents," *ASA Newsletter* 90, nos. 2–3 (2002): 1.

31. Although this case study focuses on the potential misuse of bioregulators as antipersonnel weapons, plant regulators such as Agent Orange have been used in the past for deforestation or to destroy crops. See Jeanne Mager Stellman, Steven D. Stellman, Richard Christian, et al.,

"The Extent and Patterns of Usage of Agent Orange and Other Herbicides in Vietnam," *Nature* 422 (April 17, 2000): 681–687.

32. British Medical Association, *The Use of Drugs as Weapons: The Concerns and Responsibilities of Healthcare Professionals* (London: BMA Board of Science, 2007).

33. Switzerland, "Riot Control and Incapacitating Agents under the Chemical Weapons Convention," OPCW document RC-2/NAT/12, April 9, 2008, http://www.opcw.org/documents-reports/conference-of-the-states-parties/second-review-conference.

34. OPCW Director General, "Report of the Scientific Advisory Board on Developments in Science and Technology," OPCW document RC-2/DG.1 and Corr. 1, February 28, 2008, and March 5, 2008, http://www.opcw.org/documents-reports/conference-of-the-states-parties/second-review-conference.

35. International Committee of the Red Cross, *Incapacitating Chemical Agents: Implications for International Law*, Expert Meeting, Montreux, Switzerland, March 24–26, 2010 (Geneva, Switzerland: ICRC, 2010).

36. The Netherlands, "Scientific and Technological Developments Relevant to the Biological Weapons Convention," para. 10.

37. Michele S. Garfinkel, Drew Endy, Gerald L. Epstein, and Robert M. Friedman, *Synthetic Genomics: Options for Governance* (Rockville, MD: J. Craig Venter Institute; Washington, DC: Center for Strategic and International Studies; Cambridge, MA: Massachusetts Institute of Technology, October 2007).

38. Julian Perry Robinson, "Non-lethal Warfare and the Chemical Weapons Convention," Harvard-Sussex Program, submission to the OPCW Open-Ended Working Group on Preparation for the Second CWC Review Conference, October 24, 2007.

39. Nayef R. F. Al-Rhodan, Lyubov Nazaruk, Marc Finaud, and Jenifer Mackby, *Global Biosecurity: Toward a New Governance Paradigm* (Geneva: Editions Slatkine, 2008), 200–201.

12

Immunological Modulation

Nancy Connell

Antimicrobial drugs have produced dramatic victories against infectious diseases, but many pathogens have developed resistance mechanisms that render these drugs ineffective. In contrast, vaccination remains an efficient and cost-effective approach for preventing infectious diseases and controlling their spread. Although early vaccines were developed by trial and error, today the field of vaccinology is harnessing scientific insights into the operation of the human immune system to design vaccines that induce optimal immune defenses. There is also a new emphasis on developing vaccines for the treatment of noninfectious diseases, such as autoimmune conditions, neurological disorders, cancer, heart disease, allergies, and Alzheimer's disease.[1]

The growing ability to manipulate the immune response has a potential for misuse because it might be combined with new delivery methods to yield lethal biological warfare agents. Although delivery technologies also provide fertile ground for dual-use analysis, this case study focuses on immunological modulation. Because the benefits of this technology are inseparable from the risks, an outright ban is not a realistic option. Nevertheless, existing oversight mechanisms and soft-law and informal governance measures could be adapted to mitigate the dual-use risks.

Overview of the Technology

For more than two centuries, vaccines have protected against infectious disease. A vaccine works by directing the immune system to recognize specific molecules called antigens (made up of proteins, lipids, and/or carbohydrates) on the surface of an infectious agent such as a bacterium or a virus. Based on the characteristics of the antigens, the vaccine induces the host's immune system to mount an adaptive response. This response involves the activation of white blood cells and the production of antibodies to defend against a subsequent infection by the same agent.

A parallel defensive system called the innate immune response is activated within minutes of an invasion by a foreign pathogen. In this case, several mechanisms, including a different set of white blood cells (macrophages and natural killer cells), provide immediate, nonspecific defenses. The innate immune response influences the subsequent adaptive response by triggering complex cellular signaling pathways. One analysis estimated that the stimulation of a single receptor in the innate immune system called TLR4 results in 2,531 interactions involving 1,346 different genes or proteins.[2]

The adaptive immune response, which follows from and is influenced by the innate immune response, can persist for decades through the creation of "memory immunity" and provides a strong, rapid, and highly specific defense against a subsequent infection by the same organism. Adaptive immunity in mammals has two separate components: the humoral response, characterized by antibody-producing B cells, and the cell-mediated response, involving two types of T cells. "Helper" T cells produce signaling molecules called cytokines, whereas "cytotoxic" T cells kill infected cells directly. Scientists have learned which subtypes of T cells are required to combat different kinds of infections and how cytokines organize, increase, and decrease the activity of these cells.

T cells are also involved in two types of immunological responses, known as Th1 (the inflammatory arm) and Th2 (the anti-inflammatory arm). Th1 is characterized by the production of cytotoxic T cells and cytokines that induce inflammation and is mainly responsible for protection against viral or intracellular bacterial infections. In contrast, Th2 signals B cells to produce antibodies and plays a greater role in combating extracellular bacterial and parasitic diseases. Furthermore, many of the cytokines expressed during the Th1 response suppress the Th2 response, and vice versa. This reciprocal relationship between the two types of T cell response prevents the immune system from overreacting to infection in a harmful manner. At the same time, the slightest imbalance between the two types of T cell response can lead to a potentially disastrous outcome. In addition to the reciprocal Th1/Th2 relationship, other complex interactions exist between T and B cells and among the immune, nervous, and endocrine systems.[3]

The earliest vaccines for smallpox, rabies, cholera, and other infectious diseases consisted of avirulent (non-disease-causing) forms of the infectious agent and were developed by trial and error. As the operations of the immune system have gradually been elucidated, however, it has become possible to design vaccines that activate specific elements of the immune response for effective protection. For example, the viral vectors used by gene-therapy researchers to insert foreign genetic material into cells can also serve as vehicles for delivering antigens to immune cells. Such viral vectors can also be engineered to carry genes that encode proteins called immuno-modulators, which control the immune response.

History of the Technology

The first documented vaccine was developed in 1796 by the English country doctor Edward Jenner, who inoculated a thirteen-year-old boy with cowpox virus and observed that the boy was protected from infection when subsequently challenged with the closely related—but far more deadly—smallpox virus. Today dozens of infectious diseases are preventable by vaccination.[4]

The timeline of major discoveries since Jenner suggests that advances in the understanding of immune mechanisms have driven progress in vaccine design. During the nineteenth century, Louis Pasteur confirmed that the host is capable of an effective defense against infectious agents. He coined the general term "vaccine" (derived from the Latin word for cow) as a tribute to Jenner's use of cowpox to protect against smallpox. By the early twentieth century, several mechanisms and cell types involved in innate immunity had been identified, and the 1940s witnessed a number of breakthroughs in the study of cellular immunity. Over the following decades, the cooperative interactions between T and B cells came into focus.

During the 1980s, the discovery of two immune-system proteins, interleukins 1 and 2, led to an explosion of research into the field of immune signaling. The Th1/Th2 paradigm of the T cell response was delineated in 1986. With respect to the innate immune response, Charles A. Janeway Jr. at Yale Medical School predicted in 1989 that pattern-recognition receptors mediate the body's innate ability to recognize infection. Although Janeway made this striking prediction on theoretical grounds, subsequent experimental work in his laboratory demonstrated the existence of two key sets of signaling molecules in the innate immune system: pathogen-associated molecular patterns (PAMPs) and Toll-like receptors (TLRs). In 2001, Ralph M. Steinman and his coworkers at Rockefeller University in New York showed that dendritic cells—the first cells of the innate immune system that interact with an invading pathogen—play a key role in collecting, processing, and presenting antigenic material to T cells.[5] Steinman's work led to the revolutionary idea that receptor signaling in the innate immune system serves as a trigger for the adaptive immune response.

Utility of the Technology

Insights into the innate and adaptive immune responses have opened new avenues for manipulating the immune system to prevent and treat disease.[6] New vaccines designed to modulate the immune response are under development in a variety of institutional environments, including academic, pharmaceutical, and military research organizations. Vaccine vectors have been created that carry genes encoding specific cytokines, which direct the immune system to respond in a desired manner. In addition, Toll-like receptors have been used to modulate and direct subsequent immune responses.

Several laboratories have developed experimental viral vaccines that deliver antigen genes to dendritic cells to increase the effectiveness of immunotherapy for cancer.[7] For example, dendritic cells derived from patients suffering from brain tumors called glioblastomas can be genetically engineered to express specific antigens associated with this form of cancer and then reintroduced into the patients to combat the disease.[8] (Unfortunately, preliminary clinical trials of dendritic-cell immunotherapy have encountered a number of setbacks, including the deletion of key effector cells and the development of autoimmunity.)[9]

Vaccinology has also moved beyond infectious disease into several other areas of medical therapeutics. So-called lifestyle vaccines are being developed to treat otherwise healthy people for obesity, addiction to nicotine and other drugs, and dental caries, as well as for contraception.[10] One lifestyle vaccine targets ghrelin, a weight-gain-signaling protein that was first identified in 1999.[11] An antighrelin vaccine has been tested in rats, with the goal of inducing antibodies against the protein and blocking its access to the brain, where it stimulates appetite.[12] A recent review paper described the promise of antighrelin vaccines to combat the epidemic of obesity.[13]

On a more speculative note, genetic analyses have identified specific genes that appear to be associated with criminality, raising the possibility of manipulating human behavior by immunological means. In 2008 an editorial in the journal *Vaccine* discussed the prospect of vaccines that could down-regulate specific neurotransmitter systems in the brain to "achieve the regulation of the emotionality of humans who are physically incapable of controlling their emotions. Such individuals, when abused in childhood, make up a significant proportion of the criminally inclined. It may therefore be possible to make an anti-criminal vaccine to protect society against those whose natural monoamine oxidase [a brain enzyme] is not capable of sufficiently deactivating the neurotransmitters dopamine, norepinephrine, and epinephrine."[14]

Potential for Misuse

The dual-use potential of immunological modulation is exemplified by two experiments that produced unexpectedly adverse results. The first experiment was conducted in 2000 by a team of Australian researchers who sought to develop a contraceptive vaccine for the control of wild mouse populations. Their plan was to induce female mice to produce antibodies against antigens on the surface of their own egg cells, thereby destroying the eggs and rendering the mice infertile. To this end, the scientists inserted genes coding for the egg antigens into the mousepox virus, which was used to infect the mice. The researchers also inserted into the engineered mousepox virus a mouse gene coding for interleukin-4 (IL-4), an immune regulatory protein that they expected would enhance the production of mouse antibodies against the egg antigens.

Before conducting the experiment, the scientists sought approval from their local Institutional Biosafety Committee and the Australian government's Genetic Manipulation Advisory Committee.[15] Although the researchers considered the possibility that the inserted gene for IL-4 might increase the virulence of the mousepox virus, this outcome was judged unlikely because the strain of mouse used in the experiment was genetically resistant to mousepox infection. As it turned out, however, the IL-4 gene not only stimulated antibody production in the experimental animals but had the unintended effect of shutting down the cellular arm of the immune response, which plays a key role in defending against viral infection. As a result, the inserted IL-4 gene rendered the mousepox virus highly lethal in the treated mice, even in animals that were genetically resistant to the virus or had been vaccinated against it.

In retrospect, it was clear what had happened: the inserted IL-4 gene simulated the Th2 (antibody) response and simultaneously suppressed the Th1 (cellular immune) response, which was essential to protect the host. Enough preliminary data were available at the time about the ability of IL-4 to down-regulate the cellular immune response for the investigators to have predicted the result of the experiment, but apparently they did not.[16] Once the results of the IL-4/mousepox experiment had been confirmed, the Australian authors debated whether or not to publish. The concern was that actors with nefarious intent might seek to repeat the experiment with a poxvirus that infects humans, such as smallpox virus or monkeypox virus, potentially creating a highly lethal strain that could defeat the standard protective vaccine. If such an agent could be produced, it would pose a serious threat of biological warfare or terrorism.

In September 2000, the U.S. poxvirologists Peter Jahrling and Richard Moyer learned of the IL-4/mousepox experiment and warned that its publication would provide "a blueprint for the biological equivalent of a nuclear bomb."[17] Yet when the Australian authors consulted other leading experts on smallpox, such as D. A. Henderson and Frank Fenner, they argued that publication was warranted in view of the previously published papers describing similar results.[18] The Australian government agreed, and the IL-4/mousepox paper was scheduled for publication in the February 2001 issue of the *Journal of Virology*.[19] On January 10, however, one month before the scientific paper was due to appear, the British popular science magazine *New Scientist* published an article describing the IL-4/mousepox experiment with the sensational title "Killer Virus: An Engineered Mouse Virus Leaves Us One Step Away from the Ultimate Bioweapon."[20] Prepared with the cooperation of the Australian researchers and peppered with quotes from experts on biological weapons and smallpox, the article triggered an explosion of concern in the scientific and lay press.

After the publication of the Australian paper, experimentation with IL-4 poxviruses continued. In October 2003, *New Scientist* published a second article titled "U.S. Develops Lethal New Viruses," describing the work of Mark Buller, a virologist at the University of St. Louis.[21] The article quoted Buller as saying that the construction of an "optimized" IL-4/mousepox virus was needed to test antiviral drugs as potential defenses against bioterrorist weapons based on recombinant poxviruses. Buller also reportedly stated that a similar construct, created by inserting the gene for mouse IL-4 into cowpox virus (which can infect humans), would be tested at the U.S. Army Medical Research Institute of Infectious Diseases at Fort Detrick, Maryland. Nevertheless no paper on the IL-4/cowpox experiment ever appeared, suggesting that the findings may have been too sensitive to publish.

To date, the IL-4/mousepox experiment remains the most widely cited example of the dual-use dilemma in biomedical research. It is also a classic example of modifying a virus-based vaccine with the aim of directly enhancing the immunological response. Although IL-4 is one of hundreds of cytokines and other immune-regulatory molecules whose function could contribute to vaccine effectiveness, there was—and still is—no formal process to evaluate such potentially risky experiments.

A second ill-fated experiment also demonstrated the risks and dual-use potential of immunological modulation. In March 2006, a research group in Britain began a clinical trial of a monoclonal antibody (designated TGN1412) directed against a cell-surface marker on T cells known as CD28.[22] Previous studies in animals, including primates, had shown that the binding of antibodies to CD28 resulted in a modest level of cytokine production, followed by a reversible increase in the number of T cells. Accordingly, the monoclonal antibody was developed to treat a specific type of leukemia that causes a severe deficiency of T cells.

The clinical trial, however, had unexpectedly adverse effects. Within ninety minutes after the intravenous infusion of the monoclonal antibodies into six healthy young volunteers, all of them developed a systemic inflammatory response with high levels of circulating cytokines. Over the next twelve hours, the six subjects became critically ill and suffered multiple organ failure. Although all six ultimately survived, they appear to have suffered permanent immune-system damage that could render them vulnerable to cancer and other diseases. Many questions remain about the study design, the qualifications of the investigators, and the immune mechanisms that led to this unfortunate result.[23]

Both the IL-4/mousepox experiment and the clinical trial of TGN1412 suggest that immunological modulation has a potential for deliberate misuse. The main areas of dual-use concern are briefly discussed in the following paragraphs.

First, *manipulating cytokine levels can have serious harmful effects.* IL-4 is just one of many cytokines that regulate the innate immune response. Whenever the Th1 cellular immune response is required for protection against certain pathogens, the increased expression of Th2 cytokines such as IL-4 can dramatically increase the host's susceptibility to infection. In addition to the IL-4/mousepox experiment, support for this finding comes from experiments that examined the molecular and genetic basis of susceptibility to mousepox infection in various mouse strains. These studies showed that purebred mice with high levels of Th2 cytokines and/or low levels of Th1 cytokines are highly sensitive to poxvirus infection.[24] Another example of a dangerous immunological manipulation is the inappropriate stimulation of Toll-like receptors with pathogen-associated molecular patterns (PAMPs), leading to the extreme overexpression of inflammatory cytokines. The resulting "cytokine storm" can cause autoimmunity, shock, multiple organ failure, and death.

Second, *vaccines can be used to modify neural circuitry*. Developing a vaccine for the treatment or prevention of Alzheimer's disease is an active and promising area of neuroscience research.[25] The primary target of vaccine therapy is the protein amyloid beta, the major constituent of the brain plaques associated with this type of dementia. When the molecular mechanism of plaque formation is better understood, it may be possible to develop "neurotropic" vaccines that can inhibit plaque formation. Nevertheless the knowledge and approaches developed to vaccinate against Alzheimer's disease might be misused for harmful purposes by attacking key neural circuits in the brain.

Third, *vaccines could be designed to interfere with the interactions of the nervous, immune, and endocrine systems*. Extensive study of the neuro-endocrine-immune axis has increased the understanding of how the human body maintains homeostasis in the face of external and internal stress. This understanding could potentially be misused to create vaccines that interfere with vital regulatory systems.

Accessibility

Designer vaccines that have shown unexpected harmful effects during testing in animal models (e.g., the IL-4/mousepox experiment) or in humans (e.g., the TNG1412 clinical trial) illustrate that the line between beneficial and harmful research is defined largely by intent. Once dual-use knowledge has been created, it has a potential for misuse regardless of the motivation of those who produced it.

Ease of Misuse

Although basic information about immunological modulation is freely accessible, actually designing novel vaccines for hostile purposes would require a high level of expertise and tacit knowledge, along with the resources of an academic or government research laboratory.

Magnitude and Imminence of Risk

Although a state biological warfare program might have the resources and expertise to exploit immunological modulation for weapons purposes, it is highly unlikely that a terrorist organization would have the resources do so. Taking these factors into account, the magnitude and imminence of dual-use risk associated with this technology appear to be moderate.

Awareness of Dual-Use Potential

Beyond the IL-4/mousepox experiment, which received a great deal of publicity, most researchers appear to have little awareness of the dual-use risks associated with immunological modulation. Indeed, surveys have shown that most practicing life scientists have little or no awareness of the potential harmful applications of their work.[26]

Characteristics of the Technology Relevant to Governance

Embodiment Immunological modulation is based almost entirely on intangible information and is not associated with specific hardware.

Maturity The field of immunological modulation is at the stage of advanced research and development, with relatively few commercial applications.

Convergence Immunological modulation draws on several areas of science and technology, including bioinformatics, systems biology, and cellular immunology. Another key aspect is the development of vaccine delivery systems, which draws on fields such as nanotechnology,[27] microencapsulation,[28] DNA shuffling,[29] aerosolization and stabilization,[30] and microbiology (e.g., incorporation of vaccines into microbial spores).[31]

Rate of advance In several cases, the discovery of a specific immunological mechanism, such as dendritic cell targeting or the role of ghrelin in weight gain, has led to the rapid development of clinical applications in four to six years. Serious complications have often resulted, however, suggesting the need for a more nuanced assessment of progress.

International diffusion Although most advances in immunological modulation are published in the open scientific literature, the future of rational vaccine design will be dictated primarily by local needs. Whereas developing countries are still preoccupied with controlling infectious diseases, developed countries have recently begun to develop vaccines for chronic diseases and "lifestyle" problems, such as addiction and obesity.

Susceptibility to Governance

The rapid evolution and diffusion of new vaccine technologies based on immunological modulation, coupled with the intensity of commercial competition, make it nearly impossible to restrict such activity. Restrictions might also impede crucial advances in immunology research and the development of life-saving vaccines.

Nevertheless a number of soft-law and informal governance measures might be adopted to manage the risks of this technology.

Relevance of Existing Governance Measures

To date, there have been no attempts to govern immunological modulation beyond the stringent U.S. Food and Drug Administration (FDA) regulations that cover vaccine development and production. However, federal and private funding agencies and professional societies are pursuing awareness training as a means of alerting bench scientists and graduate students about the dual-use potential of immunological research. Providing a framework for this discussion is the U.S. National Science Advisory Board for Biosecurity (NSABB), a federal advisory body established in the wake of the 2001 anthrax letter attacks.[32] In addition, several professional scientific organizations have begun to educate life scientists about dual-use issues, including the Federation of American Scientists, the American Society for Microbiology, and the American Association for the Advancement of Science.

At the international level, the potential misuse of immunology for hostile purposes falls outside the scope of both the 1972 Biological Weapons Convention (BWC) and the 1993 Chemical Weapons Convention (CWC).

Proposed Governance Measures

At the national level, Institutional Biosafety Committees (IBCs) should be responsible for raising awareness about the risks of immunological modulation and for reviewing proposed experiments to prevent adverse outcomes. In the case of unexpected findings with dual-use implications, such as the IL-4/mousepox and TGN1412 experiments, an ongoing review process is required.

To monitor such dual-use research at the international level, Alexander Kelle and his colleagues have proposed the negotiation of a framework convention.[33] A second international policy option would be to establish a "global issues network" focusing on the dual-use aspects of immunology.[34] Such a network would monitor new developments in the field of immunological modulation and raise awareness of their dual-use potential.

Conclusions

Current research efforts are elucidating the profound complexity of the immune system, the delicate balance of its regulation, and its close integration with the nervous and endocrine systems. The time interval between the discovery of a new immune regulatory mechanism and the clinical trial of a related drug or vaccine has shrunk to as little as five years. Mouse and human trials of new therapeutic interventions have already yielded unanticipated results, some of them with catastrophic effects on host survival. These developments, which are linked to the development of increasingly effective delivery systems, have increased the risk of misuse of immunological modulation for harmful purposes. Because this field operates beyond the direct control of governments and existing international treaties, it should be subjected to a web of oversight mechanisms at the local, national, and international levels.

Notes

1. M. R. Dyer, W. A. Renner, and M. F. Bachmann, "A Second Vaccine Revolution for the New Epidemics of the 21st Century," *Drug Discovery Today* 11 (2006): 1028–1033; T. A. Roehn and M. F. Bachmann, "Vaccines against Non-communicable Diseases," *Current Opinion in Immunology* 22 (2010): 391–396.

2. J. L. Gardy, D. J. Lynn, F. S. Brinkman, and R. E. Hancock, "Enabling a Systems Biology Approach to Immunology: Focus on Innate Immunity," *Trends in Immunology* 30 (2009): 249–262; B. Pulendran, L. Shuzhao, and H. I. Nakaya, "Systems Vaccinology," *Immunity* 33 (2010): 516–529.

3. S. Bambini and R. Rappuoli, "The Use of Genomics in Microbial Vaccine Development," *Drug Discovery Today* 14 (2009): 252–260.

4. F. E. Andre, "Vaccinology: Past Achievements, Present Roadblocks, and Future Promises," *Vaccine* 21 (2003): 593–595.

5. D. Hawiger, K. Inaba, Y. Dorsett, M. Guo, K. Mahnke, M. Rivera, J. V. Ravetch, R. M. Steinman, and M. C. Nussenzweig, "Dendritic Cells Induce Peripheral T Cell Unresponsiveness under Steady State Conditions in vivo," *Journal of Experimental Medicine* 194 (2001): 769–779.

6. C.-J. Huang, A. J. Lowe, and C. A. Batt, "Recombinant Immunotherapeutics: Current State and Perspectives Regarding the Feasibility and Market," *Applied Microbiological Biotechnology* 87 (2010): 401–410.

7. P. J. Tacken, I. J. de Vries, R. Torensma, and C. G. Figdor, "Dendritic-Cell Immunotherapy: From ex vivo Loading to in vivo Targeting," *Nature Reviews Immunology* 7 (2007): 790–802.

8. E. L. Smits, S. Anguille, N. Cools, Z. N. Berneman, and V. F. Van Tendeloo, "Dendritic Cell-Based Cancer Gene Therapy," *Human Gene Therapy* 10 (2009): 1141–1146.

9. K. Shortman, M. H. Lahoud, and I. Caminschi, "Improving Vaccines by Targeting Antigens to Dendritic Cells," *Experimental Molecular Medicine* 41 (2009): 61–66; K. Paluska, J. Banchereau, and I. Mellman, "Designing Vaccines Based on Biology of Human Dendritic Cell Subsets," *Immunity* 33 (2010): 464–478.

10. P. Mettens and P. Monteyne, "Life-Style Vaccines," *British Medical Bulletin* 62 (2002): 175–186.

11. M. Kojima, H. Hosoda, Y. Date, et al., "Ghrelin Is a Growth-Hormone-Releasing Acylated Peptide from Stomach," *Nature* 402 (1999): 656–660.

12. E. P. Zorrilla, S. Iwasaki, J. A. Moss, et al., "Vaccination against Weight Gain," *Proceedings of the National Academy of Sciences USA* 103 (2006): 13226–13231.

13. H. Schellekens, T. G. Dinan, and J. F. Cryan, "Lean Mean Fat Reducing 'Ghrelin' Machine: Hypothalamic Ghrelin and Ghrelin Receptors as Therapeutic Targets in Obesity," *Neuropharmacology* 58, no. 1 (January 2010): 2–16.

14. R. Spier, "'Vaccine': 25 Years On," *Vaccine* 26 (2008): 6173–6176.

15. Federation of American Scientists, "Mousepox Case Study," *Case Studies in Dual Use*, http://www.fas.org/biosecurity/education/dualuse/index.html.

16. D. P. Sharma, A. J. Ramsay, D. J. Maguire, et al., "Interleukin-4 Mediates Down Regulation of Antiviral Cytokine Expression and Cytotoxic T-Lymphocyte Responses and Exacerbates Vaccinia Virus Infection in vivo," *Journal of Virology* 70 (1996): 7103–7107.

17. Richard Preston, *The Demon in the Freezer* (New York: Random House, 2002), 158.

18. R. J. Jackson, D. J. Maguire, L. A. Hinds, and I. A. Ramshaw, "Infertility in Mice Induced by a Recombinant Ectromelia Virus Expressing Mouse Zona Pellucida Glycoprotein 3," *Biology of Reproduction* 58 (1998): 152–159.

19. R. J. Jackson, A. J. Ramsay, C. D. Christensen, et al., "Expression of Mouse Interleukin-4 by a Recombinant Ectromelia Virus Suppresses Cytolytic Lymphocyte Responses and Overcomes Genetic Resistance to Mousepox," *Journal of Virology* 75 (2001): 1205–1210.

20. Rachel Nowak, "Disaster in the Making," *New Scientist* 169, no. 2273 (January 13, 2001): 4–5.

21. Debora MacKenzie, "U.S. Develops Lethal New Viruses," *New Scientist* 180, no. 2419 (November 1, 2003): 6–7.

22. G. Woerly, N. Roger, S. Loiseau, D. Dombrowicz, A. Capron, and M. Capron, "Expression of CD28 and CD86 by Human Eosinophils and Role in the Secretion of Type 1 Cytokines (Interleukin 2 and Interferon Gamma): Inhibition by Immunoglobulin A Complexes," *Journal of Experimental Medicine* 190 (1999): 487–495.

23. E. William St. Clair, "The Calm after the Cytokine Storm: Lessons from the TGN1412 Trial," *Journal of Clinical Investigation* 118, no. 4 (April 2008): 1344–1347.

24. G. Chaudhri, V. Panchanathan, R. M. Buller, et al., "Polarized Type 1 Cytokine Response and Cell-Mediated Immunity Determine Genetic Resistance to Mousepox," *Proceedings of the National Academy of Sciences USA* 101 (2004): 9057–9062.

25. C. A. Lemere, "Developing Novel Immunogens for a Safe and Effective Alzheimer's Disease Vaccine," *Progress in Brain Research* 175 (2009): 83–93.

26. Malcolm Dando and Brian Rappert, "Codes of Conduct for the Life Sciences: Some Insights from U.K. Academia," *Briefing Paper No. 16* (Bradford: University of Bradford, May 2005).

27. M. Foldvari and M. Bagonluri, "Carbon Nanotubes as Functional Excipients for Nanomedicines: II. Drug Delivery and Biocompatibility Issues," *Nanomedicine* 4 (2008): 183–200.

28. K. D. Wilson, S. D. de Jong, and Y. K. Tam, "Lipid-Based Delivery of CpG Oligonucleotides Enhances Immunotherapeutic Efficacy," *Advances in Drug Delivery Reviews* 61 (2009): 233–242.

29. C. P. Locher, V. Heinrichs, D. Apt, and R. G. Whalen, "Overcoming Antigenic Diversity and Improving Vaccines Using DNA Shuffling and Screening Technologies," *Expert Opinion in Biological Therapeutics* 4 (2004): 589–597.

30. J. L. Burger, S. P. Cape, C. S. Braun, et al., "Stabilizing Formulations for Inhalable Powders of Live-Attenuated Measles Virus Vaccine," *Journal of Aerosol Medicine and Pulmonary Drug Delivery* 21 (2008): 25–34.

31. N. Q. Uyen, H. A. Hong, and S. M. Cutting, "Enhanced Immunization and Expression Strategies Using Bacterial Spores as Heat-Stable Vaccine Delivery Vehicles," *Vaccine* 25 (2007): 356–365.

32. National Research Council, *Biotechnology Research in an Age of Bioterrorism: Confronting the Dual-Use Dilemma* (Washington, DC: National Academies Press 2004).

33. Alexander Kelle, Kathryn Nixdorff, and Malcolm Dando, *Controlling Biochemical Weapons: Adapting Multilateral Arms Control for the 21st Century* (Basingstoke: Palgrave Macmillan, 2006).

34. National Research Council, *Globalization, Biosecurity, and the Future of the Life Sciences* (Washington, DC: National Academies Press, 2006), 251–256.

13

Personal Genomics

Nishal Mohan

Thanks to advances in DNA sequencing technology and the discovery of genes associated with common diseases, the era of personalized medicine is dawning. Genetic information can be used for disease prevention, early detection, and targeted treatment that tailors drug regimens to a patient's genetic makeup. Given these prospective benefits, the number of companies providing direct-to-consumer genetic testing services is expanding rapidly. When DNA sequencing reaches a level of cost and accuracy at which decoding entire genomes becomes routine, it may have a revolutionary impact on clinical medicine.

Personal genomics also has potential dual-use implications. If large amounts of human genetic information become publicly available, systematic data mining might make it possible to identify genetic similarities and differences among ethnic or racial groups. Conceivably, this information could be exploited to develop biological or chemical agents that can selectively harm specific populations. Because the overall set of genomic data is still small, however, scientists have not yet identified genetic traits that could be used for targeting purposes. Thus, at least for now, the dual-use implications of personal genomics do not warrant the development of dedicated governance measures.

Overview of the Technology

The main technology underlying the field of personalized medicine is genotyping, or determining genetic differences among individuals. One approach involves identifying single-nucleotide polymorphisms (SNPs, pronounced "snips"), which are substitutions, deletions, or insertions in the sequence of DNA nucleotides that make up the human genome. These subtle changes distinguish each individual from the other members of the species and may affect the body's susceptibility to disease and its response to infections and drugs.[1]

There are two approaches to personal genotyping. In the first method, DNA is extracted from a small sample of an individual's cells, purified, and exposed to a

"DNA chip," a small piece of silicon to which hundreds of thousands of single-stranded DNA fragments with sequences complementary to known SNPs have been attached. Through a hybridization process, fragments of the individual's DNA bind to complementary sequences on the chip, making it possible to determine which SNPs are present.[2] As research continues to identify additional SNPs, more complex DNA chips will become available for purchase. The main drawback of current chips is that they are expensive and not easily reusable. The second approach to personal genotyping involves determining all or part of an individual's genome with an automated DNA sequencer. Although this technique is more time-consuming and costly, it offers the advantage that it can identify multiple, sequential, and rare SNPs. Using an individual's full or partial genomic sequence, bioinformatics software tools can correlate known SNPs with disease risks.

The main drawback of the whole-genome approach is that the sequencing and annotation of an entire human genome currently cost about $250,000. Because of technological improvements, however, the cost of whole-genome sequencing is declining rapidly. Several next-generation DNA sequencing machines are under development. IBM Corporation, for example, recently joined the race with a new technology that it expects will eventually permit the sequencing of an entire human genome in a matter of hours.[3] Another important factor is accuracy. For clinical genetics applications, DNA sequences must be decoded with an accuracy of only one error for every 10,000 to 100,000 DNA units.[4] The U.S. National Human Genome Research Institute has projected that by 2014, it will be possible to sequence an entire human genome for about $1,000—the cost threshold at which genome sequencing could start to become a routine part of medical practice.[5]

Regardless of the method used to determine a personal genotype, the utility of the information depends on the availability of public and private databases that support epidemiological correlations between SNPs and the susceptibility to certain diseases.[6] At present, the limited pool of human genetic data limits the usefulness of personal genomics for diagnosis and prevention. As the price of whole-genome sequencing drops, however, the amount of epidemiological data will grow exponentially.[7]

History of the Technology

In 2003, after thirteen years of effort by scientists worldwide, the Human Genome Project finished sequencing the three billion chemical base pairs that make up the human genome. This sequence has since been used to advance medicine, human biology, and the knowledge of human origins. Although decoding the entire human genome cost an estimated $2.7 billion, rapid advances in high-throughput DNA sequencing have since reduced the cost of full-genome sequencing by several orders of magnitude.

In 2006, George Church, a geneticist at Harvard Medical School, founded the Personal Genomics Project, which is developing "a broad vision for how personal genomes may be used to improve the understanding and management of human health and disease."[8] The project has the long-term goal of sequencing the genomes of some 100,000 persons. In 2007, taking advantage of improved DNA sequencing technology, start-up companies began offering direct-to-customer personal genomics services. Firms such as 23andMe (Mountain View, California) and Navigenics (Redwood Shores, California) identify SNPs in an individual's genome that have known links to disease and sell this information to the customer.[9] The profitability of such companies is not guaranteed, however, and an Iceland-based personal genomics company, DeCode Genetics, filed for bankruptcy protection in 2009.

Utility of the Technology

One of the major benefits of personal genomics is its use in preventative medicine to assess an individual's risk of developing certain diseases and to permit early detection and intervention, such as lifestyle changes that reduce the odds that the disease will develop.[10] Combining personal genomics with family history provides a more accurate and complete means of predicting disease risk. Personal genetic information can also help physicians select the best drugs to treat their patients with maximum effectiveness. Information generated by personal genomics techniques can also be a powerful tool for predicting adverse drug reactions in individuals, leading to changes in pharmacotherapy that can save lives.[11] This application of personal genomics is particularly powerful because adverse drug reactions cause approximately 100,000 deaths each year in the United States alone.[12]

Because the human genomics data set is currently limited, much more information must be acquired on the relationships between specific genes and disease. An international research consortium called the 1000 Genomes Project plans to sequence the genomes of over a thousand people around the world to create a detailed, publicly available map of human genetic variation and its relevance to health and disease.[13] Another resource is SNPedia, an open-source database that draws on information from peer-reviewed scientific publications to map the effects of genetic variation.[14] The SNPedia database can be accessed with Promethease, a free informatics tool that compares and analyzes personal genomic sequences.[15] As additional SNPs are identified, it will become possible to determine an individual's predisposition to a wide range of diseases and drug-response patterns.

Potential for Misuse

Although no one has yet tried to exploit personal genomics for hostile purposes, the theoretical possibility exists. For example, it may be possible to exploit pharmacogenomics to identify and develop drugs that can cause significant harm to a subset of the population that is genetically vulnerable. In February 2010, for example, an international team announced that it had sequenced the genomes of South African archbishop Desmond Tutu and an indigenous bushman from Namibia as part of an effort to enable researchers and drug companies to bring the benefits of personalized medicine to people in developing countries. This analysis identified 1.3 million genetic variations that had not previously been identified, potentially making it possible to tailor drug therapies for people living in southern Africa. Certain drugs for treating AIDS, for example, are less effective in Africans than in Europeans. But critics of the research, including the ETC Group in Canada, suggested that the information might be used to create drugs for profit or even to design biological weapons capable of targeting specific ethnic groups.[16]

Accessibility

DNA sequencing technology for the identification of SNPs is available to most biological research laboratories, either in-house or from commercial suppliers. DNA sequencing machines have been developed and marketed by several biotechnology companies in Europe and North America, including Roche, Illumina, and Applied Biosystems Inc. The use of these machines relies on standardized protocols, so that they can be operated by trained technicians lacking an advanced degree in molecular biology. To date, the major factor limiting the spread of DNA sequencing technology has been cost, but it is only a matter of time before the machines become more affordable. Private firms in various parts of the world also provide commercial DNA sequencing and genotyping services for customers and institutions that lack their own hardware.

Some genomic databases are available free of charge, while others are controlled by private companies. In response to growing concerns about the sensitivity of personal genomic data, the United States and Germany recently adopted legislation limiting the use and availability of such information. In response to these developments, public SNP databases may cease to be updated and may even disappear. In 2008, for example, the U.S. National Institutes of Health removed aggregate genomic data from the public domain for privacy reasons after research showed that individuals could be identified from the seemingly anonymous pool.[17]

Ease of Misuse

At present, identifying the pharmacogenetic vulnerabilities of a particular population would be difficult and prohibitively expensive, for a number of reasons. First, DNA samples would have to be collected from the selected population, and a great deal of time and effort would be required to generate personal genomics data using

DNA sequencing techniques. Second, few verified correlations exist between genetics and adverse drug reactions, and none of them could easily be exploited to cause large-scale harm. Even when such correlations have been identified, the expertise needed to translate the genetic data into harmful drugs would require the skills of a multidisciplinary team of experts. Such an undertaking might conceivably be launched by a country with a sophisticated biotechnology industry, but it is clearly beyond the capacity of terrorist organizations. Even in the unlikely event that a genetically targetable weapon could be developed, delivering it effectively would depend on several other technologies (such as nanotechnology) that are themselves at an early stage of development.

Magnitude and Imminence of Risk

Given the high cost of personal genomic technologies and the limited number of known correlations between genetic variations and disease, personal genomics technology does not pose an imminent threat of misuse. Even if a proliferant state or terrorist group managed to gain access to a large database of human genetic data and identified a drug likely to cause harm in a subset of the population, it would still be extremely difficult to develop a genetically targetable weapon and mate it with an appropriate delivery system for effective dissemination. Moreover, in the unlikely event that these technical hurdles could be overcome, the genetically vulnerable population would not necessarily be concentrated in a single geographic area where it could be targeted in a selective manner.[18]

Awareness of Dual-Use Potential

To date, the scientific and policy communities have given little attention to the possibility that personal genomics could be misused to cause physical harm, and certainly not in the form of ethnic weapons. Instead government policy makers have been concerned with other issues: the protection of privacy rights, the risk that individuals could misinterpret genetic information, and the potential misuse of personal genomic information by employers and health insurers to discriminate against individuals with a genetic predisposition to certain diseases. In addition, the scientific community has warned of the potential harm to clients of direct-to-consumer companies that could result from inaccuracies and false positives in genetic risk predictions.[19]

The exponential growth in personal genomics has not gone unnoticed by the U.S. Department of Defense for applications for determining phenotypes relevant to military performance and medical cost containment.[20] The DOD is extremely interested in uses for both offensive and defensive military operations but is aware of the limitations created by the lack of data linking genotypes and phenotypes.

Characteristics of the Technology Relevant to Governance

Embodiment   Personal genomics is based largely on intangible information, namely, genomic sequences stored in large databases. Although DNA sequencing hardware is an enabling technology, it is advancing independently.

Maturity   Personal genomics services are available from commercial, direct-to-customer suppliers. At present, the chief limitation on the usefulness of personal genomics is the lack of accurate databases of known SNPs and their correlations with disease risks. Because only a few such databases are publicly available, the technology is still emerging and remains somewhat unpredictable.

Convergence   To reach its potential to enhance human health, personal genomics will require the systematic integration of several disciplines, including high-throughput DNA sequencing, epidemiology, systems biology, bioinformatics, and genetic counseling.

Rate of advance   Since 2007, the speed, accuracy, and cost parameters of DNA sequencing have advanced exponentially, helping to drive the emergence of personal genomics.

International diffusion   Most direct-to-consumer personal genomics companies are based in North America and Europe but extend their services to other parts of the world. At present, cost is the major constraint on the diffusion of such services.

Susceptibility to Governance

Governance of personal genomics is difficult because DNA sequencing technology is commercially available and plays a vital role in biomedical research in both academia and industry worldwide. It is clearly too late to regulate the sale of benchtop DNA sequencers, which are widely available from commercial sources. In addition, many for-profit personal genomics companies offer genetic data to anyone who can afford their services. Imposing strict controls on the use of data generated by personal genomics tests would impede efforts to identify correlations between the frequency of certain SNPs in a population and various diseases, limiting the potential for misuse of this information. But such regulation would also constrain the benefits of using genomic data for assessing disease risk and creating personalized therapies.

The DOD has been advised by the JASON defense science advisory panel to establish policies "that result in the collection of genotype and phenotype data, the application of bioinformatics tools to support the health and effectiveness of military personnel, and the resolution of ethical and social issues that arise from these activities."[21] These policies will apply only to the collection of genomic sequences from military personnel.

Relevance of Existing Governance Measures

In the mid-1990s, when it became clear that advances in DNA sequencing technology would lead to the ability to sequence whole human genomes quickly and cheaply, the U.S. Congress began to develop legislation to protect Americans from genetically based discrimination by employers and health-insurance companies. The resulting law, the Genetics Information Nondiscrimination Act (GINA), was debated for thirteen years before it finally passed in 2008.[22] GINA prohibits insurance companies and other service providers from using genetic information to deny coverage or determine payment rates, and it also prevents employers from purchasing genetic information about current or prospective employees from third parties. Thanks to these protections, Americans will be able to enjoy the benefits of personalized medicine without fear of discrimination.

In the United Kingdom, the House of Lords Report on Genomic Medicine did not recommend legislation against genetic discrimination but advised that direct-to-consumer genetic-testing companies adopt a unified code of conduct for assessing the medical utility of such services and the need for genetic counseling of customers.[23] Both the United States and the United Kingdom believe that personal-genomics services are evolving so rapidly and unpredictably that it would be premature to adopt specific regulations. Germany, in contrast, has taken a more restrictive approach to genetic testing in an attempt to prevent misuse. The Human Genetic Examination Act, which went into effect on February 1, 2010, permits genetic testing only when performed by a doctor with adequate informed consent, and imposes specific penalties for violations.[24] The German legislation also limits genetic testing on fetuses, prohibits genetic testing on individuals for diseases that appear later in life, and prevents insurance companies and employers from demanding or using genetic information. In effect, the Human Genetic Examination Act has made the services of direct-to-consumer personal genomics companies illegal in Germany.

Proposed Governance Measures

With the continuing decline in DNA sequencing costs, the rapid production of genomic data, and the increasing availability of direct-to-consumer genomic services, now is the time to start thinking about oversight mechanisms and regulations to prevent the deliberate misuse of personal genomics. One strategy for governance would be to add regulatory groups for personal genomics to existing oversight committees, such as the Advisory Committee on Genetics in the United States.

A more direct approach would be for the U.S. Congress to pass legislation regulating personal genomics testing and data. Such legislation would be similar to the regulations governing the privacy of medical records: it would clearly define the ownership of genetic data and its acceptable uses, protect the privacy of an individual's genetic data, and require stricter security and screening procedures for submitting samples to direct-to-consumer companies and similar institutions. Any such legislation should also address privacy issues related to the DNA sequencing of individual genomic material obtained without the permission of the owner.

Conclusions

The age of personal genomics has dawned, and personalized drug therapy is poised to be the next step in the evolution of medicine. Although the potential benefits of personal genomics are clear, the main downside risk at present is that personal genetic data could be used for genetic discrimination. Both the United States and Germany have passed legislation to address this concern. Far more speculative is the possibility of using pharmacogenetic data to create "ethnic weapons," such as biochemical agents that could selectively cause physical harm to a genetic subgroup of the population. Because such a scenario will remain extremely unlikely for the foreseeable future, the costs of taking preventive measures would outweigh the benefits.

Notes

1. J. Y. Hehir-Kwa, M. Egmont-Petersen, I. M. Janssen, et al., "Genome-Wide Copy Number Profiling on High-Density Bacterial Artificial Chromosomes, Single-Nucleotide Polymorphisms, and Oligonucleotide Microarrays: A Platform Comparison Based on Statistical Power Analysis," *DNA Research* 14, no. 1 (February 28, 2007): 1–11.

2. P. Yue and J. Moult, "Identification and Analysis of Deleterious Human SNPs," *Journal of Molecular Biology* 356, no. 5 (March 2006): 1263–1274; U. Väli, M. Brandstrom, M. Johansson, et al., "Insertion-Deletion Polymorphisms as Genetic Markers in Natural Populations," *BMC Genetics* 9 (January 22, 2008): 8; A. Vignal, D. Milan, M. SanCristobal, et al., "A Review on SNP and Other Types of Molecular Markers and Their Use in Animal Genetics," *Genetics, Selection, Evolution* 34, no. 3 (May–June 2002): 275–305.

3. "IBM Joins Pursuit of $1,000 Personal Genome," *New York Times*, October 6, 2009, D2.

4. Nicholas Wade, "Cost of Decoding a Genome Is Lowered," *New York Times*, August 10, 2009.

5. "NIH Promises Funds for Cheaper DNA Sequencing," *Nature* 454 (2008): 1041.

6. D. L. Wheeler, T. Barrett, D. A. Benson, et al., "Database Resources of the National Center for Biotechnology Information," *Nucleic Acids Research* 35 (January 2007): D5–D12; Michael Cariaso, "SNPedia: A Wiki for Personal Genomics," *Bio-IT World*, December 2006–January 2007, 12–17; G. A. Thorisson, O. Lancaster, R. C. Free, R. K. Hastings, P. Sarmath, D. Dash, S. K. Brahmachari, and A. J. Brookes, "HGVbaseG2P: A Central Genetic Association Database," *Nucleic Acids Research* 37 (January 2009): D797–D802; A. Hamosh, A. F. Scott, J. S. Amberger, et al., "Online Mendelian Inheritance in Man (OMIM), a Knowledge Base of Human Genes and Genetic Disorders," *Nucleic Acids Research* 33 (2005): D514–D517.

7. Nicholas Wade, "A Decade Later, Genetic Map Yields Few New Cures," *New York Times*, June 12, 2010, A1.

8. Personal Genomics Project, http://www.personalgenomes.org.

9. Erick Check Hayden, "Personal Genomes Go Mainstream," *Nature* 450 (October 30, 2007): 11.

10. M. J. Khoury, C. McBride, S. D. Schully, et al., "The Scientific Foundation for Personal Genomics: Recommendations from a National Institutes of Health–Centers for Disease Control and Prevention Multidisciplinary Workshop," *Genetic Medicine* 8 (August 11, 2009): 559–567.

11. P. C. Ng, Q. Zhao, S. Levy, et al., "Individual Genomes instead of Race for Personalized Medicine," *Clinical Pharmacology Theory* 85, no. 2 (February 2009): 306–309; D. Ge, J. Fellay, A. J. Thompson, et al., "Genetic Variation in IL28B Predicts Hepatitis C Treatment-Induced Viral Clearance," *Nature* 461 (2009): 399–401; S. J. Gardiner and E. J. Begg, "Pharmacogenetics, Drug-Metabolizing Enzymes, and Clinical Practice," *Pharmacology Review* 58 (2006): 521–590.

12. J. Lazarou, B. H. Pomeranz, and P. N. Corey, "Incidence of Adverse Drug Reactions in Hospitalized Patients: A Meta-analysis of Prospective Studies," *Journal of the American Medical Association* 279 (1998): 1200–1205.

13. The 1000 Genomes Project, http://www.1000genomes.org.

14. SNPedia, http://www.snpedia.com/index.php/SNPedia.

15. Promethease, http://www.snpedia.com/index.php/Promethease.

16. Rob Stein, "Genomes for Tutu, Bushman Decoded; Researchers Hope to Customize Health Care in Developing World," *Washington Post*, February 18, 2010, A4.

17. E. A. Zerhouni and E. G. Nabel, "Protecting Aggregate Genomic Data," *Science* 322, no. 5898 (October 3, 2008): 44.

18. Ng et al., "Individual Genomes instead of Race for Personalized Medicine," 306–309.

19. P. C. Ng, S. S. Murray, and S. Levy, "An Agenda for Personalized Medicine," *Nature* 461 (2009): 724–726.

20. "The $100 Genome: Implications for the DoD," *JASON* JSR-10-100, December 2010.

21. Ibid.

22. Genetic Information Nondiscrimination Act of 2008, Pub. L. no. 110-343 (October 3, 2008).

23. House of Lords, HL Paper 107-I, July 7, 2009.

24. German Federal Parliament (Bundestag), "Human Genetic Examination Act," April 24, 2009.

14

RNA Interference

Matthew Metz

RNA interference is a rapidly expanding field of biomedical research that holds great promise for curing and preventing disease. At the same time, the technology has some potential for misuse, including the creation of pathogens with enhanced virulence and the targeted disruption of genes that serve vital functions in the human body. Because several broadly enabling biotechnologies are integral to RNA interference, it is not suited to formal regulation, but informal approaches to governance may be useful.[1]

Overview of the Technology

RNA interference is an innate cellular mechanism for controlling the expression of genes. Its best-characterized function is to defend against invading viruses, which reproduce inside the cells of the host by commandeering their biochemical machinery. During the life cycle of a virus, its viral genome is duplicated, and genes are expressed in the medium of the host cell. The sequence of nucleotide bases in the viral genome is transcribed into complementary molecules of messenger RNA, which are then translated into viral proteins. Many viral genomes consist of RNA rather than DNA, so during the process of transcription, the genomic RNA template and a complementary strand of messenger RNA are temporarily paired with each other in a double-stranded RNA duplex. In addition, many viruses with RNA genomes replicate their entire genomic sequence by creating a complementary strand of RNA, a process that also generates double-stranded RNA while the two molecules are paired.

Although double-stranded RNA is part of the life cycle of many viruses, it is foreign to the infected host cell. The cell responds to this anomalous molecular structure by using the viral RNA duplex as a template to generate "small interfering" RNA (siRNA) molecules that are complementary to specific messenger RNA sequences. A small stretch of twenty-two RNA bases defines the target gene.[2] The host cell then uses the siRNA molecules to recognize strands of viral messenger

RNA with high specificity and mark them for destruction by the cellular machinery. A wide variety of invertebrates, fungi, and plants employ RNA interference as a mechanism for innate cellular immunity against viruses. (This mechanism is also believed to occur naturally in vertebrates but has yet to be observed.) Accordingly, it appears that RNA interference evolved as a defense against viral infection. Strong evidence also suggests that RNA interference helps to regulate the expression of the organism's own genes: many genomes contain micro-RNA sequences that regulate gene expression.

History of the Technology

The biotechnology research community's first brush with RNA interference came in 1990, when Richard Jorgensen and his colleagues published the unexpected results of their attempt to increase the intensity of purple color in petunias. The researchers, then with the company DNA Plant Technology Inc., genetically modified the plants by inserting a gene to increase the level of a key enzyme involved in the synthesis of the purple pigment. Paradoxically, many of the genetically modified petunias produced little or no purple pigment in their flowers because the expression of the enzyme was reduced. Further study showed that the inserted gene had been transcribed into messenger RNA as expected, but the transcripts had then been destroyed.[3] Jorgensen and his colleagues termed this phenomenon "co-suppression" because the messenger RNA transcripts of both the endogenous plant gene (coding for the purple pigment) and the engineered gene (designed to enhance the expression of the pigment gene) had been eliminated.

In the years following this discovery, researchers found that by genetically engineering animal, plant, and fungal cells to express RNA molecules that were complementary ("antisense") to the messenger RNA for a particular gene, they could suppress the expression of the targeted gene. The mechanism behind antisense suppression remained unsolved for many years. Even so, researchers frequently used the phenomenon itself to disrupt specific genes to gain insights into their function. As researchers explored the uses of antisense RNA, they found that sense RNA could also disrupt the expression of a gene with the same sequence.[4] This finding was essentially a reiteration of the phenomenon reported by Jorgensen, in which an inserted gene designed to increase the expression of a flower pigment had blocked it instead.

Antisense technology also became a platform for pharmaceutical development. In 1998, Isis Pharmaceuticals and Novartis marketed a therapeutic based on antisense RNA to treat AIDS-associated retinitis caused by cytomegalovirus infection. It is generally believed that antisense RNA stimulates the production of siRNA, albeit less effectively than double-stranded RNA. Thus the treatment for cytomegalovirus retinitis can be considered the first drug based on RNA interference.

Meanwhile, Andrew Fire of Stanford University and Craig Mello of the University of Massachusetts Medical School, working with the roundworm *Caenorhabditis elegans*, found that injecting double-stranded RNA into cells had a far more potent gene-silencing effect than using either antisense RNA or sense RNA. The effect was so powerful that only a few molecules of double-stranded RNA were required to silence a gene, and the effect was passed on to the next generation.[5] This observation led Fire and Mello to conclude that something catalyzed by double-stranded RNA was targeting the messenger RNA for destruction, and they termed this process "RNA interference." Subsequent research began to elucidate the cellular systems that recognize double-stranded RNA, use it as a template to generate siRNA molecules, and then destroy the messenger RNA molecules that pair with the siRNAs. In recognition of this breakthrough, Fire and Mello were awarded the Nobel Prize in Physiology or Medicine in 2006.

The breadth of potential applications of RNA interference quickly led to its recognition as a revolutionary discovery and fueled intense interest from the scientific community. In 2004, Acuity Pharmaceuticals (a subsidiary of Opko Health Inc.) began Phase I clinical trials of an RNA-interference-based drug for the treatment of age-related macular degeneration. Since then more than a dozen pharmaceutical applications based on RNA interference have entered clinical trials.[6] In addition, numerous biotechnology companies offer contract services for testing and optimizing the design of siRNAs.

Utility of the Technology

The powerful biomedical applications of RNA interference stem from its ability to target precise gene sequences. Numerous experiments have shown that the mismatch of even one nucleotide base among the thousands of bases making up a gene will dramatically reduce the effectiveness of gene silencing.[7] As a result, siRNA molecules can distinguish not only between the genes of an invading virus and those of the host but also between two nearly identical host genes. Recent discoveries have shown how to optimize the specificity of siRNA molecules, stabilize them as they travel through the bloodstream, and deliver them to the appropriate target cells.

Because RNA interference disrupts gene expression in a targeted manner, it provides a valuable tool for basic research in molecular genetics. Researchers can silence unknown genes and infer their function from the effects of the disruption. For example, if silencing a particular gene leads to albinism, then the gene in question probably plays a role in pigmentation. The broad appeal of RNA interference for basic research is unlikely to diminish anytime soon.

The ability of RNA interference to disrupt the life cycle of viruses is also being tested in various experimental systems as a means of combating viral disease. Researchers have made progress in developing therapeutics against important human pathogens such as HIV,[8] herpes viruses,[9] hepatitis C,[10] and Ebola virus.[11] Antiviral drugs often have side effects because of the unintended interactions of the drug with the metabolism of the host. In contrast, the sequence-specific nature of RNA interference makes it unlikely to interact with metabolic pathways or other drugs.

Another promising biomedical application of RNA interference is for regulating harmful gene expression in genetic disorders. Genes define the inherited characteristics of all organisms, and different versions of each gene, called alleles, confer variation within species. Each individual of a species (excepting identical twins) carries a unique set of alleles. RNA-interference-based drugs are being developed to block the expression of alleles that are responsible for inherited diseases or have mutated to cause cancer, while leaving the healthy alleles of genes untouched due to sequence differences. In experimental systems, RNA interference selectively disrupts the expression of an allele that causes the severe neurological disorder amyotrophic lateral sclerosis (Lou Gehrig's disease).[12] Clinical trials have also shown that the use of RNA interference to silence cancer-associated genes can inhibit several types of human cancer cells or sensitize them to chemotherapy.[13]

Other promising applications of RNA interference include the treatment of burns, allergies, and autoimmune disorders. Part of the injury caused by burns—ranging from simple sunburn to severe burns caused by heat or radiation exposure—results from the expression of genes associated with inflammation. RNA interference might help to moderate the inflammation-associated aspects of an injury. Drugs have long been used to suppress the immune system's overreaction to allergens in severely allergic individuals, as well as the inappropriate recognition of the body's own tissues as foreign that occurs in autoimmune diseases such as lupus. RNA interference is being examined as a new and potentially more specific method for switching off the genes responsible for allergic and autoimmune disorders.[14]

Finally, the ability of RNA interference to disrupt gene expression in a targeted manner may lead to many other applications. Genes associated with obesity, for example, are potential targets for weight-loss therapies, and it might also be possible to block genes that restrict muscle mass to enhance performance in competitive sports. RNA interference might even be used to alter the expression of genes that control hair, eye, and skin color or play a role in aging. Despite the ethical concerns surrounding such lifestyle and cosmetic applications, strong consumer interest is likely to drive research in these areas.

Each biomedical application of RNA interference will require innovation and a unique product development cycle. Current hurdles for the development of drugs based on RNA interference include delivering double-stranded RNA templates into cells and ensuring that the templates persist long enough to stimulate sufficient siRNA disruption of gene expression. Two possible delivery mechanisms exist. One approach is to use genetically engineered viral "vectors" (developed for gene-therapy purposes) to deliver siRNA, double-stranded RNA templates, or genes that express siRNA into a patient's cells. A second approach is to administer bulk quantities of double-stranded template RNA or siRNA in a formulation that enables the molecules to survive the trip through the patient's bloodstream, lungs, muscle tissue, or other route of administration and reach the target cells. Because the latter method would not confer a permanent genetic source of RNA interference, sustained administration would be required to provide ongoing effectiveness. Such administration is being rapidly improved by the use of nanoparticles.[15]

Potential for Misuse

In addition to its promising medical applications, the ability of RNA interference to block gene expression has potential for misuse. Just as the mechanism can prevent the expression of disease-related genes from inherited disorders or invading viruses, so it could be manipulated to disrupt healthy metabolism and immunity. In theory, RNA interference might be used to cause harm in two ways: by blocking vital genes in the host or by interfering with the host's defenses against infection (see table 14.1).

For example, genetic engineering methods could be used to insert template RNA into the genome of a viral pathogen. After infecting the host, the virus would express interfering RNAs that disrupt the host's metabolism and produce effects

Table 14.1
Potential misuses of RNA interference

Manifestation	Target	Application method		Effect
		Genetic engineering	Formulation	
RNA interference is used to disrupt normal host metabolism or immunity	Metabolism	Toxin-like gene engineered into a virus *(more likely)*	Toxic formulation *(less likely)*	Toxicity
	Immunity	Virulence gene engineered into a virus *(more likely)*	Addition to a pathogen weapon formulation *(less likely)*	Immune suppression
Disruption of host RNA interference mechanisms	Host defenses against viral infection	Virulence gene engineered into a virus *(more likely)*	Addition to a pathogen weapon formulation *(theoretical)*	Immune suppression

similar to those of a toxin. It might also be possible to target the RNA interference genes of the host to increase susceptibility to viral infection. If the viral vector was contagious and spread from person to person, it would have the potential to inflict mass casualties.

Ethnic Weapons

The precision of RNA interference has troubling implications for the potential misuse of the technology. Anyone who has access to human DNA sequence data can identify RNA interference targets specific to a particular gene allele, and the ongoing public release of additional genomic information makes it possible to identify new targets. Accordingly, it may eventually be possible to use RNA interference as an "ethnic weapon" by targeting an allele for a critical metabolic or immune function that is present at high frequency in a particular human population. Although it is politically contentious to discuss whether ethnic groups have genetic features that clearly separate them from other populations, the distinct ancestry of ethnic groups does correlate to instances of unique genetics.[16]

Of course, DNA sequence alone is not sufficient to identify a target gene for silencing with RNA interference; the function of the gene must also be understood. Today scientists understand the function of a small but significant minority of human genes and can infer the functions of others by their relatedness to genes in other organisms whose function is known. Determining both the sequence and function of a gene creates the opportunity to disrupt it through RNA interference. Given the vast quantity of data generated by the Human Genome Project and other efforts to map human genetic diversity,[17] it may eventually become possible to design weapons based on RNA interference that target gene alleles characteristic of certain human populations.

Because of the genetic diversity that exists within ethnic groups, it is rare that all the individuals within an ethnic group have the identical allele for a given gene. (Such a universal allele is said to be "monomorphic.") In some cases, attempts to identify genetic markers for the forensic identification of individuals based on DNA fingerprinting have been confounded by the presence of monomorphic alleles.[18] Although monomorphic alleles are rare, in many cases certain alleles are found only in a given ethnic group or lineage. Such "private" alleles are defined as occurring with a prevalence of 5 percent to 20 percent or more in a particular population and less than 0.1 percent to 1.0 percent in other populations.

Recent efforts to trace the relatedness and ancestry of different ethnicities (African Americans, Asians, Hispanics, and Europeans) have analyzed large quantities of genomic data for patterns of shared and distinctive genetic markers. These studies have borne out the existence of thousands of private alleles at frequencies of 10 percent to 12 percent within the ethnic group.[19] Cases of private alleles with even higher frequencies are less common but do occur. For example, the discovery of a private allele in Native Americans supports the hypothesis that a single founding population colonized the Americas.[20] This study examined 1,249 individuals from twenty-one Native American and Western Beringian populations and fifty-four other populations worldwide and found that the private allele was present more than 35 percent of the time in Native Americans and Western Beringians but was absent from all other subjects. Although a private allele would not provide a genetic target that could be used to attack all members of a particular ethnic group, it might be possible to target a subset of that group exclusively.

Accessibility

Although basic and applied research on RNA interference provides a roadmap for the possible development of a biological weapon, only a state actor with an advanced biological warfare program could hope to do so successfully. Publicly available research could help inform the design of such a weapon, but the development process would require a multidisciplinary team with expertise in genomics, molecular biology, cell biology, tissue culture, and drug formulation. In addition, sophisticated biotechnological capabilities would be needed to generate, test, and optimize the effectiveness of specific siRNA designs. Finally, translating RNA interference into pharmacologically active products would require access to gene-therapy techniques or methods for stabilizing and delivering siRNA molecules.

Some of the skills required for performing RNA interference might be outsourced because commercial companies offer DNA and RNA synthesis, gene expression analysis, and DNA sequencing on a fee-for-service basis. The popularity of RNA interference as a research tool has also spawned the emergence of numerous providers of molecule design and synthesis specific to this application.

Ease of Misuse

The design of siRNA molecules and their double-stranded RNA templates draws on several fields of research, each of which involves a skill set not fully codified in the scientific literature. As a result, applications of RNA interference require tacit knowledge associated with genomics, genetic engineering, cell culture, microbiology, and medicinal chemistry. The requirement of hands-on experience raises the threshold that an offensive biological warfare program would have to reach to succeed, although the skill sets are not so rare as to preclude trained individuals from being recruited or coerced. Even under the best of circumstances, developing RNA interference into a weapon would require a sustained program of development and testing lasting several years.

Magnitude and Imminence of Risk

As mentioned earlier, the possible hostile applications of RNA interference technology include genetically engineering a virus for enhanced virulence or interfering with the immune defenses of the host. There is also the possibility of developing so-called ethnic weapons that would selectively target particular races or ethnic groups. Although the large amount of genetic variation within ethnic groups makes it unlikely that a critical allele will be both private and monomorphic within a population, that scenario is not essential for an ethnic weapon to be effective. A biological weapon that was capable of causing exclusive harm to 20 percent or even 10 percent of a specific ethnic group would have devastating social, cultural, and public health consequences and would certainly suffice for terrorists or dictators seeking social engineering or vengeance. Moreover, those diabolical enough to develop and use an ethnic weapon would probably not be deterred by the potential collateral damage caused by the presence of the targeted allele in other populations at a frequency of 1 percent or less. Accordingly, the theoretical possibility of developing an ethnic weapon warrants concern. At the same time, given the lengthy and technologically advanced efforts that would be required to develop such a weapon, the risk remains remote at this time.

Awareness of Dual-Use Potential

National policy makers have not treated the malicious use of RNA interference as an immediate concern but have focused instead on the unintended safety hazards of the technology. For example, the June 2007 oversight framework proposed by the National Science Advisory Board for Biosecurity (NSABB) assessed RNA interference as unlikely to pose a risk of misuse.[21] Similarly, the Federation of American Scientists included a case study of RNA interference in its set of online tutorials on dual-use biological research but characterized the technology as mainly posing a biosafety concern.[22]

Nevertheless, a few policy analyses concerned with dual-use biotechnologies have recognized the potential offensive applications of RNA interference.[23] In November 2006, the United Kingdom presented a report to the Sixth Review Conference of the Biological Weapons Convention (BWC) that mentioned the potential hostile application of RNA interference, noting that "theoretically, the technology now exists for the long-term, efficient silencing of an allele that segregates with ethnicity."[24] The British report did not discuss the implications of this possibility, however, for the potential harm to society or the attractiveness of such a weapon to malicious actors.

An earlier report by the Sunshine Project provided a more in-depth analysis of the potential misuse of RNA interference to target gene alleles present exclusively in specific ethnic groups. This report reviewed two databases containing point mutations, or single-nucleotide polymorphisms (SNPs). Of the roughly 300 SNPs found within genes, 3.3 percent were absent from one ethnic group but present in another group at a frequency of greater than 20 percent.[25] Extrapolating from this finding, the report concluded that a large number of potential genetic targets exist for ethnic weapons. Although questions remain about the sample size and how the SNPs were selected, these results support the possibility that a significant—and insufficiently understood—risk of misuse may exist.

A report in 2006 by the U.S. National Research Council (NRC) came to the opposite conclusion, dismissing the likelihood that ethnic-specific SNPs could serve as a target for biological weapons based on RNA interference. The NRC report argued that "the hugely large number of point mutations and other polymorphisms within the genome are not likely to lead to any selective targeting in the near future. . . . The proportion of such mutations lying in functionally important areas of the genome is small and the technical difficulties associated with exploiting them are real."[26] The report provided no references or analysis to justify extrapolating from a small number of potential targets to an insignificant risk.

Characteristics of the Technology Relevant to Governance

Embodiment   RNA interference technology is based not on novel or specialized hardware but on intangible knowledge about the molecular mechanisms that control gene expression. Only a handful of reagents and laboratory resources are specific to the technique, and these can be derived from more common materials or purchased commercially. In addition, the key principles that underlie RNA interference can be understood from information in the open scientific literature.

Maturity   RNA interference is a popular tool in basic biomedical research. The pharmaceutical industry is also applying the technology to develop new therapeutic drugs, more than a dozen of which are in clinical trials. Commercial firms provide a number of services related to RNA interference, including the design, production, and testing of reagents, greatly reducing the amount of in-house expertise and laboratory capability needed to pursue such research. At the same time, each application of RNA interference is unique and requires substantial technical knowledge and laboratory resources to develop, test, and produce.

Convergence   The development of RNA interference tools draws on several enabling technologies in the biological and chemical sciences. Bioinformatics provides the sequence data needed to design the siRNAs that silence the target genes. The capability to synthesize short pieces of RNA is required to generate the double-stranded RNA that serves as the template for siRNA. In addition, genetic-engineering techniques are required to design a pathogen to exploit RNA interference, while drug formulation is essential to using siRNAs directly as a weapon.

Rate of advance   Few biomedical technologies have progressed as rapidly as RNA interference from discovery to applied research and clinical trials. The development of the technology has been fueled by advances in genomic science and gene-therapy techniques and vectors. The selection of appropriate targets for RNA interference is limited only by the scientific understanding of gene structure and function. In addition, the effectiveness of siRNA as a therapeutic depends on the ability to deliver these molecules into cells, where they act to silence genes. Methods for improving the specificity, potency, and stability of siRNA molecules in drug formulations are advancing rapidly. The genetic engineering techniques needed to incorporate siRNA-generating sequences into viral vectors are well established, and the vectors themselves are on the verge of broad clinical application for gene therapy. In general, the development of new products based on RNA interference is constrained less by a lack of effective methods than by patents and trade secrets that limit their availability.

International diffusion   The advantages of pharmaceutical products based on RNA interference have contributed to the rapid international diffusion of the technology. Academic and corporate laboratories in every advanced industrial country are conducting research involving the technique, and a review of the scientific literature indicates that RNA interference research is also taking place in developing countries such as Pakistan and Iran. Not only is expertise in this field widely distributed, but a large number of companies provide reagents and research services. While many of these companies are based in the United States, several of them have worldwide distribution networks for their products and services.

Susceptibility to Governance

Any attempt to impose legally binding regulations on an enabling technology like RNA interference would be confounded by its global diffusion and the fact that it draws on much of the same information, equipment, reagents, and procedures as other areas of biotechnology. The genomic sequences that inform the design of specific RNA interference tools, for example, are publicly available and play a pivotal role in biomedical research. Although a few specialized algorithms have been developed for the optimized design of the activating molecules, standard molecular biology techniques and reagents would be sufficient to produce a weapon based on RNA interference.

As a result of these factors, even narrowly focused governance measures would be expensive, burdensome to unrelated areas of research, and of little value in deterring those who wish to do harm. Mandatory declarations and inspections, personnel screening or licensing, registration or certification of labs and individuals, and scrutiny of equipment and reagent purchases would all be ineffective in limiting the risk of misuse. Similarly, restricting access to research laboratories or educational settings where RNA interference skills could be acquired would have a chilling effect on basic research, slow pharmaceutical advances, and impede international scientific collaboration. Restrictions on access to RNA interference materials or training would also hamper innovation in several nonmedical fields, such as agriculture and renewable energy. Nevertheless, although the costs of formal regulation would outweigh the benefits, informal approaches to governance may still be practicable.

Relevance of Existing Governance Measures

To date, the United States has not attempted to regulate RNA interference technology. The NSABB has dismissed the likelihood of RNA interference being used for hostile purposes, and the Recombinant DNA Advisory Committee has addressed the technique only from the standpoint of biosafety.[27] Regulation of RNA interference is also absent at the international level. The Australia Group focuses exclusively on harmonizing national export controls on lists of dangerous pathogens and dual-use production equipment. Finally, the Biological Weapons Convention lacks any formal verification or enforcement measures, although some parties to the treaty have raised RNA interference as a potential security concern.

Proposed Governance Measures

Because the technical and financial resources of a nation-state would probably be required to develop an RNA-interference weapon, governance efforts should focus on the state level, including efforts to make scientists involved in legitimate research less susceptible to exploitation by those who would misuse their knowledge and skills. Such approaches include training programs to increase scientists' awareness of the dual-use implications of their research and channels for reporting suspicions or concerns to the proper authorities. Governments, professional societies, and trade associations could help to formalize these steps and provide resources to implement them.

To empower the scientific community to detect and report suspicious research activities at home and abroad, it will also be essential to expand professional networks. Hosting foreign scientists in U.S. and other Western laboratories and establishing collegial relationships helps to build an international cohort of researchers who are sympathetic to Western security needs and less likely to transfer dual-use expertise to potential adversaries. Such professional collaborations also encourage members of the international scientific community to act as sentinels with respect to the legitimacy of research being conducted in countries of concern. Given these potential benefits, the U.S. government should roll back post-9/11 policies that have made it more difficult for foreign scientists to obtain visas to work in U.S. research institutions.

Conclusions

The discovery of RNA interference, a powerful and precise natural mechanism for disrupting gene expression, has spawned a multitude of basic and applied research activities. Although the development of biological weapons based on RNA interference would be a technically difficult task requiring the resources of a state, the potential to target specific ethnic groups could provide a strong incentive for actors with malicious or genocidal intent. Attempts at governance should focus on raising the awareness of scientists about dual-use issues and encouraging them to report suspicious activities. Enhancing international scientific exchanges to build enduring collegial networks would help to empower the scientific community in this regard.

Notes

1. This chapter is the author's personal analysis and does not necessarily reflect the position of the U.S. Government or the Department of Health and Human Services.

2. Yusuke Ohnishi, Yoshiko Tamura, Mariko Yoshida, et al., "Enhancement of Allele Discrimination by Introduction of Nucleotide Mismatches into siRNA in Allele-Specific Gene Silencing by RNAi," *PLoS ONE* 3 (May 2008): e2248, DOI 10.1371/journal.pone.0002248.

3. Carolyn Napoli, Christine Lemieux, and Richard Jorgensen, "Introduction of a Chimeric Chalcone Synthase Gene into Petunia Results in Reversible Co-suppression of Homologous Gene in Trans," *Plant Cell* 2 (April 1990): 279–289.

4. Su Guo and Kenneth J. Kemphues, "Par-1, a Gene Required for Establishing Polarity in *C. elegans* Embryos, Encodes a Putative Ser/Thr Kinase That Is Asymmetrically Distributed," *Cell* 81 (May 19, 1995): 611–620.

5. Andrew Fire, SiQun Xu, Mary K. Montgomery, et al., "Potent and Specific Genetic Interference by Double-Stranded RNA in *Caenorhabditis elegans*," *Nature* 391 (February 19, 1998): 306–311.

6. Akshay K. Vaishnaw, Jared Gollob, Christina Gamba-Vitalo, et al., "A Status Report on RNAi Therapeutics," *Silence* 1 (2010): 14.

7. S. Patrick Walton, Ming Wu, Joseph A. Gredell, et al., "Designing Highly Active siRNAs for Therapeutic Applications," *FEBS Journal* 277 (December 2010): 4806–4813.

8. Shuo Gu, Jianfei Ji, James D. Kim, et al., "Inhibition of Infectious Human Immunodeficiency Virus Type 1 Virions via Lentiviral Vector Encoded Short Antisense RNAs," *Oligonucleotides* 16 (2006): 287–295; Carola Pongratz, Benjamin Yazdanpanah, Hamid Kashkar, et al., "Selection of Potent Non-toxic Inhibitory Sequences from a Randomized HIV-1 Specific Lentiviral Short Hairpin RNA Library," *PLoS ONE* 5 (October 2010): e13172.

9. Yichao Wu, Francisco Navarro, Ashish Lal, et al., "Durable Protection from Herpes Simplex Virus-2 Transmission Following Intravaginal Application of siRNAs Targeting Both a Viral and Host Gene," *Cell Host and Microbe* 5 (January 22, 2009): 84–94.

10. Saba Khaliq, Shah Jahan, Bushra Ijaz, et al., "Inhibition of Hepatitis C Virus Genotype 3a by siRNAs Targeting Envelope Genes," *Archives of Virology* [online ahead of print] (December 2010), DOI 10.1007/s00705-010-0887-6.

11. Kevin B. Spurgers, Lynn S. Silvestri, Kelly L. Warfield, and Sina Bavari, "Toward RNA Interference-Based Therapy for Filovirus Infections," *Drug Development Research* 70 (2009): 246–254.

12. Dianne S. Schwarz, Hongliu Ding, Lori Kennington, et al., "Designing siRNA That Distinguish between Genes That Differ by a Single Nucleotide," *PLoS Genetics* 2 (September 2006): 1307–1318.

13. Jonathan E. Zuckerman, Teli Hsueh, Richard C. Koya, et al., "siRNA Knockdown of Ribonucleotide Reductase Inhibits Melanoma Cell Line Proliferation Alone or Synergistically with Temozolomide," *Journal of Investigative Dermatology* 131 (February 2011): 453–460; additional clinical trials not yet noted in published literature at the time this chapter was completed were registered by multiple companies at http://www.clinicaltrials.gov.

14. G. Courties, M. Baron, J. Presumey, et al., "Cytosolic phospholipase A2α gene silencing in the myeloid lineage alters development of Th1 responses and reduces autoimmune arthritis," *Arthritis and Rheumatology* [online ahead of print] (December 2010), DOI 10.1002/art.30174; J. R. Woska Jr. and M. E. Gillespie, "Small-Interfering RNA-Mediated Identification and Regulation of the Ternary SNARE Complex Mediating RBL-2H3 Mast Cell Degranulation," *Scandinavian Journal of Immunology* 73 (January 2011): 8–17.

15. Qi Yin, Yu Gao, Zhiwen Zhang, et al., "Bioreducible Poly (β-amino Esters)/shRNA Complex Nanoparticles for Efficient RNA Delivery," *Journal of Controlled Release* [online ahead of print] (January 2011), DOI 10.1016/j.jconrel.2010.12.014; Frederico Pittella, Mingzhen Zhang, Yan Lee, et al., "Enhanced Endosomal Escape of siRNA-Incorporating Hybrid Nanoparticles from Calcium Phosphate and PEG-Block Charge-Conversional Polymer for Efficient Gene Knockdown with Negligible Cytotoxicity," *Biomaterials* [online ahead of print] (January 2011), DOI 10.1016/j.biomaterials.2010.12.057.

16. Debra Robertson, "Racially Defined Haplotype Project Debated," *Nature Biotechnology* 19 (2001): 795–796.

17. For examples see the International HapMap Project, http://hapmap.ncbi.nlm.nih.gov; the Human Genome Diversity Project, http://hsblogs.stanford.edu/morrison; and the 1,000 Genomes Project, http://www.1000genomes.org.

18. Cintia Alvesa, Leonor Gusmaoa, Joselina Barbosa, et al., "Evaluating the Informative Power of Y-STRs: A Comparative Study Using European and New African Haplotype Data," *Forensic Science International* 134 (2003): 126–133; Neil Leat, Liezle Ehrenreich, Mongi Benjeddou, et al., "Properties of Novel and Widely Studied Y-STR Loci in Three South African Populations," *Forensic Science International* 168 (2007): 154–161.

19. Raymond D. Miller, Michael S. Phillips, Inho Jo, et al., "High-Density Single-Nucleotide Polymorphism Maps of the Human Genome," *Genomics* 86 (2005): 117–126; Stephen L. Guthery, Benjamin A. Salisbury, Manish S. Pungliya, et al., "The Structure of Common

Genetic Variation in United States Populations," *American Journal of Human Genetics* 81 (2007): 1221–1231; Jinchuan Xing, W. Scott Watkins, David J. Witherspoon, et al., "Fine-Scaled Human Genetic Structure Revealed by SNP Microarrays," *Genome Research* 19 (2009): 815–825.

20. Kari B. Schroeder, Mattias Jakobsson, Michael H. Crawford, et al., "Haplotypic Background of a Private Allele at High Frequency in the Americas," *Molecular Biology and Evolution* 26 (2009): 995–1016.

21. National Science Advisory Board for Biosecurity, "Proposed Framework for the Oversight of Dual Use Life Sciences Research: Strategies for Minimizing the Potential Misuse of Research Information" (June 2007), http://oba.od.nih.gov/biosecurity/pdf/Framework%20 for%20transmittal%200807_Sept07.pdf.

22. Federation of American Scientists, "Case Studies in Dual-Use Biological Research," Module 6, http://www.fas.org/biosecurity/education/dualuse/index.html.

23. Allison Chamberlain and Gigi Kwik Gronvall, "The Science of Biodefense: RNAi," *Biosecurity and Bioterrorism* 5 (2007): 104–106. National Research Council, Committee on Advances in Technology and the Prevention of Their Application to Next Generation Biowarfare Threats, *Globalization, Biosecurity, and the Future of the Life Sciences* (Washington, DC: National Academies Press, 2006), 165–169.

24. United Kingdom, "Scientific and Technological Developments Relevant to the Biological Weapons Convention," presented to the Sixth Review Conference of the Parties to the Convention on the Prohibition of the Development, Production, and Stockpiling of Bacteriological (Biological) and Toxin Weapons (BTWC), Geneva, Switzerland, November 2006.

25. Sunshine Project, "Emerging Technologies, Genetic Engineering, and Biological Weapons," Background Paper No. 12 (November 2003).

26. National Research Council, *Globalization, Biosecurity, and the Future of the Life Sciences*, 177.

27. National Institutes of Health Public Consultation, "Synthetic Nucleic Acids and the NIH Guidelines for Research Involving Recombinant DNA Molecules," Arlington, VA, June 23, 2009.

15

Transcranial Magnetic Stimulation

Jonathan D. Moreno

Transcranial magnetic stimulation (TMS), invented in 1985, uses electromagnetic induction to penetrate the skull and modulate the electrical activity of the cerebral cortex.[1] Because TMS is a relatively inexpensive technology that can modify cognition and behavior, it is certain to attract attention in the coming years for a variety of applications. As a therapeutic tool, it offers hope for individuals suffering from major depression, Parkinson's disease, and treatment-resistant migraine headaches, and it is under investigation for the treatment of post-traumatic stress disorder (PTSD). As a mind-enhancement technique, it may suppress the effects of sleep deprivation and enable individuals to perform above their baseline capability at specialized tasks.[2]

Although TMS has some potential for deliberate misuse by state and nonstate actors, the scale and scope of the resulting harm would be limited. For this reason, the technology does not warrant hard-law governance measures, although soft-law and informal approaches could be useful.

Overview of the Technology

Originally developed as a diagnostic aid for neurologists, TMS has since helped to map brain circuitry and connectivity, and offers therapeutic benefits as well.[3] To perform the technique, a technician holds an iron-core insulate coil embedded in plastic on one side of a patient's head while a large, brief current is passed through the coil. The current generates a magnetic pulse that painlessly penetrates the layers of skin, muscle, and bone covering the brain and induces weak, localized electrical currents in the cerebral cortex. Although the mechanisms by which the localized currents modulate neuronal activity are not fully understood, it is believed that the induced electrical field triggers the flow of ions across neuronal membranes and causes the cells to discharge, resulting in a chain reaction of neuronal interactions.[4]

Two basic types of TMS exist today. In single-pulse TMS, which is used for diagnostic purposes, the magnetic pulse is delivered in a nonrepetitive way, and the currents induced in the cerebral cortex do not persist beyond the period of stimulation.[5] An improved method called repetitive TMS was developed for therapeutic purposes. It involves the use of a high-speed magnetic stimulator to produce short magnetic pulses of an appropriate frequency. The repetitive pulses induce longer-lasting electrical currents that result in enduring cognitive effects.

History of the Technology

The idea of treating ailments with electricity dates back to antiquity. Around 50 BC, Scribonius Largus, the physician to the Roman emperor Claudius, advised that "to immediately remove and permanently cure a headache, however long-lasting and intolerable, a live black torpedo [electric eel] is put on the place which is in pain, until the pain ceases and the part grows numb."[6] The first attempt to stimulate the brain through the skull occurred in 1755, when Charles Le Roy tried to cure a twenty-one-year old man of his blindness by applying electrical impulses to his head. The impulses generated phosphenes (glowing spots) on the patient's retina but were not strong enough to affect the brain. Although Le Roy failed to cure blindness, he proved that nervous tissue responds to electricity.[7] His work, along with the study of electromagnetism, raised the possibility that strong magnetic fields could stimulate brain tissue.

In 1831 the renowned English chemist and physicist Michael Faraday demonstrated that when an electric current is passed through a coil of wire, the fluctuating magnetic field generated around the primary coil will induce a current in a second, neighboring coil.[8] By the beginning of the twentieth century, scientists had established that magnetic fields could modify neural activity, but it was not yet possible to generate large electrical currents using a magnet.[9] In 1980 the British neurophysiologists P. A. Merton and H. B. Morton reported stimulating the cerebral cortex of an intact human subject with "brief but very high-voltage shocks . . . without undue discomfort."[10]

In 1985 Anthony Barker and colleagues at the University of Sheffield in England succeeded in using transcranial electromagnetic induction to stimulate the human motor cortex, thereby inventing TMS. In some ways, TMS is similar to functional magnetic resonance imaging (fMRI) in that both employ intense magnetic fields, but there are important differences. First, whereas fMRI is an imaging technique, TMS is a stimulation and therapeutic technique. Second, the magnetic field plays a central role in fMRI, while the induced electrical current is paramount in TMS. Third, fMRI machines are expensive and bulky and require extensive technical knowledge for safe and effective operation, but TMS equipment is much smaller and can easily be used in a doctor's office.

Utility of the Technology

As a therapeutic tool, repetitive TMS (rTMS) can be customized to treat different illnesses. In most stroke victims, for example, one brain hemisphere is retarded while the other is largely unaffected. This asymmetry decreases motor-cortex activity in the affected hemisphere and increases activity in the unaffected hemisphere. Restoring function in both hemispheres is essential if the stroke victim is to recover. Using localized magnetic pulses, repetitive TMS can help to balance the neurological activity of the two hemispheres by enhancing the excitability of the cortical neurons on the injured side of the brain and suppressing activity on the unaffected side.[11]

Repetitive TMS is often compared to electroconvulsive therapy (ECT), which delivers a direct electrical shock to the brain rather than a magnetic pulse. ECT, developed long before TMS, is still the standard treatment for adults suffering from treatment-resistant major depressive disorder, but there are indications that this situation could soon change.[12] On January 26, 2007, the U.S. Food and Drug Administration (FDA) convened an expert panel to determine the risk-benefit profile of the Neurostar Transcranial Magnetic Stimulator (manufactured by Neuronetics Inc.) compared to standard ECT therapy. In a letter to the FDA panel, a psychiatrist stated that in her practice, ECT produced lingering side effects: patients were unable to work or drive for two or three weeks after an ECT session, and individuals of modest means could not afford to take leave for that amount of time. The psychiatrist found that repetitive TMS produced the same therapeutic benefit as ECT with fewer long-term side effects.[13]

Along similar lines, the psychiatry department at Rush University Medical Center in Chicago reported that it had been "contacted by thousands of patients interested in rTMS." Many of those who enrolled in clinical trials "had often failed multiple antidepressant trials, had few treatment options remaining, and were hesitant to risk the cognitive side effects associated with ECT." After treatment with repetitive TMS, "many subjects reported feeling like themselves for the first time in years."[14]

On October 8, 2008, the FDA granted Neuronetics approval to manufacture a TMS device for the therapy of treatment-resistant major depressive disorder in adult patients.[15] Although the FDA has not yet approved TMS for other medical purposes, several academic institutions are using the technique experimentally in clinical research settings for the "off-label" treatment of other brain disorders. For example, researchers at Columbia University have studied the effect of TMS on the memory of students after an extended period of sleep deprivation.[16]

Some evidence suggests that TMS could be employed clinically to suppress traumatic memories. According to a U.S. Army mental health survey in 2004 of troops who had fought in the Iraq War, about one in eight reported symptoms of post-traumatic stress disorder, but fewer than half of the affected individuals had sought treatment.[17] It is possible that TMS could provide an effective therapy for PTSD by helping to suppress traumatic memories and the negative emotions associated with them, or by preventing memory formation in the first place. If TMS turns out to be useful for memory suppression, however, there is a risk that soldiers could be returned to combat too soon after suffering psychological trauma. Such treatments may also have unintended long-term consequences that are not immediately apparent.

Finally, a few studies have explored the use of fMRI and TMS together for the purpose of lie detection as an alternative to polygraph use. Scientific evidence for the validity of polygraph data is lacking, and much evidence suggests that the predictive value of the technique is poor in many screening and investigative situations. According to a patent application for an fMRI/TMS "deception inhibitor," an fMRI scan would first indicate whether or not an individual was attempting to deceive the interrogator, after which TMS would be used to block the deception by inhibiting the relevant part of the cerebral cortex.[18]

Because of its relatively noninvasive nature, TMS is generally considered to be a low-risk technology.[19] Although safety studies in healthy, normal subjects show few if any side effects, TMS may pose greater risks in individuals suffering from conditions such as Parkinson's disease or major depressive disorder. A bioethical analysis concluded, "While it may be safe to stimulate healthy brain tissue, we have less information about the effects of TMS on abnormal brain tissue."[20] Among the risks associated with TMS, the most troubling is the potential induction of seizures. Seizure activity may occur when the induced neuronal excitability spreads beyond the site of stimulation, and it typically involves involuntary hand or arm movements.[21] The most important risk factor for seizures is the overall health of the subject's brain. One study found that "seizures are far less likely to occur in normal, healthy subjects than in subjects with neurologic disease such as stroke, brain tumor, or multiple sclerosis."[22]

From 1985 to 1995, seven cases of unintentionally induced seizures occurred during clinical research on repetitive TMS.[23] In 1996 the International Workshop on the Safety of Repetitive Transcranial Magnetic Stimulation reviewed these cases and developed safety guidelines. Since the guidelines were introduced in 1998, the risks associated with the therapeutic use of TMS have diminished considerably. Even so, the risks of nontherapeutic use remain substantial. Absent appropriate screening to evaluate the state of an individual's brain before TMS is performed, there is a possibility of serious side effects, including seizures. Sustained treatment sessions could also pose longer-term risks that are not well understood. Like other applied medical technologies, TMS should be studied under controlled conditions with full informed consent.

Potential for Misuse

Repetitive TMS is a clear case of a dual-use technology. Under strictly controlled conditions, it offers a promising and apparently safe intervention for persons suffering from serious mental illnesses. Given the limited expertise needed to employ the technique, however, states or terrorist organizations could potentially misuse it for harmful purposes. Aside from the obvious misuse of TMS to induce permanent brain damage, other possible scenarios warrant more nuanced ethical scrutiny. Some malign applications would be *intrinsically* unethical, such as "erasing" the memory of a highly trained operative who carried out an assassination to shield the operative from PTSD and render him immune to interrogation in the event of capture or compromise. Other applications of TMS would be *extrinsically* unethical, depending on the larger context and purpose for which the technique was used. For example, it would be unethical to enhance the ability of a terrorist to carry out an attack in a highly stressful environment.

Although TMS has existed since 1985, the U.S. Department of Defense has only recently recognized its potential for "warfighter enhancement." A report in 2009 by the U.S. National Academies titled *Opportunities in Neuroscience for Future Army Applications* recommended that the army increase its investment in TMS research. According to this report,

It is possible that TMS can be employed to enhance rather than suppress activation. One recent study showed enhancement of top-down visuospatial attention using combined fMRI/TMS stimulation (Blankenburg et al., 2008). The ability to target smaller areas is an objective sought by the TMS research community in general, but making such a device deployable in the field would require Army investment. Making this technology available in-vehicle is achievable in the medium term. The committee believes that in-helmet TMS technology would not be a useful investment until definitive applications, enhancing or inhibiting, are identified in the laboratory.[24]

The committee estimated the development time frame for using TMS to enhance attention at five to ten years, and for in-vehicle deployment at ten to twenty years. Repetitive TMS might also serve to improve learning and memory, for example, increasing the ability of an operative to speak a native dialect or to recall complicated instructions.[25] Nevertheless human experimentation with TMS in the national security context poses significant ethical challenges.[26] If any of the proposed military applications prove to be successful and cost-effective, they could lead to an arms race in the neural enhancement of combat troops. In that case, how much and what types of cognitive modification would soldiers be required to accept? Although neural enhancement through TMS might improve the combat performance of soldiers and their ability to protect one another, it might not provide as clear a benefit as a superior weapon.

Recent evidence also suggests that repetitive TMS could be misapplied to manipulate moral judgment or beliefs about right and wrong. In a laboratory simulation in which subjects were told about a failed attempt to kill an intended victim, subjects in whom TMS was used to disrupt the activity of a small region of the brain called the right temporoparietal junction (TPJ) were more likely to forgive the unsuccessful murder attempt than those subjects in whom the right TPJ was not stimulated. These findings suggest that the right TPJ is involved in the application of moral principles, and when its activity is disrupted with TMS, the brain relies more on actual outcomes.[27] Although studies with TMS could be important for elucidating the role of the brain in moral development, there is a real potential for abuse of such information. It is possible, for example, that applying TMS to the appropriate brain region could cause subjects to suspend their moral judgment and behave according to some grossly utilitarian calculus, such as "If you carry out a suicide attack, you will send a message to our foes and save the lives of many others."

Accessibility

The basic components of TMS technology—an iron-core coil and a magnetic stimulator—are commercially available at reasonable cost and are relatively easy to obtain. Major manufacturers of TMS equipment exist in Canada, China, Germany, Israel, South Korea, Switzerland, and the United States. Although the pool of manufacturers is limited, the relative simplicity of the technology will permit a rapid increase in supply in response to demand.

Ease of Misuse

Given the relatively modest nature of the equipment involved in repetitive TMS, nonstate actors and terrorists might seek to exploit the technology for its actual or perceived effects. Nevertheless, although the technique may seem straightforward, using it safely requires both training and experience. The angle at which the coil is held against the head, and which side of the coil is closer to the skull, can substantially affect the outcome of the stimulation. In addition, without proper training to pinpoint the intended region of the cerebral cortex to be activated, operators risk causing significant and potentially irreversible brain damage.

Magnitude and Imminence of Risk

The misuse of TMS could be extremely detrimental to individual subjects by suppressing the neuronal activity of localized brain regions. Accordingly, the technology might be misused to disable enemy combatants. Because TMS can be applied to only one individual at a time, however, it is incapable of harming entire groups or populations and hence poses a lower magnitude of potential risk than many of the other dual-use technologies discussed in this volume.

Characteristics of the Technology Relevant to Governance

Embodiment TMS is primarily a hardware-based technology, although it requires functional know-how to operate.

Maturity Over the past twenty years, repetitive TMS has evolved from an experimental diagnostic technique into a mature technology that has been approved for the treatment of major depressive disorder and has other promising clinical applications.

Convergence TMS is a convergent technology to the extent that it can be used in conjunction with fMRI and other brain-imaging techniques.

Rate of advance Although repetitive TMS is in advanced development for the treatment of several mental and neurological disorders for which standard therapies are ineffective, such as Parkinson's disease, stroke, and PTSD, the current rate of advance of this technology is incremental.

International diffusion The number of vendors of TMS equipment today is limited, although interest in the therapeutic use of the technology is growing among advanced industrialized countries, primarily in academic settings such as medical schools and research institutes. Several countries outside the United States have also approved TMS for the treatment of major depression, including Australia, Canada, China, and Israel. Once the safety and efficacy of TMS have been demonstrated, it is likely to spread to additional countries.

Susceptibility to Governance

Governance of TMS could mean two things: (1) regulating clinical research and legitimate therapeutic applications, and (2) regulating the sale of TMS equipment. The first approach would seek to prevent the development of hostile applications of the technology, and the second approach would seek to prevent TMS equipment from falling into the wrong hands. The feasibility of the two strategies differs. Because TMS has not yet diffused widely, it can still be controlled at the national level through the regulatory approval process. With respect to sales of TMS equipment to customers overseas, however, verifying that the technology is being used for legitimate therapeutic purposes (or, alternatively, for experiments aimed at harmful applications) would be a difficult task. It is also important to weigh the potential costs of overly stringent governance, which could impede the wider use of TMS for treating a variety of neuropsychiatric disorders.

Relevance of Existing Governance Measures

Possible governance options for TMS include control through the existing regulatory approval process for medical devices, as well as regulations on experimentation with human subjects. Because the treatment of major depressive disorder is the only indication for which TMS has so far been approved by the FDA, a window of opportunity exists to regulate other medical applications.

Beginning in 1996, the International Workshop on the Safety of Repetitive Transcranial Magnetic Stimulation adopted safety regulations to limit the risks of TMS to patients.[28] These and other rules for experimentation on human subjects may provide an opportunity to regulate the military applications of TMS at an early stage. A key determining factor will be how the first soldiers equipped with this technology are characterized. If they are designated as human subjects in an experiment, they will be subject to far greater bioethical protections than if they are seen as trainees such as test pilots, for whom a different calculus of risk and benefit applies.

Proposed Governance Measures

At least in the near future, the limited number of manufacturers and vendors of TMS machines should make it possible to track international sales. Even so, it is difficult to determine the end use of a TMS device after it has been exported, particularly in countries that, for whatever reason, are not transparent in their laboratory and hospital practices. Other options for governance include soft-law and informal approaches, such as awareness raising and education about the dual-use risks of TMS, and voluntary guidelines or codes of conduct adopted by the manufacturers and users of the technology.

Conclusions

In little more than a quarter century, TMS has evolved from a diagnostic tool into a therapeutic technique with the potential to improve the quality of life of patients with severe neuropsychiatric disorders such as treatment-resistant major depression. In addition, repetitive TMS can suppress the operation of a specific brain region to block unwanted activity and could potentially erase traumatic memories from the psyche of patients suffering from PTSD. In conjunction with other neurotechnologies, TMS may also enhance brain networks associated with attention, learning, and other cognitive processes. Although TMS could be misused to manipulate or harm individual human subjects, it has no capacity to kill or injure large numbers of people, limiting the magnitude of the potential consequences. As researchers continue to explore the applications of the technology, governance measures to prevent misuse will need to evolve as well.

Acknowledgments

The author is grateful to Amanda Foote, a former intern at the University of Pennsylvania Center for Bioethics, for assistance in preparing this paper. Anna C.

Merzagora of Drexel University provided valuable technical comments on an earlier draft. Any errors or omissions are solely the author's responsibility.

Notes

1. Vincent Walsh and Alvaro Pascual-Leone, with John E. Desmond, "Editorial: Manipulating Brains," *Behavioral Neurology* 17 (2006): 132.

2. National Research Council, Committee on Opportunities in Neuroscience for Future Army Applications, Board on Army Science and Technology, and Division on Engineering and Physical Sciences, *Opportunities in Neuroscience for Future Army Applications* (Washington, DC: National Academies Press, 2009).

3. "Transcranial Magnetic Stimulation: Safety," *Brookhaven National Laboratories*, February 1, 2008, http://www.bnl.gov/medical/TMS/safety.asp.

4. Alan Cowey and Vincent Walsh, "Tickling the Brain: Studying Visual Sensation, Perception, and Cognition by Transcranial Magnetic Stimulation," in *Progress in Brain Research*, vol. 134: *Vision: From Neurons to Cognition*, ed. C. Casanova and M. Ptito (Amsterdam: Elsevier Science, 2001), 411–425.

5. Dhwani B. Shah, Laurel Weaver, and John P. O'Reardon, "Transcranial Magnetic Stimulation: A Device Intended for the Psychiatrist's Office, but What Is Its Future Clinical Role?" *Expert Reviews* 5 (2008): 559.

6. Amanda Schaffer, "It May Come as a Shock: Can Electricity Block Migraines?" *New York Times*, November 7, 2006, F1.

7. Walsh and Pascual-Leone, "Editorial: Manipulating Brains," 132.

8. Cowey and Walsh, "Tickling the Brain," 411.

9. Several other researchers made important contributions, including the Germans Gustav Fritsch (1838–1927) and Eduard Hitzig (1838–1907), the Scotsman Sir David Ferrier (1843–1928), the Briton Sir Charles Scott Sherrington (1856–1952), and the Canadian Wilder Penfield (1891–1976). See Roland Sparing and Felix Mottaghy, "Noninvasive Brain Stimulation with Transcranial Magnetic or Direct Stimulation (TMS/tDCS)—from Insights into Human Memory to Therapy of Its Dysfunction," *Methods* 44, no. 4 (2008): 329–337.

10. P. A. Merton and H. B. Morton, "Letters: Stimulation of the Cerebral Cortex in the Intact Human Subject," *Nature* 285 (May 22, 1980): 227.

11. Felipe Fregni and Alvaro Pascual-Leone, "Technology Insight: Noninvasive Brain Stimulation in Neurology—Perspectives on the Therapeutic Potential of rTMS and tDCS," *Nature Clinical Practice Neurology* 3, no. 4 (2007): 383–393.

12. Jeffrey Rado, et al., "To the FDA Neurological Devices Panel," January 19, 2007, http://www.fda.gov/ohrms/dockets.

13. Ibid.

14. Ibid.

15. U.S. Food and Drug Administration, "510(k) Premarket Notification Database: Neuro-Star® TMS Therapy System," No. K083538 (Rockville, MD: FDA, December 16, 2008).

16. B. Luber, A. D. Stanford, P. Bulow, et al., "Remediation of Sleep-Deprivation-Induced Working Memory Impairment with fMRI-Guided Transcranial Magnetic Stimulation," *Cerebral Cortex* 18, no. 9 (2008): 2077–2085.

17. Associated Press, "One in Eight Returning Soldiers Suffers from PTSD; but Less than Half with Problems Seek Help, Report Finds," *MSNBC: Health*, June 30, 2004.

18. WO/2004/006750, Functional Magnetic Resonance Imaging Guided Transcranial Magnetic Stimulation Deception Inhibitor, publication date January 22, 2004. The author is grateful to Nita Farahany for informing him of this patent application.

19. Cowey and Walsh, "Tickling the Brain," 416.

20. Judy Illes and Marisa Gallo, with Matthew P. Kirschen, "An Ethics Perspective on Transcranial Magnetic Stimulation (TMS) and Human Neuromodulation," *Behavioral Neurology* 17 (2006): 151.

21. H. Branch Coslett, "Transcranial Magnetic Stimulation: Safety Considerations," Department of Neurology, University of Pennsylvania, http://www.ncrrn.org/papers/methodology_papers/tms.pdf.

22. Ibid.

23. "Transcranial Magnetic Stimulation: Safety," *Brookhaven National Laboratories*, February 1, 2008, http://www.bnl.gov/medical/RCIBI/Neuroimaging.asp.

24. National Research Council, *Opportunities in Neuroscience for Future Army Applications*, 85.

25. Ibid.

26. Jonathan D. Moreno, *Undue Risk: Secret State Experiments on Humans* (New York: W. H. Freeman, 1999).

27. Liane Young, Joan Albert Camprodon, Marc Hauser, Alvaro Pascual-Leone, and Rebecca Saxe, "Disruption of the Right Temporo-parietal Junction with Transcranial Magnetic Stimulation Reduces the Role of Beliefs in Moral Judgments," *Proceedings of the National Academy of Sciences* 107, no. 15 (April 27, 2010): 6753–6758.

28. Eric M. Wassermann, "Risk and Safety of Repetitive Transcranial Magnetic Stimulation: Report and Suggested Guidelines from the International Workshop on the Safety of Repetitive Transcranial Magnetic Stimulation, June 5–7, 1996," *Electroencephalography and Clinical Neurophysiology* 108 (1998): 1–16.

D

Technologies for Production, Packaging, or Delivery

16

Chemical Micro Process Devices

Amy E. Smithson

Changes are afoot in the chemical and related industries. In addition to their drive for efficiency and flexibility, chemical companies have more recently sought to reduce their environmental footprint and achieve greater process safety. Chemical micro process technology, initially developed in the 1980s, is proving that it can respond to the industry's needs. Compared to standard chemical reactors, miniaturized devices are safer, faster, more selective, and more energy efficient. Moreover, micro process devices produce higher and more uniform product yields, have greatly reduced waste streams, and are more cost-effective. These advantages explain why the chemical industry has investigated multiple applications of this technology and begun to adopt it for research and development (R&D), process development, product scale-up, and actual commercial production. Although legitimate companies are embracing micro process technology, there is a risk that state and substate actors could divert it for harmful purposes.

The hijacking of a civilian technology is hardly a new phenomenon. World War I ushered in the militarization of chemistry, and the use of poison gas was a hallmark of that conflict. Although the entry into force of the Chemical Weapons Convention in 1997 has eased concerns about the state-level pursuit of chemical weapons, Aum Shinrikyo's infamous attack in March 1995 on the Tokyo subway with the nerve agent sarin, and the use of chlorine-spiked bombs by Islamic fundamentalists in Iraq in 2007, have shown that crude forms of chemical warfare are within the reach of terrorists.

Amid this mixture of positive and negative proliferation trends, the coming-of-age of chemical micro process technology portends additional uncertainty. Such devices are ideal for the sustained processing of corrosive chemicals, a characteristic of poison gas production. Moreover, a micro plant could manufacture substantial quantities of chemical warfare (CW) agents with few of the telltale indicators commonly associated with chemical weapons factories, such as pollution abatement equipment. Thus the handful of states with ongoing CW programs could exploit micro production devices to advance and conceal these activities. Some states might even consider reentering the chemical weapons business if they were confident that an illicit weapons program would go undetected. For terrorists, micro process devices could ease the technical obstacles involved in scaling up the production of CW agents. The challenge for the international community is to find a way to allow micro process technology to flourish for legitimate commercial and scientific purposes while preventing its acquisition by those with intent to harm.

Overview of the Technology

Chemical micro process devices can be strikingly compact, and some are as small as credit cards, dice, or coins. Made of materials such as ceramic, stainless steel, glass, silicon, or the nickel-steel alloy known as Hastelloy, they are well suited for highly exothermic reactions and the long-term processing of highly corrosive chemicals. With inner channels ranging from submicrometer to submillimeter in size, chemical micro devices have a high ratio of reactor surface to reactant volume that promotes efficient surface catalysis and heat exchange. These characteristics in turn allow the precise regulation of chemical reactions and reduce the formation of unwanted by-products. Chemical micro devices also operate continuously, using miniature sensors and computers to maintain tight control over mixing, temperature, pH, flow rate, and other reaction conditions.[1]

Precision injection of chemicals into the channels of a micro device allows tiny drops to merge and react, often within seconds. To further enhance reaction efficiency, the channel walls can be seeded with catalysts, constructed in various shapes, and internally structured with etched patterns to enhance mixing. A variety of chemical micro process devices have been developed, including several different types of reactors, heat exchangers, and mixers. Such devices have successfully performed inorganic, biochemical, and organic chemical reactions and have been applied to the combinatorial production of molecular structures for high-throughput screening. Microreactors also have the potential to use chemistries that are not possible in standard industrial equipment, such as the high-yield production of hydrogen cyanide from formamide.[2]

History of the Technology

Scientists discussed the possibility of chemical nanoprocessing for over seventy years before the development of the requisite manufacturing technologies made it possible to turn theory into reality. Another significant factor that led to the emergence of the technology was the German government's response during the 1980s to the green movement's demand for policies and technologies to reduce environmental pollution. Although Germany's large chemical industry was already subject to regulation, it nonetheless became a focus of efforts to foster more environmentally friendly technologies. The German government called on leading research institutes to work together with chemical companies to explore the potential of micro process technologies and, if they showed promise, integrate them into commercial plants to reduce their environmental impact.[3] Thereafter Germany became a hub of microreactor research and development (R&D).[4]

Early studies demonstrated the utility of microreactors as tools for laboratory R&D because they can generate copious data on kinetics, residence time, and other reaction parameters, in contrast to the "black box" nature of standard chemical reactors. The resulting insights into chemical reactions enabled scientists to tinker with the reaction parameters and improve their performance. By the mid-1990s, developers of chemical micro process technology were exploring multiple commercial applications and demonstrating scale-up potential. In the late 1990s, chemical engineers and companies in the United States, a few other European countries, and Japan also began R&D on microreactors.[5] Japan later followed in Germany's footsteps by urging its chemical industry and university scientists to jointly develop micro process technology.[6] In recent years, most of the patents filed on micro process technology have been in Germany, the United States, China, and Japan.[7]

Utility of the Technology

Microreactors have several advantages over standard batch reactors. First, because the chemical industry produces and processes highly combustible and toxic materials, safety is a major concern. Reactions involving hazardous reagents or unstable intermediates or that generate high temperatures can be carried out much more safely in microreactors because their computerized monitoring permits the continuous adjustment of operational parameters to prevent a reaction from spiraling out of control.[8] Companies can also reduce safety risks by using microreactors to produce the hazardous chemicals needed for a specific manufacturing process on a just-in-time basis, rather than storing multi-ton quantities of such materials on-site.[9] Had the Union Carbide plant in Bhopal, India, produced methyl isocyanate on demand, it might have prevented the 1984 disaster that killed over 3,800 people and injured over 11,000.[10]

Reduced energy consumption is one reason that chemical micro devices are considered a green technology. In addition, switching a chemical manufacturing process from standard reactors to microreactors often allows the use of different solvents, reduced solvent volumes, and even solvent-free reactions that can radically reduce waste streams. Consequently, once certain chemical processes have been transitioned to microreactors, they do not require traditional pollution-abatement systems, such as air filter stacks or scrubbers.[11]

Other advantages of micro process technology concern the quality, quantity, and rapidity of reactions. Chemical companies typically discard between 10 and 20 percent of the output from standard reactors because it fails to meet product quality standards, but microreactors produce chemicals with low by-product contamination. Studies and initial industrial experience have also shown that converting a process from standard batch reactors to microreactors usually results in yields at least 20 to 30 percent higher.[12] The Xi'an Huian Industrial Group in China has installed a fully automated plant to produce nitroglycerine, a poisonous and explosive compound, using micro mixers, micro heat exchangers, and a reactor that is only 0.0021 cubic meters in a 30-cubic-meter plant. The production rate of this facility is 15 kilograms per hour, and the final product is 90 percent pure. Moreover, although the nitration reaction requires high temperatures in a standard batch reactor, it can be performed at room temperature in the microreactor system.[13]

Given the higher yields possible with micro process devices, companies can increase their profit margins because smaller quantities of feedstock chemicals are required. In some processes, chemical catalysts can be reused over a thousand times.[14] Additional cost savings result because companies do not have to buy and operate expensive industrial-scale chillers to control reaction temperatures. Moreover, a reaction that requires an hour in a batch reactor typically takes less than ten seconds in a microreactor, further increasing cost-effectiveness.

Another advantage of micro process devices is the ability to scale-up production rapidly. With standard batch reactors, the reaction kinetics vary considerably as volumes are increased, making it difficult to control reaction temperatures and other key operational parameters. In contrast, once a production process has been proved and optimized in a single-channel microreactor, it can be scaled up to industrial quantities simply by "numbering up" to hundreds or even thousands of identical micro devices operating in parallel arrays.[15] Rapid scale-up is particularly attractive to the pharmaceutical industry, where product specifications are demanding and regulatory approval of a scaled-up production process in standard reactors is extremely time-consuming. Already several micro device pilot plants available for purchase are capable of producing tons of chemicals per hour. Hitachi, for example, offers a miniplant with 20 microreactors that can produce 72 tons per year.[16]

Standard batch reactors can be used to make different chemicals, and this flexibility is increasingly true of microreactors as well. Several companies offer modular micro systems that can switch within hours from one production process to another. The modules perform different functions required for a chemical reaction (e.g., pumps and sensors, mixers, heat exchangers, reactors, filters and separators, valves), making it possible to change the micro plant configuration as needed.[17] Companies offering modular systems include the Institute for Mikrotechnik Mainz and Ehrfeld Mikrotechnik, a division of Bayer Technology Services. Although most applications of microreactors currently involve liquid and gaseous feedstocks, intermediates, and products, some solids (such as pigments) are also being processed and made in micro pilot plants. In such cases, manufacturers coat the channels of microreactors with special materials to prevent the solids from clogging and fouling them.[18]

Micro production devices can be applied to all sectors of the chemical industry. At present, roughly 33 percent of the chemical manufacturing in the pharmaceutical and fine-chemical industries is performed in microreactors, and it is expected that micro device technology will capture about 90 percent of these markets within a decade, except for processes involving solids. Early market penetration in these two market sectors can be explained by the fact that microreactors can meet the high purity standards required of pharmaceutical chemicals and can also produce small quantities of a wide variety of fine chemicals. Major companies in these sectors that use microreactors include Sigma Aldrich, Clariant, Johnson & Johnson, Novartis, AstraZeneca, Sanofi Aventis, Schering-Plough, Roche, and GlaxoSmithKline.[19]

Industry insiders also predict that over the next decade, microreactors will increase their share of the cosmetics market from about 10 percent to 33 percent, eventually becoming the dominant manufacturing technology in this sector because they can produce highly uniform emulsions. Other insiders expect a high level of use of micro process technology in the natural gas and biofuels areas, accounting in a decade for an estimated 40 percent of that market. Microreactors are also projected to make inroads into the polymer, petrochemical, food, and commodity chemical markets. Companies are even exploring the possibility of using micro systems for the bulk manufacture of commodity chemicals such as formaldehyde, methanol, ethylene, and styrene.[20] The cumulative forecast is that within a decade, 30 percent of all chemical processing will be performed in micro devices.

Potential for Misuse

The formulas for classic CW agents such as mustard, sarin, and VX and certain details of their manufacturing processes have long been available in the patent and professional literature. While such information could enable relative novices to synthesize small quantities of these agents, it does not include the operational details needed to scale-up production to militarily significant quantities,[21] a term used to indicate the quantity of chemical agent needed to saturate a battlefield. To inflict 50 percent casualties among unprotected troops over a square kilometer, the quantity of CW agent needed can range from two tons of the nerve agent VX to nineteen tons of mustard gas.[22]

Now that numbered-up microreactor arrays can produce tons of chemicals per day, technical specialists have expressed concern that this technology could be misused for the production of CW agents. Whereas chemical plants with standard reactors sprawl over many acres, a micro process plant is closet sized and fully automated, avoiding the need for a large staff to monitor operations to prevent accidents. Towering exhaust stacks and scrubbers are not present because chemical micro process plants do not generate significant hazardous waste streams. Moreover, energy consumption decreases greatly because industrial-scale chillers are not required to control exothermic reactions.[23] Accordingly, as the commercial chemical industry converts to micro plants, most of the traditional intelligence signatures associated with CW production will vanish, leaving intelligence agencies hard-pressed to locate clandestine production facilities. Thus micro process devices could enable states and sophisticated terrorist groups to amass significant stockpiles covertly, setting the stage for surprise attacks.

At the substate level, the most instructive case is that of Aum Shinrikyo, the Japanese cult that released the nerve agent sarin in the Tokyo subway on March 20, 1995. Aum's attack killed thirteen people, seriously injured several dozen, and frightened over 5,000 so badly that they inundated Tokyo hospitals. Fortunately, the low purity (roughly 30 percent) of the agent released and the crude method of dispersal—using sharpened umbrellas to puncture plastic bags filled with a dilute solution of sarin—prevented a much larger casualty toll.[24]

Another factor that prevented Aum from killing more commuters was its inability to scale up production of sarin at its dedicated $10 million chemical plant. The cult leaders sought to produce 70 tons of sarin—a militarily significant quantity—in 40 days. To this end, they acquired top-of-the-line equipment, including items made of corrosion-resistant Hastelloy and a $200,000 Swiss-built reactor with automatic temperature and injection controls.[25] Aum chemists had technical difficulties scaling up the process, however, resulting in recurring leaks. Several technicians inhaled toxic fumes and exhibited symptoms ranging from nosebleeds to convulsions. Citizens living near the cult's compound in the Mount Fuji foothills also lodged complaints with the police in July 1994 about noxious odors emanating from the site. In November 1994, an accident forced the cult to halt the production of sarin.[26] The use of microreactors might have enabled Aum Shinrikyo to avoid the safety problems and leaks that ultimately forced it to abort production and alerted Japanese law enforcement authorities to the cult's nefarious activities.

Accessibility

In 2010, roughly twenty companies were selling chemical micro devices. Because most manufacturers market their products over the Web without restrictions, would-be proliferators might be able to purchase micro process equipment without coming in direct contact with company officials. The accessibility of the technology must therefore be characterized as high.

Ease of Misuse

The main hurdle to the misuse of micro process technology by state proliferators, terrorists, and criminals is adapting the synthesis of CW agents to microreactors.

Relevant information on the use of chemical micro devices is available from text-books and more than 1,500 articles in professional and trade journals.[27] Operational know-how can also be gleaned from professional conferences, such as the International Microreaction Technology Conference or the World Congress of Chemical Engineers, where attendees can query top developers and scientists from companies that work with the equipment.

More than formal credentials, such as a Ph.D. in chemistry or chemical engineering, practical hands-on experience with chemical micro process technology is needed to translate the synthesis of CW agents from standard-sized to miniaturized equipment.[28] To acquire this tacit knowledge, governments, terrorist groups, or so-called lone wolf actors intent on proliferation could enroll in university courses on microreactors or get a job at a company that uses micro process technology in its research laboratories and production plants. Having acquired such experience, states, terrorist groups, and individuals could more readily overcome the most challenging aspect of exploiting this technology for malevolent purposes.

Magnitude and Imminence of Risk

Because microreactors are potentially capable of producing large volumes of CW agents, the imminence and magnitude of the risk of misuse are high. In addition to providing enhanced safety and ease of scale-up, the use of microreactors would eliminate key signatures of illicit production.[29]

Awareness of Dual-Use Potential

Although relatively little has been written on the security implications of microreactor technology, experts have noted that highly toxic chemicals, including CW agents, could be produced in micro devices. Reportedly, a number of toxic chemicals have already been synthesized in microreactors, including hydrogen cyanide and phosgene, both of which were used as chemical weapons during World War I, and methyl isocyanate, which was released in the Bhopal tragedy.[30]

Characteristics of the Technology Relevant to Governance

Embodiment Chemical micro process technology consists of hardware, but computer software is also needed to control the devices and their accompanying sensors.

Maturity Chemists and chemical engineers at the cutting edge of green chemistry and process intensification are exploring the potential of micro process technology. Refinements to the technology include new device configurations and channel coatings that improve throughput and other performance parameters. Recent years have seen a shift from the use of microreactors primarily as a laboratory R&D tool to their increasing application in pilot- and industrial-scale production. According to leading scientists in the industry, reactors, heat exchangers, and mixers are merely the first items of chemical production equipment to be miniaturized. Scientists are also working on shrinking components for chemical separation, reforming, and distillation.[31]

Convergence   Micro chemical process technology involves the convergence of standard chemical processing with the manufacturing technologies used to make modern microelectronic devices, such as laser micromachining and microlamination.

Rate of advance   The rate of development of microreactor technology is accelerating. Top chemical manufacturing companies around the world—including Dow, Sigma Aldrich, DuPont, BASF, Pfizer, and Johnson & Johnson—are working with micro process technology in their R&D departments, at pilot scale, and for larger commercial production.

International diffusion   In recent years, knowledge about the manufacture and operation of micro process devices has spread internationally, along with fabrication technologies for these devices, which are essentially the same as those that underpin the microelectronics industry. Though expensive, advanced machining technologies are increasingly found in all corners of the globe, including rapidly industrializing countries. Thus any plant or country that produces computer chips has the technological potential to make microreactors.

Another aspect of diffusion concerns the commercial market for chemical micro process technology. Although the chemical industry was once centered in Europe, North America, Japan, South Korea, and Taiwan, it has grown significantly in countries such as India, China, Brazil, Peru, Pakistan, Singapore, South Africa, and Thailand.[32] Chemical micro devices appeal to companies seeking to improve their competitive edge through enhanced product quality and profitability. As a matter of national policy, some countries are actively encouraging the adoption of chemical micro process technology for its environmental benefits and to promote economic competitiveness in an age of outsourced production.

Manufacturers of micro process devices currently comprise about twenty firms, concentrated in Europe (primarily Germany), with a few in the United States and Asia. Given the intensive R&D activity in Japan, India, and China, all of which have large chemical industries, it is reasonable to expect that additional micro device manufacturers will emerge in Asia in the near future. Industry insiders predict that within ten years, some 100 companies will be producing chemical micro devices worldwide.[33] As the industry grows in size and competitiveness, the price of microreactors will drop, making them even more accessible.

Susceptibility to Governance

Now is an opportune time to govern micro process technology because the number of suppliers is relatively small. Nevertheless the point at which governments will perceive that the security risks of the technology are sufficient to warrant regulation of their domestic industry remains an open question. This decision will not be taken lightly, because the development of regulations is time-consuming and resource intensive. The regulatory process typically includes meetings of interested government agencies, consultations with affected industry, drafting and revision of proposed regulations, public circulation of the draft regulations for comment, assignment of responsibility for regulatory implementation, allocation of resources for implementation, and updating of the regulations as needed. Parliamentary participation adds another layer of complexity to the regulatory process through hearings, drafting and passage of legislation, and oversight of implementation.

Given the safety, environmental, and economic benefits that will accrue from the widespread use of microreactors, draconian regulation of the industry would be ill-advised. Although the micro process technology industry would prefer to avoid regulation entirely, if a regulatory regime is imposed, industry will lobby for a balanced approach that does not hamper sales to legitimate customers or create an onerous implementation burden.

Relevance of Existing Governance Measures

Scientists and security analysts concerned with the nonproliferation of chemical weapons have watched, evaluated, and discussed micro process technology as it has become more common for companies to use such equipment for commercial production. To date, however, national policy makers have not intervened, and no barriers exist to the commercial sale of microreactors.

Since the entry into force of the Chemical Weapons Convention (CWC) in 1997, the Organization for the Prohibition of Chemical Weapons (OPCW) in The Hague has overseen the implementation of the treaty. In 2008, the OPCW's Scientific Advisory Board first raised the issue of microreactors and opted to reassess the implications of this technology for the CWC on a periodic basis.[34] An independent scientific advisory panel also recommended updating the verification provisions of the CWC in response to technical developments such as microreactors, lest the treaty become "frozen in time."[35]

Nevertheless, because CWC verification focuses primarily on the diversion of dual-use chemicals for the illicit production of warfare agents, "scheduled" chemicals—not processing equipment—are the items of accountability for inspectors. Thus a major shift in verification philosophy would be required to cover microreactors under the treaty's declaration and inspection procedures. Given the major political effort that would be necessary to bring about such a change, the CWC is an unlikely vehicle for addressing the proliferation potential of micro process devices.

The Australia Group (AG) is a cooperative of forty countries and the European Commission that harmonize their national export controls on dual-use materials and manufacturing equipment related to the production of chemical and biological weapons.[36] AG members have discussed chemical micro production equipment but have so far declined to subject them to export controls. Although the group's control list already includes reactors, heat exchangers, and other types of chemical production equipment, the internal capacities are orders of magnitude larger than those of chemical micro devices.[37] In the fall of 2009, however, the AG established a dedicated working group on chemical micro process technology.

Adding chemical micro devices to the AG control list would be a positive step but would not be sufficient to prevent the misuse of these devices because manufacturers would be required to obtain export licenses only for sales to known or suspected CW proliferators, such as North Korea, and those that abet them, such as freight forwarders and front companies. Sales to "friendly" nations and domestic customers would remain unregulated, an unsatisfactory situation in an era when telling friend from foe has become increasingly difficult. Purported friendly states might view microreactors as an opportunity to establish or upgrade a covert CW program, and jihadist groups and assorted domestic terrorists operate in far too many countries. Thus both domestic and international measures are needed to safeguard against the diversion and misuse of chemical micro devices.

Proposed Governance Measures
Although the OPCW and the Australia Group have taken no formal steps toward governance of micro process devices, this technology—like any other commercial product—is susceptible to control and regulation by individual governments. Any governance scheme should aim at the most useful intervention points and be structured so that legitimate users retain access to beneficial dual-use items while users with malign intent are denied. Governance measures for micro process devices should not target the underlying know-how, which has been in the public domain for over a decade, or the manufacturing technology, which is widely employed in the consumer electronics industry. Instead governance measures should aim to prevent the sale of microreactors to suspected or known proliferators. Regulations might include declaration and reporting requirements, mandatory screening of customers against lists of entities or individuals engaged in illicit activities, and training of company sales and marketing staff to recognize suspicious customers and purchase inquiries. A hotline to government authorities might also be established for reporting such cases in a timely fashion.

The most expeditious and comprehensive route to preventing misuse would be a self-governance initiative on the part of micro device manufacturers, modeled on the efforts of gene-synthesis companies to screen DNA orders and customers to prevent bioterrorists from assembling dangerous pathogens from scratch. Two consortia of gene-synthesis companies have established voluntary screening programs, partly in response to pressure from end users. To date, however, a dedicated global trade association has not yet been formed to represent chemical micro process device manufacturers. Alternatively, governments that support critical R&D on microreactors could place conditions on research funding to prod manufacturers to adopt responsible policies and procedures.

Conclusions

Chemical micro process devices can significantly enhance safety, increase efficiency, and reduce the environmental footprint of commercial chemical plants. Nevertheless these devices could also advance the CW programs of proliferant states, cause other nations to reconsider their decision to renounce chemical weapons, and accelerate terrorist efforts to acquire them. In addition, chemical micro plants lack nearly all the signatures that intelligence analysts use to identify suspect poison gas factories, depriving the intelligence community of the ability to detect covert CW production in a timely manner. The underlying know-how and manufacturing technologies for chemical micro process devices have already diffused so widely that a ban would be as impractical as it would be undesirable, given the clear benefits of the technology.

Unfortunately, existing nonproliferation tools are poorly suited to grapple with this proliferation challenge. The CWC does not regulate production equipment, and any new export controls agreed by the Australia Group, were they to materialize, would apply only to known states of CW proliferation concern and the entities that aid them. Accordingly, the most expeditious and effective way to plug this hole in the nonproliferation regime may lie in self-regulation by the international chemical micro process industry, such as an initiative to establish best practices and procedures for responsible sales and customer screening.

Notes

1. J. Yoshida, A. Nagaki, T. Iwasaki, et al., "Enhancement of Chemical Selectivity by Microreactors," *Chemical and Engineering News* 83, no. 22 (May 30, 2005): 43–52; C. Wille and R. Pfirmann, "Pharmaceutical Synthesis via Microreactor Technology Increasing Options for Safety, Scale-Up, and Process Control," *Chemistry Today* 22 (2004): 20–23.

2. Volker Hessel, Patrick Löb, and Holger Löwe, "Development of Reactors to Enable Chemistry rather than Subduing Chemistry around the Reactor—Potentials of Microstructured

Reactors for Organic Synthesis," *Current Organic Chemistry* 9, no. 8 (2005): 765–787; Paul Watts and Stephen J. Haswell, "The Application of Microreactors for Small Scale Organic Synthesis," *Chemical and Engineering Technology* 28, no. 3 (2005): 290–301; Patrick Löb, Holger Löwe, and Volker Hessel, "Fluorinations, Chlorinations, and Brominations of Organic Compounds in Micro Structured Reactors," *Journal of Fluorine Chemistry* 125, no. 11 (2004): 1677–1694.

3. Author interviews with scientists and senior corporate officials from the chemical micro process technology industry in New Orleans and Washington, DC, March–April 2008.

4. German entities that played leading roles in developing chemical micro process technology include Mikroglas ChemTech GmbH, the Institute for Mikrotechnik Mainz, the Karlsruhe Research Center, Cellular Process Chemistry Systems GmbH, the Fraunhofer Alliance for Modular Microreaction Systems, and Ehrfeld Mikrotechnik BTS, a company founded by one of the field's pioneers, Wolfgang Ehrfeld.

5. Early R&D work was performed at Pacific-Northwest Laboratories, the Massachusetts Institute of Technology, Oregon State University, and Microfluidics in the United States; the University of Hull and University College London in the United Kingdom; Lund Institute of Technology in Sweden; the University of Twente and TNO Science and Industry in the Netherlands; the National Center for Scientific Research in France; and Tokyo University in Japan, among others.

6. The Research Association of Micro Chemical Process Technology was founded to facilitate Japanese collaboration to bring about high-efficiency chemical plants.

7. Volker Hessel, Christoph Knobloch, and Holger Löwe, "Review on Patents in Microreactor and Micro Process Engineering," *Recent Patents in Chemical Engineering* 1, no. 1 (2008): 1–14.

8. Xini Zhang, Stephen Stefanick, and Frank J. Villani, "Application of Microreactor Technology in Process Development," *Organic Process Research and Development* 8, no. 3 (2004): 455–460; Kunio Yube and Kazuhiro Mae, "Efficient Oxidation of Aromatics with Peroxides under Severe Conditions Using a Microreaction System," *Chemical and Engineering Technology* 28, no. 3 (2005): 331–336.

9. Eero Kolehmainen et al., "Advantages of On-Site Microreactors from Safety Viewpoint," presentation 198e delivered at the 10th International Conference on Microreaction Technology, New Orleans, April 9, 2008.

10. On Bhopal, see Robert D. McFadden, "India Disaster: Chronicle of a Nightmare," *New York Times*, December 10, 1984; Jackson B. Browning, "Union Carbide: Disaster at Bhopal," in *Crisis Management: Inside Stories on Managing under Siege*, ed. Jack Gottschalk (Detroit: Visible Ink Press, 1993).

11. On solvent-free, alternate-solvent, and environmentally friendly processing in microreactors, see Patrick Löb, Holger Löwe, and Volker Hessel, "Fluorinations, Chlorinations, and Brominations of Organic Compounds in Micro Reactors," *Journal of Fluorine Chemistry* 125, no. 11 (November 2004): 1677–1694; Holger Löwe et al., "Addition of Secondary Amines to a,b-Unsaturated Carbonyl Compounds and Nitriles by Using Microstructured Reactors," *Organic Process Research and Development* 10, no. 6 (2006): 1144–1152.

12. With a bromination reaction, scientists from Hitachi reported an increase in yield of 98 percent. Myake Ryo and Togashi Shigenori, "Innovation of Chemical Process Engineering Based on Micro-Reactor," *Hitachi Hyoron* 88, no. 1 (2006): 916–921.

13. Ann M. Thayer, *Chemical and Engineering News* 83, no. 22 (May 30, 2005): 43. For other examples, see Volker Hessel, Patrick Löb, and Holger Löwe, "Industrial Microreactor Process Development Up to Production: Pilot Plants and Production" in *Microreactors in Organic Synthesis and Catalysis*, ed. Thomas Wirth (Weinheim, Germany: Wiley-VCH, 2008), 238–270.

14. Charlotte Wiles, "Investigation into the Preparation of Synthetically Useful α-Aminonitriles under Continuous Flow," presentation 46b delivered at the 10th International Conference on Microreaction Technology, New Orleans, April 7, 2008.

15. Wolfgang Ehrfeld, Volker Hessel, and Holger Löwe, eds., *Microreactors: New Technology for Modern Chemistry* (Weinheim, Germany: Wiley-VCH, 2000), 6–12; Anna Lee Y. Tonkovich and Eric A. Daymo, "Microreaction Systems for Large-Scale Production," in *Microchemical Engineering in Practice*, ed. Thomas R. Dietrich (Hoboken, NJ: John Wiley and Sons, 2009), 299–324.

16. S. Togashi, T. Miyamoto, T. Sano, et al., "Microreactor Systems Using the Concept of Numbering Up," in *New Trends in Fluid Mechanics Research*, ed. F. G. Zhuang and J. C. Li (New York: Springer, 2009), 678–681; Patricia L. Short, "Microreactors Hit the Major Leagues," *Chemical and Engineering News* 86, no. 42 (October 20, 2008): 37–38.

17. Daniel A. Snyder, Christian Noti, Peter H. Seeberger, et al., "Modular Microreaction Systems for Homogeneously and Heterogeneously Catalyzed Chemical Synthesis," *Helvetica Chimica Acta* 88, no. 1 (January 24, 2005): 1–9; Tassilo Moritz, Reinhard Lenk, Jorg Adler, et al., "Modular Micro Reaction System Including Ceramic Components," *International Journal of Applied Ceramic Technology* 2, no. 6 (November 21, 2005): 521–528.

18. For example, in 2004 Clariant International Ltd. opened a plant that makes over eighty tons per year of diketopyrrolopyrrole pigments. Ch. Wille, H. P. Gabski, Th. Haller, et al., "Synthesis of Pigments in a Three-State Microreactor Pilot Plant: An Experimental Technical Report," *Chemical Engineering Journal* 101, nos. 1–3 (August 2004): 179–185.

19. Author interviews with U.S. and European industry experts, New Orleans and Washington, DC, March–April 2008. Clay Boswell, "Microreactors Gain Wider Use as Alternative to Batch Production," *Chemical Market Reporter*, no. 266 (October 4, 2004): 8; Dominique M. Roberge, Laurent Ducry, Nikolaus Bieler, et al., "Microreactor Technology: A Revolution for the Fine Chemical and Pharmaceutical Industries?" *Chemical Engineering Technology* 28, no. 3 (2005): 318–323.

20. Author interviews with U.S. and European industry experts, New Orleans and Washington, DC, March–April 2008. Clay Boswell, "Microreactors Gain in Popularity among Producers: More than Meets the Eye," *ICIS News*, April 30, 2009.

21. For more detail on technical production challenges, see U.S. Congress, Office of Technology Assessment, *Technologies Underlying Weapons of Mass Destruction*, OTA-BP-ISC-115 (Washington, DC: U.S. Government Printing Office, 1993), 16, 26–27, 133. On specialized scale-up knowledge not found in the open literature, author's interview with Ph.D. chemist and chemical weapons expert, Washington, DC, July 14, 2000; Raymond A. Zilinskas, "Aum Shinrikyo's Chemical/Biological Terrorism as a Paradigm?" *Politics and the Life Sciences* 15, no. 2 (September 1996): 238.

22. Julian Perry Robinson, ed., *The Chemical Industry and the Projected Chemical Weapons Convention*, SIPRI Chemical and Biological Warfare Studies, no. 4 (Oxford: Oxford University Press, 1986), 89.

23. Mahdi Balali-Mood, Pieter S. Steyn, Leiv K. Sydnes, et al., "Impact of Scientific Developments on the Chemical Weapons Convention: Report of the International Union of Pure and Applied Chemistry" *Pure and Applied Chemistry* 80, no. 1 (2008): 188.

24. For more on Aum's chemical weapons program, see chap. 3 of Amy E. Smithson with Leslie-Anne Levy, *Ataxia: The Chemical and Biological Terrorism Threat and the U.S. Response*, report no. 35 (Washington, DC: Henry L. Stimson Center, October 2000), 80–111.

25. Anthony T. Tu, "Aum Shinrikyo's Chemical and Biological Weapons," *Archives of Toxicology, Kinetics, and Xenobiotic Metabolism* 7, no. 3 (Autumn 1999): 75, 79; David E. Kaplan and Andrew Marshall, *The Cult at the End of the World* (New York: Crown, 1996), 150, 211; U.S. Congress, Senate Committee on Governmental Affairs, Permanent Subcommittee on Investigations, staff statement and testimony of John F. Sopko, *Global Proliferation of Weapons of Mass Destruction*, 104th Cong., 1st sess. (Washington, DC: U.S. Government Printing Office, 1996), 21–22, 61–62, 87–88; D. W. Brackett, *Holy Terror: Armageddon in Tokyo* (New York: Weatherhill, 1996), 110, 113–114, 118, 146, 157, 175.

26. Anthony T. Tu, "Overview of Sarin Terrorist Incidents in Japan in 1994 and 1995," *Proceedings from the 6th CBW Protection Symposium* (Stockholm: May 10–15, 1995), 14–15.

27. Volker Hessel, Christoph Knobloch, and Holger Löwe, "Review on Patents in Microreactor and Micro Process Engineering," *Recent Patents in Chemical Engineering* 1, no. 1 (2008): 1–16.

28. Author interviews with scientists and senior corporate officials from the chemical micro process technology industry, March–April 2008 and August 2009.

29. *Report of the Scientific Advisory Board on Developments in Science and Technology*, Doc. RC-2/DG.1 (The Hague: February 28, 2008), 11.

30. Tuan H. Nguyen, "Microchallenges of Chemical Weapons Proliferation," *Science* 309 (August 12, 2005): 1021.

31. Author interviews with scientists and senior corporate officials from the chemical micro process technology industry in New Orleans, Washington, Montreal, and Kyoto, March–April 2008, August 2009, and March 2010.

32. For data on contemporary market developments, see the Global Production Index of the American Chemistry Council and trade publications such as *ICIS News*, which tracks industry activities worldwide.

33. Author's interviews with scientists and senior corporate officials from the chemical micro process technology industry, New Orleans and Washington, DC, March–April 2008.

34. *Report of the Scientific Advisory Board on Developments in Science and Technology*, 11.

35. Balali-Mood, Steyn, Sydnes, et al., "Impact of Scientific Developments on the Chemical Weapons Convention," 179.

36. Amy E. Smithson, *Separating Fact from Fiction: The Australia Group and the Chemical Weapons Convention*, Occasional Paper No. 34 (Washington, DC: Henry L. Stimson Center, March 1997).

37. The specified total internal volume for reactors is greater than 100 liters and less than 2,000 liters. The Australia Group's current export control list for chemical equipment is available at http://www.australiagroup.net/en/dual_chemicals.html.

17

Gene Therapy

Gail Javitt and Anya Prince

Gene therapy is the insertion of foreign genetic material into a person's cells or tissues to alter gene function and potentially treat or cure hereditary diseases. Although it was heralded in 1990 as a groundbreaking technology that would radically change the medical community's ability to fight disease, the ensuing two decades of research have been fraught with setbacks and complications. Researchers have encountered many technical barriers to inserting genes into individuals for therapeutic purposes, and they continue to face challenges during clinical trials.

Some analysts have speculated about the ability of terrorists and other malicious actors to use gene-therapy techniques to create enhanced biological weapons. At present this possibility is strictly hypothetical, although future technological advances could make the dual-use risks of gene therapy more of a concern. Accordingly, gene therapy should be monitored closely so that appropriate governance measures can be introduced if necessary as the technology matures.

Overview of the Technology

Gene therapy research combines recombinant DNA technology with virology and immunology to deliver a therapeutic gene into a target cell.[1] (Recombinant DNA technology, also known as genetic engineering, is the process of combining DNA from different species to create organisms with new traits or capabilities.) The three key elements of gene therapy are the accurate delivery of a therapeutic gene to the target cell, the successful expression of the gene, and the lack of adverse reactions.

Genome modification can take place either in somatic (nonreproductive) cells or in germ-line cells. Somatic-cell gene therapy has the potential to alter genetic function only in the individual whose genes have been altered, and it does not create changes that can be passed on to the next generation.[2] Germ-line gene therapy, in contrast, seeks to introduce genetic material directly into eggs or sperm (or their precursor cells) or into early human embryos. As a result, germ-line modification has the potential to create permanent, heritable changes in the offspring and descendants of a treated individual.[3]

A key challenge for the success of gene therapy is integrating the new genetic material into the target cell, whether somatic or germ line. Several methods have been developed for gene transfer: viral vectors, nonviral delivery systems (such as cationic polymers or lipids),[4] and artificial chromosomes. A vector is a virus that has been modified to deliver a gene of interest to the target cell without itself causing harm to the recipient. Viruses are, by their nature, highly efficient at transferring genes into foreign organisms.[5] Three types of viral vectors have been used in human gene therapy research: retrovirus, adenovirus, and adeno-associated virus.[6] At present, gene-therapy researchers take blood from a subject, use a viral vector to transfer therapeutic genes into the white blood cells, and infuse the blood back into the patient.

Despite advances in viral-vector technology, scientists have found it difficult to deliver vectors into target cells in a way that is standardized or repeatable.[7] Viral vectors have also been associated with adverse reactions in research subjects, including death.[8] Nonviral delivery systems are safer but less efficient at transferring genes.[9] Accordingly, this case study focuses on the use—and potential misuse—of viral vectors for the genetic modification of an organism.

History of the Technology
Although equivalent terms were used in academic discussions as early as 1960, the term "gene therapy" was coined in a published article in 1970.[10] Research on the genetic engineering of plants and animals began in the early 1970s and became a global, multibillion-dollar industry by the 1980s. In 1989 the first reported gene-therapy study in humans was performed on a child with an inherited metabolic disorder called severe adenosine deaminase deficiency.[11] This trial was considered a success because the gene for the missing enzyme was transferred safely to the child, and her white blood cells began to produce it. Nevertheless, because the treated cells did not give rise to healthy new cells as the researchers had hoped, the patient must continue to receive periodic gene therapy, supplemented with a medication that helps to maintain the level of the enzyme in her blood.[12]

Another common problem with gene therapy is that the recipient's immune system detects the viral vector carrying the therapeutic gene and destroys the vector before it can reach the target cells. Researchers have also had difficulty developing gene therapies that can be used repeatedly in the same individual, because the immune system attacks viral vectors it has seen before.[13] Because of these difficulties, the therapeutic benefit of gene therapy has not fulfilled its initial promise.[14] Although more than 900 clinical trials were conducted worldwide between 1989 and 2004,[15] the U.S. Food and Drug Administration (FDA) has yet to approve a gene-therapy product for clinical use. The field also experienced a major setback in 1999 when a human subject, Jesse Gelsinger, died while participating in a clinical trial.[16] This incident led to congressional hearings, which called into question the adequacy of oversight of gene-therapy research and led to increased involvement by the FDA and the National Institutes of Health (NIH).

Over the past two decades, gene therapy research has progressed from preclinical safety studies to clinical studies for more than 45 diseases.[17] Although a few clinical trials of gene therapy have resulted in correct gene expression, many of the patients have experienced adverse reactions. In 2000, for example, scientists transferred curative genes to children with severe combined immunodeficiency disorder, using as a vector the Moloney murine leukemia virus. Although the gene transfer was successful, several of the treated children subsequently developed a rare form of leukemia because the viral vector activated another gene when it integrated into the host DNA.[18]

As of early 2011, more than 1,000 human gene-transfer protocols were registered with the NIH.[19] However, because scientists have not perfected the key parameters as quickly as was initially hoped, the field is advancing slowly. In 1990, gene-therapy advocates believed that new cures were right around the corner, but today the technology is still considered experimental in the United States.[20]

Utility of the Technology

Gene therapy is of significant research interest because it could potentially be used to cure, ameliorate, or prevent inherited diseases of metabolism or the immune system.[21] Germ-line gene therapy could allow the genetic change to be passed on to the patient's offspring, thereby eradicating an inherited disorder from future generations. While the primary focus of gene therapy research has been to alleviate disease,[22] some posit that in the future, gene therapy could be used to "enhance" human characteristics by conferring socially desirable traits, such as height or intelligence, or to improve health by conferring resistance to infectious agents.[23] In some instances, it may be difficult to distinguish therapeutic applications from enhancement because they fall along a continuum.[24] Despite this somewhat blurry line, genetic enhancement raises more ethical and social concerns than does gene therapy.[25]

Potential for Misuse

To be effective, gene therapy will require additional technological advances, such as improved gene delivery and reliable gene expression in the targeted cells. Assuming that these advances materialize, the potential for misuse could arise either intentionally or inadvertently. A study in 1997 by the JASONs, a group of academic scientists who perform studies for the U.S. Department of Defense, reported six ways that genetic engineering could be used to create enhanced biological weapons, including the use of gene-therapy vectors.[26]

One analyst has noted that if vector technology is perfected, viral vectors could be used to transmit harmful genes into a target population.[27] Alternatively, viral vectors might be turned into "stealth viruses" that could be introduced surreptitiously into the genetic material of the host, where they would remain dormant for an extended period of time before causing disease. (Herpes virus is a naturally occurring example of a stealth virus.) In one scenario, a malicious actor would threaten to activate the stealth virus with an external signal unless his demands were met.[28] Such a scenario would be particularly complex because it would require exposing the targeted population twice, first to the virus and then to the activating signal.[29] It is unclear, however, how a would-be bioterrorist could introduce a stealth virus surreptitiously into a target population and then activate the virus at a later time. Moreover, even if a "stealthy" viral vector was used, some fraction of the exposed population (e.g., those suffering from an impaired immune system) might manifest modest symptoms prior to delayed activation by the malicious actor.[30]

Accessibility

Viral-vector technology is accessible, especially to scientists in industrialized nations. Even so, it would be extremely difficult for a terrorist or some other malicious actor to exploit the technology to transfer harmful genes surreptitiously to more than one person. If it were possible to deliver viral vectors in aerosol form, such a capability would greatly increase the potential for misuse.[31] Although scientists have attempted to deliver adenoviruses to a patient's lungs by the aerosol route, the technique has not yet succeeded in clinical trials.[32]

Ease of Misuse

It is implausible that a malicious actor could perform gene transfer experiments outside an established laboratory and without significant knowledge and training.[33] In using gene therapy for biowarfare purposes, the most difficult step would be to devise an effective means of delivering a harmful gene into a target population. The development of a stealthy viral weapon would probably require a multidisciplinary team, extensive funding, and a good deal of tacit knowledge. Accordingly, misuse by a state is far more likely than by a nonstate actor.[34]

Magnitude and Imminence of Risk

One analyst has argued that while the risk of a terrorist attack involving gene transfer is low, the possibility should be taken seriously because the consequences could be severe.[35] Yet although some scientists and policy makers have recognized the possibility that gene therapy could be misused for harmful purposes, they do not see it as an imminent risk. A common view is that natural pathogens are sufficiently deadly, making it unlikely that a terrorist group would undertake the complex process of engineering a viral vector as a weapon.[36]

Awareness of Dual-Use Potential

A few scientists have pointed to the potential misuse of "stealth" viral vectors, but the topic has not been widely discussed in the academic literature.

Characteristics of the Technology Relevant to Governance

Embodiment   Gene therapy relies primarily on intangible information and know-how, rather than hardware.

Maturity   Despite large increases in the level of research and investment in gene therapy, the field has not progressed beyond the clinical testing stage.[37] Serious adverse events, such as Gelsinger's death in 1999, have also raised persistent safety concerns.[38]

Convergence   Gene therapy research draws on three different disciplines: recombinant DNA technology, virology, and immunology.

Rate of advance   Researchers have made significant advances in understanding the strengths and weaknesses of particular gene-transfer vectors, including which ones are appropriate for treating specific diseases.[39] Despite this progress, however, human gene therapy remains experimental.[40] More study is needed of the immune responses to vectors and transferred genes before such techniques can be employed clinically.[41]

International diffusion   As of June 2010, twenty-nine countries had conducted clinical trials of gene therapy.[42] Approximately 63 percent of all such trials have been performed in the United States, and 94 percent in North America and Europe.[43] About 65 percent of gene-therapy trials have targeted cancers.[44] The two main factors affecting the rate of diffusion of gene therapy are its high cost and the nascent status of the technology.[45] As gene-therapy research progresses, the cost will decline, and a greater number of countries and actors will gain access to the technology.

Susceptibility to Governance

As a technology that is already heavily regulated on health and safety grounds, gene therapy has shown itself to be susceptible to governance. To date, however, U.S. government regulation and oversight of gene therapy have focused on biosafety issues. Gene therapy was overseen initially by the Recombinant DNA Advisory Committee (RAC), which the NIH established in 1974 to ensure the safety of genetic-engineering technology.[46] During the 1980s, the RAC created a Human Gene Therapy Subcommittee (HGTS), but when gene therapy began to involve human clinical trials, the FDA asserted regulatory authority.[47] During this period, the relationship between the RAC and the FDA was strained because both agencies claimed jurisdiction over gene therapy but took different approaches to regulation.[48] A major point of contention was the RAC's emphasis on public review, which conflicted with the FDA's preference for confidentiality.[49] In the mid-1980s, the FDA received the lead role for reviewing clinical gene-therapy studies, but tensions with the RAC persisted.[50]

After Gelsinger's death in 1999, both the NIH and the FDA intensified their oversight of gene therapy. Investigations by both agencies determined that clinical trials of gene therapy throughout the country had not been in compliance with federal regulations governing research on human subjects, and that adverse reactions had not been properly reported to federal authorities.[51] As a result of these revelations, the NIH and the FDA launched a new program called the Gene Therapy Clinical Trial Monitoring Plan, which sought to ensure that adverse reactions during clinical trials would be reported.[52] In addition, the FDA instituted random inspections of ongoing clinical trials[53] and modified the informed-consent documents to give research subjects a better understanding of the risks of participation.[54] Since 2000, federal regulations have required that before an Institutional Biosafety Committee can grant final approval of a new gene-therapy protocol, the RAC must agree to review the protocol or determine that a review is unnecessary.[55] The RAC also provides a forum for public discussion of ethical and safety issues arising from novel gene-therapy protocols.[56]

Today the FDA regulates all gene-therapy clinical trials performed in the United States, as well as those conducted abroad if the resulting data will be included in a licensing application to the agency. Sponsors seeking to conduct a clinical trial must submit an Investigational New Drug (IND) application.[57] The FDA can reject the application if it determines that the potential risks outweigh the potential benefits, or for other reasons. As a condition for obtaining an IND, the sponsors must submit the research protocol to an Institutional Review Board (IRB) that reviews experiments involving human subjects and comply with all FDA regulations governing such research.[58]

Although the NIH does not have direct regulatory authority over gene-therapy protocols, the RAC still provides some oversight of federally funded research.[59] In addition to a general RAC review, federally funded gene-therapy research protocols must be registered with the NIH, which maintains a publicly accessible database of information about clinical trials and adverse reactions.[60] Also, under the NIH guidelines, an Institutional Biosafety Committee (IBC) must approve federally funded clinical trials of gene therapy.[61] Privately funded researchers may submit their protocols to the NIH and the RAC on a voluntary basis.

Relevance of Existing Governance Measures

Because gene therapy is already highly regulated, one way to prevent misuse for hostile purposes would be to rely on the existing regulatory framework, including FDA oversight of clinical trials and additional NIH oversight of federally funded research. For example, one could extend the existing regulatory framework to cover all researchers in the field. At present, gene-therapy researchers whose work is supported entirely by private funds do not have to submit their research protocols to the RAC, although they are encouraged to do so voluntarily.[62] Nevertheless scientists funded by private sources must undergo RAC review if the research sponsor, or the institution where the research takes place, receives NIH money.[63] In addition, if the research uses recombinant DNA techniques that were developed with NIH funds and the institution that developed those techniques is a participant in the project, then the research protocol must be submitted to the RAC even if it is privately funded.[64] Given the broad coverage that already exists under current oversight mechanisms, requiring all privately funded researchers to submit gene-transfer protocols for RAC approval would not significantly increase the number of experiments under review.

Nevertheless this approach has three drawbacks. First, it is unlikely that malicious actors would submit their research for review, enabling them to evade government oversight. Second, the NIH guidelines for recombinant DNA research focus on minimizing the biosafety risks associated with the *unintended* creation of harmful recombinant organisms, so additional governance measures would be needed to prevent the *deliberate* use of gene transfers for malign purposes.[65] Third, it is difficult to create regulations when one does not know the identity of the actors of concern and what types of harmful applications they might pursue.[66]

At present the deliberate misuse of gene therapy by state or nonstate actors remains unlikely. Although technical breakthroughs may occur in the future, it is difficult to predict what they might be. Given these unknowns, increased regulation would tend to hamper legitimate research while doing little to reduce the risk of misuse. Moreover, focusing prematurely on the dual-use potential of gene therapy might cause the public to perceive the technology as sinister, hampering beneficial research in the field.[67]

Proposed Governance Measures

New governance measures might be considered if and when the risk of misuse of gene therapy becomes more apparent. One approach is education and awareness raising. Most life scientists have had little direct exposure to the issues of biological weapons and bioterrorism and tend not to consider the misuse potential of their own research. The British biosecurity expert Brian Rappert has suggested that educators stress the moral obligation of whistle-blowing if a scientist suspects a fellow researcher is engaging in foul play.[68]

A second option for addressing the possible security risks associated with gene therapy is to strengthen ties among academic researchers, the intelligence community, and the military.[69] To bolster these communication channels, Col. John Logan Black has proposed the development of a comprehensive, continuously updated database containing the history, genetic sequence, and physical characteristics of all viral vectors used in gene therapy. He has also called for accelerated research on detection systems for viral vectors.[70]

A third policy option for protecting against the misuse of dual-use research in the life sciences is prepublication review of security-sensitive data. The U.S. National Academies report *Biotechnology Research in an Age of Terrorism* recommended that such reviews be based on voluntary self-governance by scientific journals rather than on formal government regulation.[71] Given the current low risk that gene therapy could be misused, however, limiting the publication of beneficial results would hamper scientific progress more than it would prevent terrorists from learning how to exploit the technology for harmful purposes.[72]

Conclusions

To date, efforts to regulate gene therapy have focused on concerns about patient safety rather than on dual use, and the focus of governance measures has been almost entirely domestic rather than international. Although the deliberate misuse of gene therapy for harmful purposes is theoretically possible, it remains unlikely because of major technical hurdles, which would require a high level of scientific expertise to overcome. Until more is understood about gene therapy's potential for misuse, increased regulation is not advisable, because it would tend to hinder beneficial research. Priority should therefore be given to the governance of dual-use biotechnologies that pose a more imminent threat. Nevertheless policy makers should monitor the field of gene therapy and be prepared to intervene should the risk of deliberate misuse become more likely.

Notes

1. Richard A. Merrill and Gail H. Javitt, "Gene Therapy, Law, and FDA Role in Regulation," in *Encyclopedia of Ethical, Legal, and Policy Issues in Biotechnology*, ed. Thomas J. Murray and Maxwell J. Mehlman (Hoboken, NJ: John Wiley and Sons, 2000), 322.

2. National Human Genome Research Institute, "Germline Gene Transfer," National Institutes of Health (2006), http://www.genome.gov/10004764.

3. Susannah Baruch, Audrey Huang, Daryl Pritchard, Andrea Kalfoglou, Gail Javitt, Rick Borchelt, Joan Scott, and Kathy Hudson, "Human Germline Genetic Modification: Issues and Options for Policymakers," *Genetics and Public Policy Center* (2005): 13.

4. T. Niidome and I. Huang, "Gene Therapy Progress and Prospects: Nonviral Vectors," *Gene Therapy* 9 (2002): 1647.

5. Ana P. Cotrim and Bruce J. Baum, "Gene Therapy: Some History, Applications, Problems, and Prospects," *Toxicologic Pathology* 36 (2008): 97.

6. John Logan Black, "Genome Projects and Gene Therapy: Gateways to Next Generation Biological Weapons," *Military Medicine* 168, no. 11 (2003): 865.

7. Leroy Walters, Joseph P. Kennedy Professor of Christian Ethics at Georgetown University, telephone interview by Anya Prince, August 17, 2009.

8. Cotrim and Baum, "Gene Therapy," 101.

9. Ibid., 97.

10. Leroy Walters, "Gene Therapy, Law, Recombinant DNA Advisory Committee (RAC)," in *Encyclopedia of Ethical, Legal, and Policy Issues*, 336.

11. Cotrim and Baum, "Gene Therapy," 97.

12. "Hope for Gene Therapy," Public Broadcasting Service, October 23, 2001, http://www.pbs.org/saf/1202/features/genetherapy.htm.

13. Human Genome Program, "Gene Therapy," U.S. Department of Energy, Office of Science, Office of Biological and Environmental Research, June 11, 2009, http://www.ornl.gov/sci/techresources/Human_Genome/medicine/genetherapy.shtml.

14. Walters interview.

15. Cotrim and Baum, "Gene Therapy," 97.

16. Ibid., 101.

17. Eric Alton, Stefano Ferrari, and Uta Griesenbach, "Progress and Prospects: Gene Therapy Clinical Trials (Part 1)," *Gene Therapy* 14 (2007): 1439; Eric Alton, Stefano Ferrari, and Uta Griesenbach, "Progress and Prospects: Gene Therapy Clinical Trials (Part 2)," *Gene Therapy* 14 (2007): 1555; National Institutes of Health, "Human Gene Transfer Protocols," June 5, 2009, http://oba.od.nih.gov/rdna/rdna_resources.html.

18. Cotrim and Baum, "Gene Therapy," 102.

19. National Institutes of Health, "Human Gene Transfer Protocols."

20. Human Genome Program, "Gene Therapy."

21. Baruch et al., "Human Germline Genetic Modification," 13.

22. M. Kiuru and R. G. Crystal, "Progress and Prospects: Gene Therapy for Performance and Appearance Enhancement," *Gene Therapy* 15 (2008): 329.

23. Baruch et al., "Human Germline Genetic Modification," 13.

24. Kiuru and Crystal, "Progress and Prospects," 330.

25. Ibid.; Baruch et al., "Human Germline Genetic Modification," 14.

26. Steven M. Block, "Living Nightmares: Biological Threats Enabled by Molecular Biology," in *The New Terror: Facing the Threat of Biological and Chemical Weapons*, ed. Abraham D. Sofaer, George D. Wilson, and Sidney D. Drell (Stanford, CA: Hoover Institution Press, 1999), 17.

27. Col. Michael J. Ainscough, "Next Generation Bioweapons: The Technology of Genetic Engineering Applied to Biowarfare and Bioterrorism," Counterproliferation Paper No. 14 (Maxwell Air Force Base, AL: U.S. Air Force Counterproliferation Center, April 2002), 21.

28. James B. Petro, Theodore R. Plasse, and Jack A. McNulty, "Biotechnology: Impact on Biological Warfare and Biodefense," *Biosecurity and Bioterrorism* 1, no. 3 (2003): 164.

29. Black, "Genome Projects and Gene Therapy," 869.

30. Ibid., 869.

31. Block, "Living Nightmares," 62.

32. Walters interview.

33. Ibid.

34. Dr. Gigi Kwik Gronvall, Senior Associate, Center for Biosecurity, University of Pittsburgh Medical Center, telephone interview by Anya Prince, September 3, 2009.

35. Ainscough, "Next Generation Bioweapons," 28.

36. Petro et al., "Biotechnology," 162.

37. Cotrim and Baum, "Gene Therapy," 101.

38. Human Genome Program, "Gene Therapy."

39. Alton, Ferrari, and Griesenbach, "Progress and Prospects (Part 1)," 1439.

40. Human Genome Program, "Gene Therapy."

41. Ibid.

42. The twenty-nine countries in question are Australia, Austria, Belgium, Canada, China, the Czech Republic, Denmark, Egypt, Finland, France, Germany, Ireland, Israel, Italy, Japan, Mexico, the Netherlands, New Zealand, Norway, Poland, Russia, Singapore, South Korea, Spain, Sweden, Switzerland, Taiwan, the United Kingdom, and the United States. See Michael L. Edelstein, "Gene Therapy Clinical Trials Worldwide," *Journal of Gene Medicine* (updated June 2010), http://www.wiley.com/legacy/wileychi/genmed/clinical/.

43. Ibid.

44. Ibid.

45. Gronvall interview.

46. Theodore Friedmann, Philip Noguchi, and Claudia Mickelson, "The Evolution of Public Review and Oversight Mechanisms in Human Gene Transfer Research: Joint Roles of the FDA and NIH," *Current Opinion in Biotechnology* 12, no. 3 (June 2001): 304.

47. Merrill and Javitt, "Gene Therapy, Law, and FDA Role in Regulation," 322.

48. Evan Diamond, "Reverse-FOIA Limitations on Agency Actions to Disclose Human Gene Therapy Clinical Trial Data," *Food and Drug Law Journal* 63 (2008): 330; Merrill and Javitt, "Gene Therapy, Law, and FDA Role in Regulation," 322.

49. Diamond, "Reverse-FOIA Limitations," 330; Walters interview.

50. Friedmann et al., "The Evolution of Public Review," 304.

51. Ibid., 305.

52. Larry Thompson, "Human Gene Therapy: Harsh Lessons, High Hopes," *FDA Consumer* (September 2000), 2.

53. Ibid.

54. NIH Report, "Assessment of Adenoviral Vector Safety and Toxicity: Report of the National Institutes of Health Recombinant DNA Advisory Committee," *Human Gene Therapy* 13 (January 1, 2002): 8.

55. Kenneth Cornetta and Franklin O. Smith, "Regulatory Issues for Clinical Gene Therapy Trials," *Human Gene Therapy* 13 (July 1, 2002): 1145.

56. Office of Biotechnology Activities, National Institutes of Health, "About Recombinant DNA Advisory Committee (RAC)," http://oba.od.nih.gov/rdna_rac/rac_about.html.

57. Leroy Walters, "The Oversight of Human Gene Transfer Research," *Kennedy Institute of Ethics Journal* 10, no. 2 (2000): 171; Cotrim and Baum, "Gene Therapy," 101.

58. The relevant law is 21 C.F.R. Pt. 50.

59. Diamond, "Reverse-FOIA Limitations," 332.

60. Office of Biotechnology Activities, National Institutes of Health, "About Recombinant DNA Advisory Committee (RAC)."

61. Stacy M. Okutani, "Federal Regulation of Scientific Research," Center for International and Security Studies at Maryland, University of Maryland (August 2001), p. 8, http://www .cissm.umd.edu/papers/display.php?id=361.

62. Office of Biotechnology Activities, National Institutes of Health, "Frequently Asked Questions (FAQs) about the NIH Review Process for Human Gene Transfer Trials."

63. Ibid.

64. Ibid.

65. National Research Council of the U.S. National Academies, Committee on Research Standards and Practices to Prevent the Destructive Application of Biotechnology, *Biotechnology Research in an Age of Terrorism: Confronting the Dual Use Dilemma* (Washington, DC: National Academies Press, 2003), vii.

66. Gronvall interview.

67. Ibid.

68. Brian Rappert, "Biological Weapons, Genetics, and Social Analysis: Emerging Responses, Emerging Issues—Part II," *New Genetics and Society* 22, no. 3 (December 2003): 304.

69. Robert M. Frederickson, "The Third Cabo Gene Therapy Focus Panel: On the Offensive for Biodefense," *Molecular Therapy* 8, no. 2 (August 2003): 178.

70. Black, "Genome Projects and Gene Therapy," 869.

71. National Research Council, *Biotechnology Research in an Age of Terrorism*, 6.

72. Walters interview.

18

Aerosol Vaccines

Raymond A. Zilinskas and Hussein Alramini

An aerosol vaccine is a preparation of living but attenuated (nonvirulent) bacteria or viruses that is delivered in the form of an airborne suspension of microscopic particles or droplets. Studies in animals and humans have shown that delivering a vaccine as an aerosol can be more effective in inducing an immune response than administering it orally or by injection. Aerosol vaccines can be designed for delivery into either the nasal cavity or the deep regions of the lungs. Although one human aerosol vaccine for intranasal delivery has been developed and marketed and a few others are in advanced clinical trials, no vaccine for deep-lung delivery has as yet been approved for human use. This chapter assesses the potential for the misuse of this technology and concludes that although aerosol generators pose significant dual-use risks, aerosol vaccines and related delivery technologies do not.

Overview of the Technology

The challenge facing developers of aerosol vaccines is to introduce living microbes into the recipient's tissues in a way that stimulates immunity against the target pathogen without causing harm to the host. Preparation of an aerosol vaccine involves several steps. First, an attenuated bacterium or virus is cultivated, separated from the growth medium, and suspended in a special solution containing chemical preservatives and nutrients. This mixture, called a formulation, stabilizes the microbes during storage and protects them from environmental stresses after their release as an aerosol.

Each bacterial or viral species used as a vaccine requires a tailored formulation to ensure that it is delivered as a cloud of correctly sized aerosol particles and survives the stresses of delivery. Aerosol vaccines can be prepared as either a wet or dry formulation: some microbes function best as part of a liquid, while others are more effective when dispersed as a dry powder. In general, a dry formulation is more difficult to produce. The biomass must be dried in a special piece of equipment such as a spray-dryer and then milled into a fine powder. Chemicals may then be added to prevent the dry particles from clumping due to static electricity.

Developing formulations for aerosol vaccines is more of an art than a science. No known scientific method can predict which combination of chemicals will interact with an attenuated microbe to stabilize and protect it. Instead developers work by trial and error, testing various combinations of chemicals in the laboratory and then in clinical and field trials. Microencapsulation, which involves coating particles or droplets with an inert polymer, was once believe to offer possibilities for protecting microbes that were too fragile for aerosol delivery. Current microencapsulation technologies are now known not to enhance microbial survival or persistence, although in some cases they can protect against the harmful effects of ultraviolet radiation.[1]

In the case of aerosol vaccines designed for intranasal delivery, the aerosol particles must remain in the nasopharynx until they are absorbed through the nasal mucosa. This feature implies that vaccine developers must satisfy two requirements. First, the aerosol particles must be relatively large, between 20 and 50 microns in diameter, so they settle onto the nasal mucosa and thus are prevented from being inhaled into the lungs.[2] Second, the particles must be properly formulated with adhesive chemicals to enhance their residence time on the nasal mucosa. Nebulizers and inhalers designed for individual delivery of aerosol vaccines are portable devices that generate a large-particle aerosol containing a measured dose of a drug or vaccine.

The second type of aerosol vaccine is designed for delivery to the deep regions of the lungs, where maximum absorption occurs in the tiny air sacs called alveoli. To achieve deep-lung penetration, the vaccine particles must be less than 10 microns in diameter. An aerosol vaccine intended for deep-lung delivery must overcome the physiological and immunological defense systems of the host that normally prevent airborne particles from reaching the alveoli.[3] Epithelial cells lining the respiratory tract secrete mucus to trap particles and are equipped with cilia that propel the trapped particles into the esophagus and from there to the stomach, where the acid environment destroys them.

Vaccine particles that evade the defenses in the airways of the head travel through the nasal valve and into the pharynx, larynx, trachea, and bronchi before reaching the deep regions of the lungs. Along the way, the inhaled microbes are subject to "mucosal immunity," meaning entrapment and destruction by secretions containing antimicrobial chemicals and by immune-system cells that populate the lining of the respiratory tract. Because of these obstacles, fewer than 10 percent of the inhaled particles actually reach the lungs.[4] Of those that do, 50 percent are retained in the alveoli, and the other 50 percent are exhaled.[5]

Aerosol generators for large-scale immunization must be carefully standardized by a well-trained technician according to the vaccine formulation to be dispensed, the operating conditions, and the intended aerosol output. Moreover, because some liquid aerosol vaccines lose potency after sitting in an aerosol generator for more than a few minutes, it may be necessary to keep the device and its contents at a near-freezing temperature with crushed ice.

History of the Technology

The first published account of an aerosol vaccine designed for deep-lung delivery appeared in 1910, but intensive research did not begin until after World War II, primarily by Soviet military scientists.[6] Because the most effective method for delivering biological warfare agents is in the form of an aerosol, it made sense to administer biodefense vaccines by the same route. The earliest mention of such work in the Soviet military medical literature dates from the mid-1950s.

One article describes how a research team led by K. I. Aleksandrov at the Kirov Military Medical Academy in Leningrad began to develop and test "aerogenic" vaccines to protect against diseases of biological warfare concern, including brucellosis, plague, and tularemia. The vaccines consisted of live attenuated bacteria in dry formulations. After initial testing in animal models suggested that the vaccines were safe and efficacious, in 1957 and 1958 the scientists exposed 487 human subjects in sealed test chambers. The recipients experienced minimal side effects, and clinical and serological tests indicated that they had acquired protective immunity against plague, tularemia, and brucellosis.[7]

Aleksandrov and his colleagues subsequently developed a dry aerosol vaccine against anthrax consisting of spores of a nonpathogenic strain of *Bacillus anthracis*.[8] Clinical trials demonstrated that aerosol vaccination could immunize a large number of people simultaneously. In a 40-cubic-meter room, up to 300 persons received five-minute exposures over the course of an hour. Using three small rooms or tents, each with a volume of 40 to 50 cubic meters, five or six men could vaccinate more than one thousand persons per hour.[9]

U.S. military scientists working in the pre-1969 offensive biological warfare program read the Soviet publications on aerosol vaccination and tried to emulate them. In 1962 a team from the U.S. Army Biological Laboratories at Fort Detrick, Maryland, reported on efforts to develop an aerosol vaccine against *Franciscella tularensis*, which causes tularemia. Unlike their Soviet counterparts, the U.S. scientists received permission from higher authorities to conduct human trials involving live pathogens.[10]

During Operation Whitecoat (1954–1973), several thousand volunteers were exposed to Q fever and tularemia bacteria in closed chambers and open-air tests.[11] In one such experiment, aerosolized tularemia vaccine was delivered through breathing masks to 253 subjects divided into five groups, each of which received concentrations of live vaccine ranging from 10^4 to 10^8 bacteria. The five groups, plus a control group, were then challenged with an aerosol of a virulent strain of

F. tularensis. When the recipients developed symptoms of the disease, they were treated with antibiotics. This trial demonstrated that the aerosol vaccine was more effective in protecting against aerosol challenge than was the existing injected vaccine.[12] However, no finished vaccine resulted from this effort.

After the Soviet Union's dissolution in 1991, Russian research on aerosol vaccines continued; a review paper in 1999 reported that dry or rehydrated live vaccines against plague, tularemia, and anthrax were being developed for aerosol administration.[13] Elsewhere, in 2008 a multinational team of investigators from European research institutions announced the development of an aerosol vaccine against smallpox, consisting of two attenuated strains of live vaccinia virus. After testing the candidate vaccine on six rhesus monkeys, the investigators concluded that the aerosol vaccine was safe, induced long-lasting systemic and mucosal immune responses, and could offer a viable solution for future mass-vaccination campaigns.[14] Other reports have described the development of aerosol vaccines against measles and tuberculosis.[15]

In 2003 the U.S. Food and Drug Administration (FDA) approved the first live-virus vaccine designed for nasopharyngeal delivery, the FluMist influenza vaccine manufactured by MedImmune LLC of Gaithersburg, Maryland. This vaccine is marketed as a nasal spray for use in healthy children and adults aged 5 through 49. It contains three strains of live influenza virus in a formulation designed for delivery as a large-particle spray into the nasal cavity, where it stimulates mucosal and systemic immunity.[16] FluMist is packaged in individual disposable sprayers, each of which contains a single dose. Several studies have demonstrated the safety and efficacy of the vaccine.[17] A dry formulation (using inactivated virus) is being developed that, unlike FluMist, would not have to be refrigerated.[18] Before the dry vaccine is ready for clinical testing, however, researchers will need to overcome numerous technical problems with respect to the reproducibility of particle size, distribution, stability, and performance characteristics.[19]

Aerosol vaccines have also been prepared against a number of livestock diseases. During the 1960s, East German veterinary scientists developed a vaccine against a bacterial infection of pigs called swine erysipelas.[20] Ten years later, another East German team developed an aerosol vaccine against erysipelas septicemia in ducks raised in large coops. The researchers estimated that it took only ten man-hours to vaccinate 10,000 ducks, significantly lowering production costs.[21] More recently, a team of U.S. Army scientists at Fort Detrick developed an aerosol vaccine to protect horses against glanders, a disease caused by the bacterium *Burkholderia mallei*, a putative biowarfare agent.[22]

The most widely used veterinary aerosol vaccine immunizes chickens against Newcastle disease, caused by a virus that affects poultry. Because factory farms raise chickens in coops containing up to 45,000 birds, it is impractical to vaccinate them individually. Aerosol vaccination is particularly effective because it stimulates a twofold immune response: production of antibodies that circulate in the blood-stream and a local mucosal response in the respiratory tract.[23] The aerosol vaccine is administered twice, first when the chicks are a day old and again when they are several weeks old.[24] Intervet/Schering-Plough Animal Health, the world's largest supplier of veterinary vaccines, offers seven different poultry vaccines that, when diluted with distilled water and disseminated by sprayer, create aerosol particles in the 50- to 100-micron range.[25]

Utility of the Technology

Aerosol vaccines for deep-lung delivery offer several advantages over traditional vaccines. They are rapidly absorbed through the alveoli of the lung and protect better than traditional vaccines because they stimulate both mucosal and systemic immunity.[26] In addition, aerosol delivery methods can be used by small numbers of support personnel to immunize large groups of people or animals simultaneously.

Nevertheless the aerosol vaccines for deep-lung delivery that have been developed to date have drawbacks that outweigh their benefits. Major safety concerns exist because of the potential of these vaccines to exacerbate respiratory diseases such as bronchitis, pneumonia, chronic obstructive pulmonary disease, emphysema, and allergic asthma, and because the excipients (carriers) used in formulations may be allergenic or irritating in some individuals. Other drawbacks of aerosol vaccines for deep-lung delivery are that they require a higher concentration of live microorganisms than conventional vaccines, as well as adjuvants to enhance the immune response. Clinical testing of aerosol vaccines is also more difficult and expensive than for other types of vaccines.

Today the pharmaceutical industry is making a significant effort to develop small-molecule drugs for nasopharyngeal delivery but is doing little work on aerosol vaccines. Although the success of FluMist may inspire companies to develop additional aerosol vaccines against influenza and possibly the common cold for naso-pharyngeal delivery, it is unclear whether other diseases could be prevented by this method. If, however, a safe and efficacious live vaccine for nasopharyngeal delivery was developed that could cross the blood-brain barrier, it would change the situation dramatically.

Potential for Misuse

Aerosol vaccines designed for nasopharyngeal delivery have several characteristics that minimize the risk of misuse for hostile purposes: they are packaged for individual use and dispersed into the nasal cavity in particles greater than 20 microns in diameter, usually with a handheld inhaler. Such large-particle aerosols are not suitable for delivering biological warfare agents because they are retained in the upper airways and do not travel into the deep regions of the lungs, where maximum absorption occurs. Moreover, individual inhalers are not capable of generating a large aerosol cloud that could expose many people simultaneously. These factors exclude nasopharyngeal vaccine technology from the risk of misuse for hostile purposes.

The main biosecurity concern associated with aerosol vaccines for deep-lung delivery does not involve the vaccine per se but rather the aerosol generators that are used to disperse particles small enough to reach the alveoli. These devices are dual use because they could deliver virulent bacteria or viruses as easily as attenuated ones. Even so, the misuse potential of generators for the delivery of aerosol vaccines is limited because they are designed to disseminate small quantities of liquid under controlled circumstances, rather than in the open air.

Accessibility

Classical aerosol vaccines are described extensively in the scientific literature. In the past, states with biological warfare programs, such as the United States and the Soviet Union, have assembled and supported development teams for aerosol vaccines. The development of such vaccines is also within the means of large pharmaceutical companies.

Ease of Misuse

The development of aerosol vaccines requires an advanced research infrastructure and an interdisciplinary scientific team that possesses the explicit and tacit knowledge needed to overcome complex technical problems during the development process. Such a team must include trained scientists and engineers with years of practical experience in aerobiology, microbiology, biochemistry, fermentation processes, formulation, materials, and downstream industrial processing.

Magnitude and Imminence of Risk

In view of the considerations discussed in the previous sections, the magnitude and imminence of the risk associated with aerosol vaccines appear to be low.

Characteristics of the Technology Relevant to Governance

Embodiment Aerosol vaccines consist of a live agent, a formulation, and an aerosol generator and are thus a hybrid of hardware and intangible information.

Maturity Only one human vaccine for nasopharyngeal delivery (FluMist) is commercially available, as well as a few deep-lung aerosol vaccines for veterinary use. Despite extensive research and development, no human aerosol vaccines for deep-lung delivery have yet been approved or marketed.

Convergence   The development of aerosol vaccines depends on innovations in both microbiology and vaccinology. Aerobiology techniques are also required to deliver a sufficient number of living organisms into the nasal cavity or the alveoli to stimulate the immune defenses of the host.

Rate of advance   Ever since aerosol-vaccine technology was introduced in the 1950s, progress has been slow. Methods for dispersing live microorganisms in aerosol form have changed little since Soviet times. MedImmune's development of FluMist in the early 2000s was a significant advance, but it hardly represented a paradigm shift.

International diffusion   At present, the development of aerosol vaccines appears to be limited to Europe, Russia, and the United States.

Susceptibility to Governance

Because aerosol vaccine technology is based on hardware as well as intangible information and is advancing slowly, it is moderately susceptible to governance. Nevertheless the technologies used to formulate aerosol vaccines entail little risk of misuse. Although the aerosol generators used to deliver aerosol vaccines might be modified to make them more useful for biological warfare or bioterrorism, such an effort appears unnecessary because generators with a much greater potential for misuse are commercially available from uncontrolled sources, such as suppliers that sell their wares over the Internet.[27]

For example, equipment for the aerosol dispersal of biopesticides such as *Bacillus thuringiensis* (Bt) is far more amenable to misuse for biowarfare purposes than is aerosol vaccine technology.[28] The Danish company GEA Niro manufactures an atomizer that is described as "particularly well suited for the production of particles with a sub-five micron mean particle size, such as particles required for inhalation products within the pharmaceutical industry."[29] Despite the clear dual-use potential of this device, it and similar devices are available for purchase with no restrictions.

Relevance of Existing Governance Measures

Although aerosol generators have been subject to regulation in the past, significant gaps in the nonproliferation regime still exist. For example, countries participating in the Australia Group, an informal forum of like-minded states that harmonize their national export controls, agree to deny sales to suspected biological proliferators of "complete spraying or fogging systems, specially designed or modified for fitting to aircraft, lighter than air vehicles or UAVs [unmanned aerial vehicles], capable of delivering, from a liquid suspension, an initial droplet [with a median diameter] of less than 50 microns at a flow rate of greater than two liters per minute."[30]

Because of this narrow definition, the Australia Group control list covers only a small fraction of commercially available aerosol generators that are potentially suitable for biological warfare or bioterrorism. To help close this gap, the control list should be revised and strengthened. Even so, several manufacturers and suppliers of aerosol generators are located in countries that are not members of the Australia Group and lack effective national export controls on dual-use biological equipment.

Proposed Governance Measures

The dual-use risk associated with aerosol vaccines per se is low and does not warrant the development of dedicated governance measures. Although the aerosol generators used to deliver aerosol vaccines provide a more plausible target for regulation, similar devices intended for other applications pose a much greater risk of misuse. For this reason, the security community should concentrate on controlling access to aerosol generators that are commercially available and could be readily adapted for biological warfare purposes.

Conclusions

For the foreseeable future, human aerosol vaccines for deep-lung delivery will have limited commercial prospects. The difficulty and expense of conducting clinical trials and the unacceptably high risks to test subjects have caused pharmaceutical companies to conclude that aerosol vaccines, despite their advantages, are not worth the cost and effort to develop. Given the lack of deep-lung aerosol vaccines on the market, there is little reason for concern that the technology could be misused. Accordingly, while urgent steps should be taken to reduce the risk that commercially available aerosol generators could be exploited for biological warfare and terrorism, no immediate action is required with respect to the governance of aerosol vaccines per se.

Acknowledgments

The authors are grateful to Dawn Verdugo for comments on drafts of this chapter.

Notes

1. V. K. Rastogi and K. P. O'Connell, *Studies on the Microencapsulation of Bacteria with Polymer, DL-Lactide Co-glycolide (PLGA), and Starch/Gelatin* (Edgewood Chemical and Biological Center, Research and Technology Directorate, ECBC-TR-550, 2007).

2. A micron is one-billionth of a meter. For comparison, the diameter of an average human hair is approximately 100 microns.

3. Bruce Lighthart and Alan J. Mohr, *Atmospheric Microbial Aerosols: Theory and Applications* (New York: Chapman & Hall, 1994). See also Robert F. Phalen, *Inhalation Studies: Foundations and Techniques*, 2nd ed. (New York: Informa Healthcare USA, 2009).

4. John Rees, "ABC of Asthma: Methods of Delivering Drugs," *British Medical Journal* 331, no. 7515 (September 3, 2005): 504.

5. S. A. Shoyele and A. Slowey, "Prospects of Formulating Proteins/Peptides as Aerosols for Pulmonary Drug Delivery," *International Journal of Pharmaceutics* 314 (2006): 1–8.

6. K. Petzoldt, C. von Benton, and W. Floer, *Study, Based on Published Sources, of the Applications and Limitations of Mass Aerogenous Immunization against Bacterial Infections under Field Conditions* [in German] (Bonn: Forschungsbericht aus der Wehrmedizin, Bundesministerium der Verteidigung, 1976).

7. N. I. Aleksandrov, N. Y. Gefen, N. S. Garin, et al., "Reactogenicity and Effectiveness of Aerogenic Vaccination against Certain Zoonoses" [in Russian], *Voyenno-Meditsinskiy Zhurnal*, no. 12 (1958): 51–59.

8. N. I. Aleksandrov, N. Y. Gefen, N. S. Garin, et al., "Experiment of Mass Aerogenic Vaccination of Humans against Anthrax" [in Russian], *Voyenno-Meditsinskiy Zhurnal*, no. 8 (1959): 27–32.

9. Ibid., 32.

10. R. B. Hornick and H. T. Eiglsbach, "Aerogenic Immunization of Man with Live Tularemia Vaccine," *Bacteriological Reviews* 30 (1966): 532–537.

11. U.S. Senate, *Is Military Research Hazardous to Veterans' Health? Lessons Spanning Half a Century*, Staff Report Prepared for the Committee on Veterans' Affairs, 103rd Congress, 2nd sess., S. Prt. 103-97, chap. C, December 8, 1994.

12. Hornick and Eiglsbach, "Aerogenic Immunization of Man with Live Tularemia Vaccine," 537.

13. A. V. Stepanov, L. I. Marinin, and A. A. Vorobyev, "Aerosol Vaccination against Dangerous Infectious Diseases" [in Russian], *Vestnik Rossiiskoi Akademii Meditsinsikh Nauk*, no. 8 (1999): 47–54.

14. M. Corbett, W. M. Bogers, J. L. Heeney, et al., "Aerosol Immunization with NYVAC and MVA Vectored Vaccines Is Safe, Simple, and Immunogenic," *Proceedings of the National Academy of Sciences* 105, no. 6 (2008): 2046–2051.

15. N. Low, S. Kraemer, M. Schneider, and A. M. Restrepo, "Immunogenicity and Safety of Aerosolized Measles Vaccine: Systematic Review and Meta-analysis," *Vaccine* 26, no. 3 (2008): 383–398; L. Garcia-Contreras, Yun-Ling Wong, P. Muttil, et al., "Immunization by a Bacterial Aerosol," *Proceedings of the National Academy of Sciences* 105 (2008): 4656–4660.

16. The manufacturer states: "Immune mechanisms conferring protection against influenza following receipt of FluMist vaccine are not fully understood." See MedImmune, "FluMist Influenza Vaccine, Live, Intranasal Spray," 2009–2010 formula, package insert of June 2009, par. 12.1.

17. M. J. Gagliani, "Direct and Total Effectiveness of the Intranasal, Live-Attenuated, Trivalent Cold-Adapted Influenza Virus Vaccine against the 2000–2001 Influenza A(H1N1) and B Epidemic in Healthy Children," *Archives of Pediatric and Adolescent Medicine* 158 (2004): 65–73. See also Z. Wang et al., "Live Attenuated or Inactivated Influenza Vaccines and

Medical Encounters for Respiratory Illnesses among U.S. Military Personnel," *Journal of the American Medical Association* 301, no. 9 (2009): 945–953.

18. R. J. Garmise, K. Mar, T. M. Crowder, et al., "Formulation of a Dry Powder Influenza Vaccine for Nasal Delivery," *AAPS Pharmaceutical Science and Technology* 7, no. 1 (March 2006), http://www.aapspharmscitech.org/articles/pt0701/pt070119/pt070119.pdf.

19. Ibid.

20. H. Möhlman, Margot Meese, P. Stohr, and V. Schultz, "Technology of Aerogenic Immunization against Swine Erysipelas under Conditions of Actual Practice" [in German], *Monatshefte für Veterinärmedizin* 25, no. 21 (1970): 829–832.

21. H. Müller and G. Reetz, "Aerosol Immunization of Ducks with Spirovak Erysipelas Vaccine 'Dessau'" [in German], *Archiv für experimentelle Veterinärmedizin* 34, no. 1 (1980): 55–57.

22. R. L. Ulrich, Kei Amemiya, David M. Waag, et al., "Aerogenic Vaccination with a *Burkholderia mallei* Auxotroph Protects against Aerosol-Initiated Glanders in Mice," *Vaccine* 23 (2005): 1986–1992.

23. P. W. Cargill and J. Johnston, "Vaccine Administration to Poultry Flocks," Merial Avian Business Unit, December 2006, http://www.poultry-health.com/library/cargilljohnson0612.htm; DeKalb Veterinary Service Bulletin, "Vaccination by Spray," DeKalb University.

24. D. Lu and A. J. Hickey, "Pulmonary Vaccine Delivery," *Expert Review of Vaccines* 6 (2007): 213–226.

25. Intervet/Schering-Plough Animal Health, "All Vaccines," 2009, http://www.intervet.com/ipg/vaccines.aspx.

26. G. Scheuch, Martin J. Kohlhaeufl, Peter Brand, et al., "Clinical Perspectives on Pulmonary Systemic and Macromolecular Delivery," *Advanced Drug Delivery Reviews* 58 (2006): 996–1008.

27. Micron, "View Our Complete Product Range," 2010, http://www.micron.co.uk/all_products.

28. David B. Levin and Giovana Valadares de Amorim, "Potential for Aerosol Dissemination of Biological Weapons: The Science of Biological Control," *Medscape Today*, 2003, http://www.medscape.com/viewarticle/452339_2.

29. GEA Niro, "Atomizers," http://www.niro.com/niro/cmsdoc.nsf/WebDoc/webb7ezgp8.

30. Australia Group, "Control List of Dual-Use Biological Equipment and Related Technology and Software," September 2009, http://www.australiagroup.net/en/dual_biological.html.

III
Historical Case Studies

19

Development of the V-Series Nerve Agents

Caitríona McLeish and Brian Balmer

During the twentieth century, many beneficial dual-use technologies were transferred from the military to the civilian sector, including atomic energy, microelectronic chips, FM radio frequencies, and systems analysis techniques. Less well known and understood are historical transfers of technology from the civilian to the military sector. Such civil-to-military transfers are particularly relevant when discussing chemical warfare (CW) because the production of chemical weapons requires access to a "civilian chemical industry capable of manufacturing organic chemicals on a substantial scale."[1] Thus a CW capability is inherently characterized by dual-use technologies.

Drawing on recently released documents from the National Archives in Britain, this case study explains how technology transfer from civilian pesticide research to the British army during the Cold War resulted in a new generation of chemical weapons called the "V-agents." By tracing the process through which this technology moved from peaceful to military application, the case study makes clear that the weapons application did not arise automatically from the inherent properties of the artifact itself (such as high toxicity) but required the active intervention of defense officials. Using slightly outdated terminology from science policy, the technology transfer resulted as much from "pull" from the military customer as "push" from the civilian developer. This observation suggests that the governance of dual-use technologies requires a nuanced and historically informed understanding of dual-use technology transfer.

Characteristics of the Technology

The V-series compounds belong to the class of chemical warfare agents known as "nerve agents" because they interfere with the transmission of nerve signals at synapses—the microscopic gaps between nerve cells (neurons) in the central nervous system or between a neuron and an effector organ, such as a muscle or gland. The arrival of an electrical nerve impulse at a nerve ending triggers the release of neurotransmitter molecules, which diffuse across the synapse and bind to receptor sites on the postsynaptic cell to initiate a physiological response. Acetylcholine is a major neurotransmitter at synapses in the brain and the peripheral nervous system. Under normal conditions, an enzyme called cholinesterase breaks down acetylcholine after its release into the synapse, rapidly resetting the electrochemical switch to the "off" position so that a new chemical signal can be transmitted.

Organophosphorus nerve agents (so called because they contain both carbon and phosphorus atoms) work by blocking the action of cholinesterase, preventing the enzyme from breaking down acetylcholine and resetting the electrochemical switch. The result is an excessive buildup of acetylcholine in the synapse that causes the postsynaptic cells to remain in a state of continual excitement until the nervous system seizes up. The symptoms of nerve-agent poisoning include major convulsions followed by a flaccid paralysis in which the breathing muscles cease to function, leading to death by asphyxiation within several minutes.

In the civilian sector, organophosphorus compounds are used primarily as pesticides, as well as flame retardants and lubricants. A successful pesticide is highly toxic to insects but much less so to humans, whereas a successful CW agent is highly toxic to humans. During the Cold War, pesticide manufacturers and CW establishments both became interested in organophosphorus chemistry because of the benefits it offered to their respective pursuits. Between 1952 and 1953, at least three companies independently discovered a new family of organophosphate compounds with strong anticholinergic activity that proved to be effective insecticides, especially against mites.[2] One such compound was marketed under the trade name Amiton, but it proved to be too toxic for agricultural use.[3] In the early 1950s, Amiton was transferred from the pesticide industry to the British defense establishment, where it led to the development of a new generation of nerve agents.

The Policy Context

For a short period after the end of World War II, British defense policy gave equal priority to chemical, biological, and atomic weapons. Even before the decision in 1947 to proceed with an atomic bomb, the Labour government of prime minister Clement Atlee decided that it was imperative that Britain be "in a position to wage chemical warfare from the start of hostilities."[4] Scientists at the Defense Research Policy Committee (DRPC), which advised the minister of defense and the chiefs of staff on defense research priorities, supported this policy, as did the chiefs of staff.[5] Whitehall's enthusiasm for CW research and development was relatively short-lived, however, and the prioritization of the nuclear weapons program soon drew attention away from chemical warfare.[6] The interest of British military planners in chemical warfare also waned when they realized that the 1925 Geneva Protocol, to which Britain was a party, banned the first use of these weapons.[7]

For a brief period, however, the discovery of the German nerve agents, which were far more toxic and fast acting than any known chemical warfare agents, stimulated the British government's interest in CW research.[8] The staff of the Chemical Defense Experimental Station at Porton Down, Wiltshire—renamed in 1948 the Chemical Defense Experimental Establishment (CDEE)—considered the German nerve agents to be a development of major importance that had to be properly assessed. Adding to the sense of urgency were intelligence reports that the Soviet Union had dismantled one of the German nerve-agent factories and rebuilt it in Stalingrad.[9] After World War II, Britain confiscated substantial stocks of German nerve agents and weapons until it could build its own production capacity.[10]

The technological opportunities offered by the German nerve agents gave Porton Down a new mission and probably saved it from closure in the postwar environment of financial austerity. Much of the work at Porton focused on assessing the properties and effectiveness of tabun and sarin as lethal weapons. Scientists also determined that low-dose exposures caused a severe blurring and darkening of vision that could incapacitate soldiers temporarily.[11] A Porton report in 1947 concluded that "the advent of the nerve gases makes obsolete the previous lethal gases CG [phosgene], AC [hydrogen cyanide] and CK [cyanogen chloride] except for very special purposes."[12] The research and development program also yielded a number of new candidate CW agents, including some that were comparable in potency to the German nerve agents.[13]

In November 1951, however, Rudolph Peters, the Whitley Professor of Biochemistry at Oxford University, reported that nerve-agent research at Porton "had reached a transition stage with past work largely finalized and new fields envisaged."[14] A few months later, in a discussion titled "Further Work on G Agents," the Chemical Defense Advisory Board (CDAB), a technical advisory committee to Porton Down, concluded that "little of value was likely to arise from investigation of further changes in the groupings of known nerve gases."[15]

The Civil–Military Overlap

The diminishing returns from research on sarin, soman, and tabun, also known as the G-series of nerve agents, stimulated a quest for novel CW agents. As early as November 1950, the members of the CDAB, including A. E. Childs, director of the Chemical Defense Research Department in the Ministry of Supply, discussed how to proceed.[16] The CDAB minutes quoted hereafter indicate that the chemical industry was seen as a possible—but by no means guaranteed—source of new directions:

The chairman [Prof. Alexander R. Todd] remarked that it could not be expected that extra-mural workers would be found who were prepared to proceed along speculative lines in a search for new toxic compounds and it was, therefore, necessary for all concerned to watch the literature carefully for any guidance. Dr Barrett [Porton Down] agreed that, in the absence of sufficient information on structure/toxicity relationships, the only possible approach was empirical but he wondered if it were possible for industry to submit all likely compounds for test. Mr. Childs [Ministry of Supply] said that a scheme of this nature had been started in regard to antibiotics but had produced no worthwhile results; nothing was lost, however, in trying to get information this way.[17]

The difficulty of predicting toxicity from chemical structure had led to a process of trial and error. In a letter, Childs noted the shortcomings of this "empirical" technique:

The methods of approach at the moment are mainly empirical and, in short of a staff out of all proportion greater than we can afford, only the fringes can be touched. In consequence, success depends largely on a stroke of luck. This state of affairs is regarded by all concerned as dangerous and unsatisfactory.[18]

The advisers considered two strategies: literature scanning and soliciting assistance from private industry. Porton Down had a long history of ties with representatives of the chemical industry. During World War I, the British government had contracted with chemical companies to produce poison gases and had called on academic chemists to aid in military research and development. The links between government, industry, and academia continued after the war.[19] Several companies that the British government had contracted to produce chemical weapons during World War I merged to form Imperial Chemical Industries (ICI).[20] In 1937 the government turned to ICI to build and operate factories that would produce and fill the next generation of chemical munitions.[21]

Childs and his assistant Dr. J. McAulay led Porton Down's effort to seek assistance from the chemical industry. Childs drew on his personal contacts, and in January 1951 he wrote to James Davidson-Pratt at the Association of British Chemical Manufacturers.[22] During World War II, Davidson-Pratt had worked in the Ministry of Supply, where he had dealt with biological warfare research, and he was currently a member of the ministry's biological equivalent of the CDAB, the Biological Research Advisory Board.[23] Childs's letter outlined the haphazard and inefficient search for new toxic compounds at Porton and mentioned that the United States had established a Chemical and Biological Coordination Center in Washington, D.C., to collate information provided by industry.[24] Childs then requested that the British association circulate to its members a request for information on new toxic compounds.

Davidson-Pratt responded immediately, offering to send out a confidential letter to member companies asking them to communicate "anything which appears unusually toxic or novel" to the Ministry of Supply. On a less optimistic note, he pointed out that a proposal two years earlier to create a center similar to the one in the United States had not been well received. "The general view of British industry was to express grave doubts as to whether the results to be obtained would ever be commensurate with the efforts involved," he said. "I.C.I. felt particularly strongly on this point in the light of their past research experience."[25] In sum, while industry was a potentially useful ally, neither its cooperation nor its utility could be taken for granted.

Plant Protection Limited

One industrial field of particular interest to chemical weapons researchers at Porton Down was pesticide research and manufacture. After World War II, the pesticide and insecticide industries expanded rapidly, and many companies began working on organophosphorus compounds. Between 1952 and 1953, at least three firms identified a group of organophosphate esters with potent insecticidal activity, especially against mites.[26] After these substances were patented and their properties published in open literature, some of them were marketed as insecticides.

One such compound was a candidate miticide that was later named Amiton.[27] It was discovered by the chemists Ranajit Ghosh and J. F. Newman, working at Plant Protection Limited (PPL) in Yalding, Kent. A subsidiary of ICI and another company called Cooper, McDougall and Robertson (CMR), PPL had been established in 1937 to end the competition in pesticide production between the two firms.[28] Ghosh probably synthesized Amiton in early 1952, although one source claims that he did so in 1948.[29] PPL did not apply for a patent on the compound until November 1952, and the details were not published until 1955, when the synthesis of Amiton was protected by several patents.[30]

In 1954, ICI marketed a form of Amiton (the hydrogen-oxalate salt) as an insecticide under the trade name Tetram.[31] Three years later, the journal *Nature* reported that PPL was manufacturing a "new" pesticide under the trade names Tetram and ICI Amiton that "has a high toxicity to man, but as an insecticide it is claimed to be largely specific for red spider and other mites and for scale insects, and to have little effect on insect predators."[32] In 1955, ICI placed a production contract for Amiton with the Chemical Defense Establishment Nancekuke, a British government facility in Cornwall that also manufactured nerve agents for the military.[33]

Amiton was not a successful insecticide. It was highly toxic to humans and readily absorbed through the skin into the blood circulation, making it too dangerous for agricultural use. According to an unpublished document, "Although [Amiton] showed great promise as a systemic insecticide against sucking arthropods such as mites and scale insects, and despite the absence of accidents during trial, it was eventually decided that the intrinsic toxicity of the material was too high for commercial exploitation."[34] As a result, the product was withdrawn from the market around 1958.[35] Nevertheless, while high percutaneous toxicity is not a quality that lends itself to a successful agrochemical, it is a great asset in a CW agent.

Technology Transfer

Current restrictions on the availability of primary documents make it difficult to trace the exact process by which Amiton was transferred from PPL to Porton Down. Porton's initial request in 1951 for assistance from industry yielded little of interest.[36] The Ministry of Supply began a renewed effort in July 1953, when it sent a series of letters directly to chemical and medical firms (nineteen were named on one list) and through the Association of British Chemical Manufacturers, which contacted twenty-five firms.[37] Although some of the letters were individually tailored, the main circular is worth quoting at length:

The main duty of this Directorate is to plan protective measures against any toxic materials which might be used in war.

The rapid growth of research and development in chemistry, and particularly in the fields of biochemistry, chemotherapy and insecticides, greatly increases the chance that new and unexpected types of toxic materials may be brought to light. Some of these compounds might prove to be effective as war chemical agents against which a method of protection will be needed.

We should therefore appreciate the co-operation of industrial and other research organisations in providing us with data on the synthesis and properties of any new compounds which you prepare (or extract from natural products) and which show high toxicity or toxicity associated with new molecular structures or toxicity of a novel type.[38]

The suggestion that new toxic chemicals were required for protective purposes was clearly disingenuous, unless the retaliation-in-kind policy was interpreted at the time as a form of protection. The letter from the Ministry of Supply to the research controller of ICI, dated July 16, 1953, was more personalized and contained an intriguing addendum:

We have been very grateful for the co-operation of the I.C.I. in the past and hope very much that we can count on it in the future. For your own private information the last item received from you has now been put well within the barbed wire fence and is receiving much attention.[39]

The items provoking "much attention" in the search for new toxic compounds at the time were related to Amiton, suggesting that PPL had passed the compound to Porton Down through its parent company ICI.

An important feature of the relationship between the military and industry was secrecy. In their correspondence with industry, British government officials expressed the hope that their requests for information would be held in confidence.[40] The need for confidentiality operated in both directions, however. Porton's initial outreach efforts in 1951, rather than provoking a flood of information from industry, had met a significant barrier in companies' wish to protect their commercial secrets. Thus, before sending the second set of letters to industry in 1953, the Ministry of Supply established a system of commercial (C) codes to identify each compound submitted. A memo from Porton explained, "In the hope of inducing firms etc to bring forward toxic compounds for test more freely, it has been decided to restrict knowledge of the origin of such compounds as much as possible."[41]

Instead of revealing the names of the originating firms, the ministry would forward to Porton Down only the C number of each compound, along with information about its composition and properties. This system was explained in the letter-writing campaign and flagged as a means of protecting commercial interests. In correspondence dated April 29, 1953, the first compound transferred under the new system, R5158, was given the code number C11.[42] One year later, Dr. McAulay reported to the Chemical Defense Advisory Board that the ongoing outreach efforts to industry were "an extension of our previous vigorous efforts to maintain contact with all important industrial and academic research laboratories on matters which might have a CW interest. (Compound C11 . . . resulted.)"[43]

Compound C11 was described as a compound that was "notified to the Ministry by a commercial firm, [and] was proving to be of great importance."[44] The structure of C11 was either closely related or identical to that of compound T2274, the internal code name assigned to Amiton.[45] CDAB member Professor Ewart Jones, an organic chemist from Manchester University, summed up the excitement generated by compound C11:

There had, thus, appeared an entirely new lead in the nerve gas field, when it was thought to have been completely circumscribed, and it was inevitable that new light would be thrown on the structure/activity relationships. The compounds could be seen on inspection to be a remarkable combination of the nerve gases, mustard gas and the nitrogen mustards, and it was natural that the Committee had been able to put forward many suggestions for further work.[46]

Once C11 or Amiton had been transferred to Porton Down, it was given the military code name VG. (The "V" apparently stood for "venomous" because of the compound's skin-penetrating characteristics.) The members of the Chemical Defense Advisory Board noted that C11 and another related compound designated T2290 (later code-named Agent VE) were "by far the most dangerous of all for attack through the bare skin." Porton was therefore aware of the properties of Agent VE by the end of 1953.

Although VE had powerful insecticidal properties, it was even more toxic than Amiton to warm-blooded animals. As a result, VE superseded Amiton (VG) as a candidate agent for weaponization. PPL continued its development work in this area, and in June 1955 Ghosh applied for a patent on VE.[47] It is unclear whether the military scientists at Porton Down synthesized VE independently by modifying Amiton or whether PPL discovered VE and transferred it to Porton. Subsequently,

Porton scientists identified substances that were even more toxic than VE by modifying the basic Amiton-type molecular structure.

The historical documents available today leave unanswered many important questions about the nature of the civil-to-military technology transfer. In particular, ambiguities remain concerning how much information about Amiton and Agent VE was transferred by PPL to Porton Down and how much Porton acquired on its own.[48] Based on the available historical record, all that can be stated with confidence is that the dual-use chemical C11 (closely related if not identical to Amiton) was transferred from PPL to Porton Down sometime between 1951 and 1953, probably in late 1952 or early 1953.

By May 1954, the British government passed information about Amiton and VE to allied chemical-weapons scientists at Edgewood Arsenal in the United States and the Suffield Experimental Station in Canada. Members of the Chemical Defense Advisory Board were told that it was not known if the Soviet Union possessed similar knowledge.[49] As with the G-series nerve agents, British military chemists sought to develop more potent forms of the new Amiton-based agents. The CDAB report for 1954 noted that "a number of new phosphorus-sulphur compounds have been synthesised and much effort has been devoted to preparing each structural form in a pure state."[50] The report also noted that "many of the fascinating problems presented by these new compounds could not have been solved without the up-to-date infra-red spectrometric equipment recently purchased; even so these dangerous investigations are still impeded by lack of standard analytical techniques."[51] Porton scientists reported synthesizing over 200 organophosphorus compounds with anticholinesterase activity and noted that "at Edgewood probably more have been examined."[52]

In July 1956 the Cabinet Defense Committee, in a bid to reduce defense spending, decided not to proceed with the planned large-scale production of nerve agents and to destroy all the remaining stocks. By this time, however, U.S. military scientists had created about 50 different V-series agents, and in February 1957 the U.S. Army Research and Development Command selected Agent VX for large-scale production.[53]

A Porton report from 1957 estimated that the toxicity of V-agents had a limit and that VX approached that limit.[54] Concurrently the twelfth Tripartite Conference of U.S., British, and Canadian chemical weapons experts issued a call to "divert the maximum possible effort to research for new agents and recommended that the field of natural products should receive particular attention."[55] ("Natural products" was a reference to toxins, highly toxic substances produced by living organisms.) In parallel with the new emphasis on toxins, the conference participants recommended seeking help with the search from industry and academic research institutions.

Further Outreach to Industry

Despite the overlap between the interests of the commercial pesticide industry and the military, their agendas were not perfectly aligned. The goal of the pesticide industry was to develop chemicals that were highly toxic to insects but not to mammals, whereas the military sought chemicals that were highly toxic to mammals. This discrepancy was highlighted in May 1957, when the Ministry of Supply sent out a fresh round of letters, both directly to individual firms and indirectly through the Association of British Chemical Manufacturers. Although the industry responses to the letters generally offered cooperation, some expressed reservations. For instance, the reply from Fisons Pest Control Ltd. read:

We do of course occasionally find compounds which are exceptionally toxic to mammals, in the course of our search for the other thing, but as you appreciate this is the signal for doing no further work on the subject and usually our information at this stage is meagre. . . . I feel that a little more exchange of information between people concerned with organo phosphorus compounds as insecticides and people concerned with them as war gases would be helpful to both parties.[56]

Porton scientists followed up the letter campaign with visits to a small number of companies. After a visit to Glaxo Research Laboratories, the scientists noted optimistically, "It would seem that the possibility exists of mutual benefit in that Glaxo's failures may be Porton's successes and vice versa."[57] But Glaxo responded to Porton's request by noting, "Generally our aim is to find substances with low mammalian toxicity and high activity as therapeutic agents, insecticides, etc. It is very unusual for us, therefore, to prepare substances of very high toxicity."[58]

This response underlined the point that private firms would not, in the normal course of events, get as far as synthesizing or characterizing the highly toxic materials sought by the Porton scientists. The same point was made by Shell: "We regret we have no products of this kind, mainly by reason of the fact that the type of chemical compounds which we are synthesising as potential agricultural chemicals are based on structures which might reasonably be supposed to possess low mammalian toxicity."[59] Similarly, a pharmaceutical research director stated that time pressures prevented him from following up on the properties of toxic agents that surfaced from time to time in the company's research laboratories.[60]

More fine-grained examples gleaned from the historical record underline the mismatch between the agendas of the defense establishment and the civilian chemical industry. Porton scientists visited a company that used acute oral toxicity in rats to determine the suitability of a compound for its research program. The firm did not test toxicity by routes such as the skin, which were of greater interest to the military.[61] For their part, Porton staff resisted industry proposals for a two-way flow of information by discouraging companies from making reciprocal visits.[62]

Although the British chemical industry and Porton Down operated under conditions of strict secrecy—industry to protect proprietary information, Porton to safeguard national security—the secret of the V-agents managed to leak out. By the late 1950s it had spread to the Soviet Union and France.[63] Between 1960 and 1972, chemists in seven countries (the United States, Sweden, West Germany, the Netherlands, Yugoslavia, the United Kingdom, and Czechoslovakia) published information on V-agents in the scientific literature.[64] Other countries known or suspected to have synthesized V-agents include Iraq, Israel, and Syria.

In 1975, British journalists discovered the previously secret patent for VX in a public library, raising concerns about the possible synthesis of V-agents by terrorists.[65] Indeed, the Aum Shinrikyo cult in Japan produced small quantities of VX for assassination purposes in the mid-1990s.[66] Since the entry into force in 1997 of the Chemical Weapons Convention (CWC), any offensive development work on nerve agents has been banned.[67]

Conclusions

Although defense research and development laboratories achieved incremental improvements in chemical warfare agents, major breakthroughs such as the discovery of the G- and V-agents were spin-offs of civilian technologies. The transfer of Amiton from PPL to Porton Down demonstrates how the British defense establishment interacted with the domestic chemical industry to develop a new family of nerve agents. Nevertheless it was not preordained that Amiton (the pesticide) would become VG (the nerve gas).

Despite a degree of civil–military overlap, forging the conditions for technology transfer from the chemical industry to the military sector required an active process of outreach. ICI did not simply knock on the door of Porton Down with their new discovery. The British defense authorities mounted repeated letter-writing campaigns to industry in search of new toxic compounds, established secrecy measures to protect industrial trade secrets, and sought to translate the goals of the pesticide industry into those of the chemical warfare laboratory. Even so, these efforts failed to generate a flood of new research leads. Instead several of the solicited chemical companies stated repeatedly that they would not, in the normal course of events, be interested in the same things as the military scientists.

Focusing narrowly on technology alone generates the paradigm of the dual-use dilemma, in which both benign and malign applications are construed as inherent to the technological artifact or scientific discovery itself.[68] Examining the historical context, however, reveals that although Amiton and VG shared the same molecular structure, the pesticide Amiton was not identical to the nerve gas VG. Instead the defense authorities had to intervene actively to influence the goals of the pesticide industry.

In other words, Amiton had to be *translated* into VG through an active network of artifacts and people, including a governmental policy environment that encouraged the search for new toxic compounds. The concept of translation comes from research in the sociology of science.[69] Because the interests of different groups crucial to the success of a research project do not always align, the interests of one actor must be translated into the interests of the other for the work to move forward and succeed.

Lessons for Policy
Although the exact details of the process by which Amiton was transferred to Porton remain obscure, the history of the V-series nerve agents suggests that the governance of dual-use technologies must look beyond the particular technological artifact under consideration and understand the social and political context in which innovation and technology transfer occur.[70] Governance architecture should seek to direct technological change along socially beneficial trajectories by influencing the sociotechnical networks involved.[71]

The case of the V-agents also reminds us that the effective governance of dual-use technologies must accommodate change and innovation by moving away from governance measures based on lists of artifacts or technical characteristics and toward those that focus on intent and purpose. The prohibitions in the CWC, for example, are based on purposes rather than specific chemicals, which are used only for verification. Thus the treaty bans the development, production, stockpiling, transfer, and use of *all* toxic chemicals and their precursors regardless of their origin or method of production, *except* for "purposes not prohibited under this Convention," as long as the "types and quantities . . . are consistent with such purposes." The comprehensive nature of this prohibition means that as soon as a new chemical agent is developed for hostile purposes, it immediately falls under the purview of the treaty. Although intent is often viewed as ineffable and difficult to regulate, the case of the V-agents suggests that intent may be more susceptible to governance than a static list of compounds.

Notes

1. Julian Perry Robinson, "Supply-Side Control of the Spread of Chemical Weapons," in *Limiting the Proliferation of Weapons: The Role of Supply-Side Strategies*, ed. Jean-Francois Rioux (Ottawa: Carlton University Press, 1992), 63.

2. Julian Perry Robinson, "V-Agent Nerve Gases," in *The Problem of Chemical and Biological Warfare*, vol. 1, *The Rise of CB Weapons*, by Stockholm International Peace Research Institute (SIPRI) (New York: Humanities Press, 1971), 74.

3. Christa Fest and Karl-Julius Schmidt, *The Chemistry of Organophosphate Pesticides* (Berlin: Springer, 1982), 128.

4. The National Archives, Kew [TNA], CAB [Cabinet Office papers] 131/1. Minutes of Defense Committee of the Cabinet, June 20, 1946.

5. Jon Agar and Brian Balmer, "British Scientists and the Cold War: The Defence Research Policy Committee and Information Networks, 1947–1963," *Historical Studies in the Physical Sciences* 28 (1998): 209–252.

6. In October 1952, Britain tested its first atomic bomb, elevating nuclear weapons above the other types of unconventional weapons, and soon afterward, the defense budget was cut in the wake of rearmament for the Korean War. See Gradon Carter and Graham Pearson, "North Atlantic Chemical and Biological Research Collaboration: 1916–1995," *Journal of Strategic Studies* 19 (March 1996): 74–103.

7. TNA DEFE 10/33, Defense Research Policy Committee, 10th Memo, Review of Defense Research and Development, *Trends in Scientific and Technical Development: Weapons of Mass Destruction*, March 10, 1954, p. 27.

8. Robinson, "V-Agent Nerve Gases," 73.

9. Gradon Carter and Graham Pearson, "Past British Chemical Warfare Capabilities," *RUSI Journal* 141 (February 1996): 61.

10. Gradon Carter and Brian Balmer, "Chemical and Biological Warfare and Defence, 1945–90," in *Cold War, Hot Science: Applied Research in Britain's Defense Laboratories, 1945–1990*, ed. Robert Bud and Philip Gummett (Amsterdam: Harwood, 1999).

11. Carter and Pearson, "Past British Chemical Warfare Capabilities," 62.

12. TNA W[ar] O[ffice papers], 195/9236 Porton Report No. 2747, *Preliminary Report on the Potential Value of Nerve Gases as CW Agents,* January 18, 1947. Mustard gas continued to rank highly in the chemical arsenal. See TNA WO 195/12063 CDAB, 21st Meeting of the Board, November 6, 1953.

13. Julian Perry Robinson, director, Harvard-Sussex Program on Chemical and Biological Weapons, personal correspondence with authors, August 6, 2009.

14. TNA, WO195/11648 CDAB, 18th Meeting of the Board, November 1, 1951.

15. TNA, WO195/11754 CDAB, 19th Meeting of the Board, February 7, 1952.

16. The Ministry of Supply was responsible for the defense research establishments, such as the CDEE. See David Edgerton, "Whatever Happened to the British Warfare State? The Ministry of Supply, 1946–1951," in *Labour Governments and Private Industry: The Experience, 1945–1951*, ed. Helen Mercer, Neil Rollings, and Jim D. Tomlinson (Edinburgh: Edinburgh University Press, 1992).

17. TNA, WO195/11216 CDAB, 15th Meeting of the Board, November 9, 1950.

18. TNA, WO 188/2716, letter to James Davidson Pratt, Association of British Chemical Manufacturers from A. E. Childs, DCDRD, Ministry of Supply, January 4, 1951.

19. In 1920, *Nature* ran a series of letters to the editor commenting on the government's proposal to continue the relationship forged with universities during the interwar years. "Letters to the Editor," *Nature* 106, nos. 2662–2663 (November 4, 11, and 18, 1920).

20. Carter and Pearson, "Past British Warfare Capabilities," 60.

21. Ibid.

22. The Association of British Chemical Manufacturers (ABCM) was founded in 1916 as a trade association for the chemical industry. Today it is called the Chemical Industries Association.

23. Brian Balmer, *Britain and Biological Warfare: Expert Advice and Science Policy, 1930–65* (Basingstoke: Palgrave, 2001), 45, 101.

24. TNA, WO 188/2716, letter to James Davidson Pratt, Association of British Chemical Manufacturers (ABCM), from A. E. Childs, DCDRD, Ministry of Supply, January 4, 1951.

25. TNA, WO 188/2716, letter from J. Davidson Pratt (ABCM) to A. E. Childs (Ministry of Supply), January 5, 1951.

26. Robinson, "V-Agent Nerve Gases," 74.

27. Amiton has the chemical formula O, O-diethyl-S-2-diethylaminoethyl phosphorothiolate.

28. From early 1953, CMR began to reduce its investment in PPL, and the partnership between ICI and CMR ended in 1958. See William J. Reader, *ICI: A History*, vol. 2, *The First Quarter Century, 1927–52* (London: Oxford University Press, 1975), 335, 455–456. See also Wellcome Trust Archives, London, Wellcome Foundation, Cooper McDougall & Robertson Ltd., WF/C/6/1/302, WF/C/6/1/303, WF/C/M/PC/07.

29. A. Calderbank, "Organophosphorus Insecticides," in *Jealott's Hill: Fifty Years of Agricultural Research, 1928–1978*, ed. F. C. Peacock (Bracknell: ICI, 1978), 50. The earlier date is given in Fest and Schmidt, *Chemistry of Organophosphate Pesticides*, 128.

30. Ranajit Ghosh, *New Basic Esters of Phosphorus-Containing Acids*, British Patent Number 738839, application date November 19, 1952, complete specification published October 19, 1955; Ranajit Ghosh, *Manufacture of Basic Esters of Phosphorothiolic Acid*, British Patent Number 763,516, application date July 16, 1954, complete specification published December 12, 1956; Ranajit Ghosh, *New Pesticidal Basic Esters of Phosphorothiolothionic Acid*, British Patent Number 763,516, application date July 16, 1954, complete specification published December 12, 1956.

31. Stockholm International Peace Research Institute, *The Problem of Chemical and Biological Warfare: A Study of the Historical, Technical, Military, Legal, and Political Aspects of CBW and Possible Disarmament Measures*, vol. 1, *The Rise of CB Weapons* (New York: Humanities Press: 1971), 70–75, 280–282.

32. "A New Organophosphorus Insecticide," *Nature* 4563 (April 13, 1957): 763.

33. Graham Pearson, reply to a question by the Countess of Mar, *Hansard* (House of Lords), April 11, 1994, 2.

34. Unpublished document, Ministry of Defense, Annex C, *Properties of the Insecticide "Amiton" and Its Salts R6199 and R6200*, Sussex Harvard Information Bank (SHIB), Harvard-Sussex Program, University of Sussex.

35. Robinson, "V-Agent Nerve Gases," 74. The termination of the contract with Nancekuke to manufacture Amiton in 1958 suggests this was the date of withdrawal from the market.

36. TNA, WO 188/2716, F. Savage, Assistant Managing Director for Anchor Chemical Company (Manchester) to Ministry of Supply, Chemical Defense Research Department, January 16, 1951.

37. TNA, WO 188/2721 TA 9002, "Firms Asked to Co-operate in Providing Data on New Toxic Compounds" (n.d., but with 1953 papers); letter from Davidson-Pratt to McAulay, July 22, 1953.

38. TNA, WO 188/2721, Ministry of Supply, Directorate of Chemical Defense Research and Development, "New compounds prepared in industrial and in other research laboratories and found to be too toxic for medical or industrial use," July 1953.

39. TNA, WO 188/2721, letter from J. McAulay to R. M. Winter, Research Controller, Messrs ICI Ltd., Nobel House, Buckingham Gate, July 15, 1953.

40. TNA, WO 188/2721, letter from J. McAulay (CDR1) to Prof. Bergel, Chester Beatty Research Institute, July 15, 1953, and similar correspondence in this file.

41. TNA, WO 188/2721, Code Numbers for Toxic Compounds Received from Firms etc. by Superintendent Research Division, CD Experimental Establishment, May 27, 1953.

42. TNA, WO188/2721, letter from Superintendent Research Division, CD Experimental Establishment, Porton [signature unclear but same as on following letter] to Dr. J. McAulay (Ministry of Supply) (April 29, 1953); Code Numbers for Toxic Compounds Received from Firms etc. by Superintendent Research Division, CD Experimental Establishment, May 27, 1953.

43. TNA, WO188/2721, from J. McAulay, copy to Sir Rudolph Peters. Chemical Defense Advisory Board. Minutes of 26th Meeting, Action 1, Minute 257(d). Liaison with Commercial Interests. July 28, 1954.

44. TNA, WO195/12549, Ministry of Supply, Chemical Defense Advisory Board, Minutes 24th Meeting of the Board, November 5, 1953.

45. It is possible that C11 was a structural isomer composed of exactly the same atoms as Amiton but arranged into a different three-dimensional structure. C11 might even have been a mixture. A later report noted, "It has been established that the occasionally erratic behaviour of these compounds [organic-phosphorus-sulphur compounds] is due to their tendency to undergo internal structure change (isomerisation), one form being highly toxic and the other non-toxic." WO195/13005, CDAB Annual Review of the Work of the Board for 1954, November 11, 1954.

46. TNA, WO195/12549, Ministry of Supply, Chemical Defense Advisory Board, Minutes 24th Meeting of the Board, November 5, 1953.

47. Rajanit Ghosh (ICI Ltd.), "New Basic Ester of Thiophosphonic Acids and Salts Thereof," British Patent no. 797603 (applied June 1955).

48. Some of the still-unanswered questions include the following: Did PPL tell Porton about Amiton before or after the first patent application was filed in 1952? Was the compound C11, originally postulated to have a different structure from Amiton, transferred at the same time as Amiton? A further possibility is that PPL passed C11 but not Amiton to Porton via ICI, believing the two compounds to be different substances. In this case, one could conjecture that Porton synthesized T2274, and PPL synthesized Amiton, independently. Further uncertainties remain. Was further information passed to Porton in 1954, when Ghosh discovered Agent VE, or had CDEE independently synthesized VE? Was PPL's role in the technology transfer limited to information concerning the existence of the V-agents, or did it provide practical details about how Amiton or VE could be synthesized in high yields?

49. TNA, WO195/12802, Ministry of Supply, Chemical Defense Advisory Board, 26th Meeting of the Board, May 13, 1954.

50. TNA, WO195/13005, CDAB, Annual Review of the Work of the Board for 1954, November 11, 1954.

51. Ibid.

52. TNA, WO 188/2716, A Memorandum on Possible Increases in Intrinsic Toxicity of Organo-Phosphorus Compounds and the Case for a Search for New Agents, by D. R. Davies and A. L. Green. CDEE Porton, December 9, 1957.

53. Jonathan B. Tucker, *War of Nerves: Chemical Warfare from World War I to Al-Qaeda* (New York: Anchor Books, 2007), 158.

54. TNA, WO 188/2716, A Memorandum on Possible Increases in Intrinsic Toxicity of Organo-Phosphorus Compounds and the Case for a Search for New Agents.

55. TNA, WO 188/2716, The Search for New Agents, T. F. Watkins, December 12, 1957.

56. TNA, WO 188/2716, G. S. Hartley (Director of Research) to E. E. Haddon, Director of Chemical Defense Research & Development, Ministry of Supply, June 4, 1957.

57. TNA, WO188/2721, visit by Mr. Watkins and Mr. Callaway (CDEE) to Glaxo Research Laboratories, Greenford, Middlesex, to see Mr. Toothill and Dr. Child, April 14, 1958.

58. TNA, WO188/2721, from TG Maccrae, Executive Director of Research and Development, Glaxo Laboratories Ltd., January 1, 1958.

59. TNA, WO188/2721, from C. G. Williams, Shell Research Limited, to D. E. Woods, Ministry of Supply, June 13, 1957.

60. TNA, WO188/2721, consultation report of visit by R. W. Brimblecomb (CDEE) to Glaxo (Sefton Park, Stoke Poges, Bucks) to meet Dr. Campbell (Director of Research) and Dr. Ball, March 24, 1958.

61. TNA, WO188/2721, ASG Hill (for Director), Porton Down to DCDRD, "Search for New Toxic Compounds: Visits to Firms," February 18, 1958, "Appendix: Visit to Murphy Chemical Co. Ltd. (T. F. Watkins)," February 12, 1958.

62. Ibid.

63. Tucker, *War of Nerves*, 181, 169.

64. TNA, DEFE 13/823, Security Classification and Production of VX, R. Holmes, January 8, 1975.

65. Brian Balmer, "A Secret Formula, a Rogue Patent, and Public Knowledge about Nerve Gas: Secrecy as a Spatial-Epistemic Tool," *Social Studies of Science* 36 (2006): 691–722.

66. David E. Kaplan, "Aum Shinrikyo (1995)," in *Toxic Terror: Assessing Terrorist Use of Chemical and Biological Weapons*, ed. Jonathan B. Tucker (Cambridge, MA: MIT Press, 2000), 214.

67. In 1997, Russia declared 15,558 metric tons of a V-agent termed R-33, and the United States declared 4,032 metric tons of VX, all of which must be destroyed under the terms of the CWC.

68. Caitríona McLeish, "Reflecting on the Dual-Use Problem," in *A Web of Prevention: Biological Weapons, Life Sciences, and the Governance of Research*, ed. Brian Rappert and Caitríona McLeish (London: Earthscan, 2007).

69. Bruno Latour, *Science in Action* (Cambridge, MA: Harvard University Press, 1987).

70. Paul Nightingale, "Technological Capabilities, Invisible Infrastructure, and the Un-social Construction of Predictability: The Overlooked Fixed Costs of Useful Research," *Research Policy* 33, no. 9 (2004): 1259–1284.

71. Giovanni Dosi, "Technological Paradigms and Technological Trajectories: A Suggested Interpretation of the Determinants and Directions of Technical Change," *Research Policy* 11, no. 3 (1982): 147–162. See also Donald MacKenzie and Judy Wacjman, eds., *The Social Shaping of Technology* (Milton Keynes: Open University Press, 1999).

20

The Use and Misuse of LSD by the U.S. Army and the CIA

Mark Wheelis

The use of chemicals to modify brain function is an ancient practice. For millennia, humans have employed alcohol, marijuana, coca leaf, psychedelic fungi, and other plant extracts for ritual, therapeutic, and recreational purposes. There have also been sporadic reports of the use of psychoactive drugs for hostile ends. A wide variety of drugs have been examined for their potential to incapacitate enemy soldiers, enhance the capabilities of friendly troops, assist in interrogation, and induce psychosis in enemy leaders.[1] Chemicals studied for these purposes have been drawn largely from recreational or ritual drugs, as well as known categories of pharmaceuticals (the two categories overlap). This historical case study examines the efforts by the U.S. Army and the Central Intelligence Agency (CIA) during the 1950s to develop the hallucinogenic drug lysergic acid diethylamide (LSD) for chemical warfare and covert operations.

The army and the CIA were attracted to LSD because of its extraordinary potency, dramatic disturbance of cognition and perception, and low lethality, which suggested that the drug might have potential as a military incapacitant and an agent for intelligence use. Although the mechanism of action of LSD was unknown when the programs took place, such understanding was not needed for the proposed applications. The army's attempt to develop LSD into a battlefield weapon did not involve scientific innovation but simply extended traditional CW technology to a new agent. Ultimately the effort failed for the same reasons that have prevented many other chemicals from being developed as effective weapons, namely, the instability of the drug when dispersed as an aerosol and the difficulty and high cost of its synthesis.

The CIA's attempts to develop LSD as a mind-control agent, an interrogation aid, and a weapon to induce psychosis in enemy leaders were an extension of its previous development efforts with other psychoactive chemicals, such as mescaline, tetrahydrocannabinol (THC), scopolamine, and barbiturates. LSD was usually administered by adding it to a drink offered to an unwitting subject, an extremely low-tech delivery method. The CIA development program failed because the drug did not produce desirable effects in a reproducible manner and because of belated concerns about the legality of the program.

Although U.S. experimentation with LSD as an agent for hostile purposes ended in the 1960s, military and intelligence agencies around the world continue to be interested in the development of other psychoactive drugs for riot control, counter-terrorism, interrogation, and troop enhancement. The potential use of such chemicals raises serious ethical and legal issues about manipulating the mental function of individuals without their informed consent. Broader themes addressed in this case study include the interpretation of misuse, the importance of oversight, normative dynamics, human rights issues, and the role of individuals.

Background on LSD

LSD disrupts the perceptual and cognitive systems in the brain, leading to powerful visions and hallucinations. These effects are sometimes experienced as profoundly meaningful, creating a sense of cosmic unity. Alternatively, the hallucinations induced by LSD can be terrifying, particularly if the subject is unaware of having been drugged. It is now understood that LSD is structurally similar to the neurotransmitter serotonin and mimics its excitatory action at sites in the cerebral cortex called 5-HT_{2A} receptors. LSD is therefore a serotonin receptor "agonist." (An agonist is a compound that mimics the action of a natural signaling molecule, as opposed to an "antagonist," a compound that blocks the action of the natural transmitter.) Because other 5-HT_{2A} agonists do not all produce hallucinations, it is clear that some aspects of the mechanism of action of LSD are not fully understood. Recent research has begun to identify the specific cortical pathways that are responsible for the drug's hallucinogenic effects.[2]

LSD was first synthesized in 1934 by Albert Hoffmann, a chemist at the Swiss pharmaceutical company Sandoz who was investigating derivatives of compounds isolated from ergot, a fungus that grows on rye and related plants, as possible drugs. Because lysergic acid is present in significant amounts in ergot-infected grains, Hoffmann extracted it and systematically synthesized derivatives of the molecule, including LSD. Several years later, in 1943, Hoffmann was renewing work with some of these derivatives when he suddenly felt dizzy and intoxicated in a way he had never experienced before. He left work early, bicycled home, and lay down. Several hours of vivid hallucinations followed before he gradually returned to normal.

Hoffmann suspected that he had accidentally absorbed one of the experimental compounds he was handling. To determine if that was the case, he deliberately ingested 250 micrograms of LSD, a dose so tiny that no other drug known at that time would have had a noticeable effect. His plan was to gradually take more of the drug throughout the day until he reached a dose at which the first symptoms appeared. In fact, he had already ingested an amount that was several times the ED_{50} (effective dose 50), or the quantity that produces a specified effect in 50 percent of people taking it. Thus Hoffmann experienced the first deliberately induced acid trip.[3]

LSD remained a curiosity until the early 1950s, when the psychiatric community became interested in the compound as a "psychotomimetic" agent—a drug that mimicked mental illness, especially schizophrenia. The hope was that LSD intoxication and schizophrenia shared a common biochemical basis and that the drug would provide a reversible clinical model for the study and eventual cure of the psychosis. It was also believed that LSD would provide effective therapies for a number of mental illnesses by disrupting entrenched patterns of thought.

Numerous studies of the drug were carried out in academic laboratories, psychiatric hospitals, and prisons, mostly with financial support from the army and the CIA that was funneled through front organizations to conceal the source. Although the studies continued through the 1960s, it gradually became clear that LSD intoxication was not a valid model of schizophrenia and provided no clinical benefit for any mental illness studied. In recent years, however, there has been a resurgence of interest in LSD for treating mental illnesses that involve serotonin pathways, and it is possible that legitimate clinical uses of the drug may yet be discovered.[4]

The Army LSD Program

In May 1955 the U.S. Army officially launched Project M-1605, which sought to develop a psychochemical agent as a military weapon. The requirements were for a chemical that was as potent as sarin nerve agent, produced effects in less than an hour, was stable in storage, and could be disseminated from aircraft under all weather conditions. An absence of long-term effects was considered useful but not essential. About 45 compounds—including mescaline, LSD, THC, and related compounds—were tested in animals. In 1956 the army approved testing by the Chemical Corps of psychochemicals in human subjects.

Over the next two decades until 1975, more than 250 different chemical compounds were tested in over 2,000 experiments involving some 6,700 soldier volunteers and 1,000 civilian volunteers, mostly prisoners. In addition, the experimenters regularly subjected themselves to the agents they were testing. Most of the tests were conducted at Edgewood Arsenal, the Chemical Corps's research and development facility near Aberdeen, Maryland, on the Chesapeake Bay. This facility was already involved in human experimentation because the Chemical Corps had long conducted tests of low doses of standard CW agents, such as mustard gas and nerve agents, on human volunteers.[5]

LSD was one of the most promising candidates for a new incapacitating weapon. Human testing showed that the compound was highly potent and demonstrated its ability to disorganize small military units performing routine tasks. By 1958 the army was sufficiently enthusiastic about the potential of psychochemical agents that it mounted a major public relations campaign to solicit additional funding from Congress, including a movie that showed a cat on LSD cringing in terror before a mouse. Nevertheless army researchers encountered problems when trying to move LSD out of the laboratory and onto the battlefield. The compound was unstable in sunlight, limiting the ability to disseminate it as an aerosol cloud, the standard delivery method for military chemical agents.

LSD was also a highly complex molecule that was costly to produce. Initially the drug was prepared by chemically modifying lysergic acid extracted from ergot and was available only in small quantities from Sandoz. Even after Eli Lilly achieved the complete chemical synthesis of LSD in 1953, the multi-ton quantities needed for a chemical weapon stockpile would have been prohibitively expensive to produce. For these reasons, the army's interest in LSD waned, and it ended the research program in the early 1960s. Instead the Chemical Corps turned its attention to another hallucinogenic agent called quinuclidinyl benzilate (BZ), a plant glycolate related to atropine and scopolamine. BZ was eventually weaponized and stockpiled, although it was never used on the battlefield or even deployed to forward bases.[6]

Despite this setback, the army continued a research program to investigate the utility of LSD as an interrogation aid, similar to what the CIA was doing but on a much smaller scale. As part of this program, ninety-five volunteers were dosed with LSD and subjected to mock interrogations. This effort was followed in 1961 and 1962 by two programs, code-named Third Chance and Derby Hat, which involved the administration of LSD during interrogations in Europe and the Far East. The army interrogation programs, which ended in 1963, involved many of the same legal and ethical issues as the CIA program and are not discussed further in this case study.[7]

The CIA's LSD Program

The U.S. Office of Strategic Services (OSS), the predecessor to the CIA, tested several drugs as aids to interrogation during World War II. A "truth drug" committee studied mescaline, scopolamine, and barbiturates before turning to marijuana in 1943. Human testing was performed on employees of the Manhattan Project (the atomic bomb program), presumably because it was subject to intense secrecy. When the subjects were given cigarettes injected with an extract of marijuana, the results were encouraging: they became talkative and freely disclosed information. The OSS then tested the technique on an unwitting subject, a gangster who was cooperating with the U.S. government to recruit Mafia support for the Allied forces preparing to invade Sicily. Again, this experiment was considered a success, because the subject volunteered sensitive details about the mob's involvement in the drug trade. Additional trials on suspected Communist sympathizers were also considered successful. Ultimately, however, the OSS concluded that the drug treatment worked only on people who were predisposed to talk and not on resistant subjects.[8]

After the creation of the CIA in 1947, there was a renewed interest in enhanced interrogation techniques and the use of drugs to destroy a subject's will or to induce amnesia. These interests were inspired in part by two incidents that occurred in Hungary in 1949: the show trial of Cardinal József Mindszenty, who acted drugged and confessed to absurd charges; and the arrest later that year of Robert Vogeler, an American executive with the International Telephone and Telegraph company who was charged with spying and given unknown drugs during his interrogation and trial. (Although Vogeler was convicted and sentenced to fifteen years in prison, he was released and repatriated after seventeen months.) In 1950, shortly after the outbreak of the Korean War, captured U.S. Air Force pilots began confessing to fictitious activities, such as waging biological warfare. These events convinced the CIA that the Soviet Union and its allies had developed techniques for "mind control" and that the United States had to catch up, both to understand the interrogation methods being used against U.S. soldiers and spies and to employ them against the Communist enemy.[9]

In response to these concerns, in April 1950 the CIA approved a program code-named Bluebird, directed by Morse Allen, a polygraph expert from the agency's Office of Security. The purpose of this effort was to explore various methods of enhanced interrogation, including drugs, electroconvulsive shock treatment, lobotomy, and hypnotism. The drug component of Bluebird involved giving subjects a mixture of sedatives (the barbiturates amytal, seconal, or pentothal) and stimulants (amphetamines, caffeine, atropine, or scopolamine), together with hypnosis and occasionally marijuana, and subjecting them to polygraph interrogation. In July 1950, a CIA team went to Japan for a few months to test these techniques on suspected Communist agents and North Korean prisoners-of-war. Although the results of these studies are unknown, they were apparently not encouraging, because the search for new drugs continued.

In August 1951, Bluebird was renamed Artichoke for security reasons. Beginning in 1952, the CIA sent teams of interrogators to several countries, including West Germany, France, Japan, and South Korea, where they set up safe houses to conduct their activities. At least one safe house was established in Washington, D.C. For several years, the Artichoke teams used the new techniques to interrogate known or suspected double agents and defectors. The results were inconsistent: sometimes the interrogations produced useful information, but often the results were disappointing. The CIA also became concerned about releasing subjects who had been interrogated with Artichoke techniques, for fear that they would talk about their experiences. This concern led to studies of chemical or physical ways to induce amnesia, which ultimately failed. Artichoke also investigated whether drugs, hypnosis, or other techniques could enable the CIA to control a subject's mind and force him to carry out a command, such as to assassinate a specified target. This effort appears to have been unsuccessful, although some have claimed that the CIA had limited success in controlling the minds of a small number of subjects who suffered from preexisting psychiatric conditions, such as multiple personality disorder.[10]

Given this background and the prior use of LSD as a model for schizophrenia, it is not surprising that CIA officials became interested in the drug when they learned of it in the early 1950s. Much of the voluminous experimental work on LSD under project Artichoke was supported by the CIA but carried out in universities, prisons, and mental hospitals. To conceal the source of the funds, the CIA channeled the money through front companies and other government agencies. In some cases, the investigators failed to obtain informed consent and administered LSD to unwitting people, such as adult or pediatric mental patients, but they usually informed the subjects in general terms about the nature of the experiments. Even so, many ethically marginal experiments took place.[11]

Because the CIA planned to administer LSD to unwitting, mentally healthy, resistant individuals, informing the subjects that they were participating in a drug trial would limit the value of the data being gathered. Given that the experiments that the CIA wished to conduct were clearly illegal and unethical, they could not be performed by outside agencies. In 1953, CIA scientists began a series of projects, including one code-named MK-Ultra, in which they administered LSD to unwitting subjects.[12] MK-Ultra had two sister projects. The first, called MK-Naomi, was a joint program with the U.S. Army Chemical Corps's Special Operations Division (SOD) at Fort Detrick, Maryland, to develop delivery devices and tactics for the covert use of chemical and biological agents, including LSD. The second sister project, called MK-Delta, aimed to use MK-Ultra products in field trials overseas, taking over from Artichoke. All three projects were run by Sidney Gottlieb, a Ph.D. chemist who headed the Chemical Division of the CIA's Technical Services Staff.

In April 1953, Director of Central Intelligence Allen Dulles approved MK-Ultra, and Richard Helms, the head of the CIA's Directorate of Operations, gave the project an initial budget of $300,000. Due to its sensitivity, MK-Ultra was exempted from the usual internal financial controls and requirements for written contracts. The initial project team consisted of six Technical Services Staff professionals. At first the experimental subjects were CIA agents who knew that they might be dosed with LSD at any time, but did not know when. These experiments, however, did not provide useful information about the responses of individuals who were unaware of being drugged.[13]

Beginning in May 1953, the CIA began testing LSD and other drugs on completely unwitting subjects. This field program, called MK-Ultra Subproject 3, was run by George White, a narcotics agent who had been seconded on a part-time basis to the CIA from the Federal Bureau of Narcotics. White had previously worked for the Office of Strategic Services and had been part of the truth drug development program during World War II. The testing of LSD on unwitting subjects began at a CIA-rented safe house in New York, but because of concerns that the location was vulnerable to exposure, the program was moved to San Francisco in 1955. The CIA opened a second safe house in Marin County, across the Golden Gate Bridge from San Francisco, and a third back in New York in 1961.

At all three locations, CIA operatives picked up prostitutes, petty criminals, and drug dealers in bars and on the streets and brought them to the safe house or used cooperating prostitutes to lure clients there. Once at the safe house, the unwitting subjects were given drinks spiked with LSD or other drugs, and CIA scientists monitored their reactions by observing them through one-way mirrors. George White also administered LSD to suspects he arrested as a narcotics agent, to serve as an interrogation aid. The logic behind the choice of subjects was that their illicit professions and marginal social status made it unlikely that they would talk or protest afterward, and this assumption proved to be correct.

It is unclear whether the safe-house experiments provided any useful information about drugs and interrogations, but they did yield a great deal of information about the practices of prostitutes and the proclivities of their clients. Although LSD was the focus of much of MK-Ultra's efforts, many other drugs were tested on unwitting subjects in the safe houses, including drugs considered too dangerous for CIA staff to experiment with on themselves. Deaths resulting from such experiments were rumored within the agency, and at least one hospitalization occurred. There were also claims of long-term mental health consequences, although such cases are not well documented.[14]

One incident proved to be highly problematic. In late 1953, near the start of the MK-Ultra project, LSD was administered to a group of CIA agents and members of the Army Chemical Corps's SOD unit who collaborated with the agency on the covert use of chemical and biological agents. They had gathered for a retreat of SOD and MK-Ultra staff at Deep Creek Lake in rural Maryland. One SOD member, Frank Olson, had a bad LSD trip that left him suffering from severe depression, paranoia, and anxiety. About a week later, during a visit to New York City to consult with a CIA psychiatrist, Olson crashed through the tenth floor window of a hotel and fell to his death. Another CIA agent, who was acting as his escort, was the only other person known to have been with him at the time. Although Olson's death was ruled a suicide, suspicions persisted that it was murder. In any event, the CIA covered up the connection of the incident to the LSD program until a congressional investigation in 1975.[15]

After Olson's death, the CIA briefly suspended the testing of LSD on unwitting suspects, but the experiments soon resumed. The testing continued until a 1963 oversight investigation of the Technical Services Staff conducted by the CIA's inspector general uncovered MK-Ultra Subproject 3 and brought it to the attention of the senior CIA leadership. Although the use of unwitting subjects was discontinued, the program remained officially in existence. In 1973, Gottlieb destroyed most of the records of MK-Ultra, MK-Naomi, and MK-Delta, with the permission of the CIA's director at the time, Richard Helms.[16]

Despite the CIA's failure to identify a drug that could serve as a truth serum, a 1957 report suggests that at least six drugs were moved out of the experimental category and into operational use against at least thirty-three targets. The goals of these operations remain unknown, but in some cases the objective may have been to induce symptoms of mental illness so that the subject would be committed to a psychiatric hospital. Some of these efforts were apparently successful. It is not known if the drugs employed for this purpose included LSD, but it is likely.[17]

Use or Misuse?

When the U.S. Army attempted to develop LSD as a battlefield chemical weapon, no legal barriers regulated the development, production, or stockpiling of chemical weapons. Thus the army program did not violate any treaties and was not considered misuse in the context of the time, although the release forms for human experimentation were later judged inadequate by a Senate investigative committee.[18]

In contrast, the CIA's LSD program clearly went beyond the bounds of what was legal or ethical and constituted a gross misuse of pharmacological technology. The trials of Nazi doctors at the Nuremberg War Crimes Tribunal after World War II had firmly established the principle of "informed consent," including the requirements that human subjects be fully informed of the nature of an experiment and its potential risks, and that participation be strictly voluntary and not coerced. Yet the CIA routinely ignored these restrictions, and its abuses grew progressively worse over time. The agency's willingness to violate the Nuremberg Code is a major blot on its history. Further, the use of LSD to augment the interrogation of enemy POWs during the Korean War was a violation of the Geneva Conventions,[19] and the CIA's experiments with LSD on unwitting civilians and enemy defectors were violations of criminal law. Last but not least, the physicians who participated in the LSD program were guilty of gross violations of medical ethics.[20]

The goals of the CIA program included controlling the mind of an unwilling subject, inducing psychosis, and forcing unwilling subjects to reveal information they did not choose to release. The abuses were thus embedded in the program's goals from the outset, rather than being an unintended consequence. Although this account focuses on U.S. efforts to use LSD for hostile purposes, it is likely that several other countries have conducted similar programs, perhaps involving abuses that equaled or exceeded those of the CIA program.

Lessons for Governance

The abuses committed by the CIA in its efforts to develop LSD and other drugs as an interrogation aid, for "mind control," and to induce psychosis, were a product of the intense paranoia of the times and the lack of effective internal and external oversight at the agency. Throughout the programs, the United States and its allies saw themselves as engaged in an existential struggle with the Soviet Union. This perception of acute threat undoubtedly made it easier to condone violations of legal and ethical norms as permissible or necessary.[21] Even in extraordinary times, however, many individuals and organizations behave with integrity, and it is not clear why the CIA programs were so egregiously abusive. In fact, a few individuals within the agency did raise moral or legal concerns. In 1953 a member of the informal advisory committee to Artichoke wrote:

What in God's name are we proposing here? Does it not strike anyone but a few that these projects may be immoral and unethical, and that they may fly in the face of international laws? What really are we attempting to accomplish? Where does respect for life and human dignity come into play?[22]

Clearly the agency had inadequate mechanisms to allow such expressions of concern to reach higher levels of the bureaucracy. Moreover, a formal legal review of the proposed programs never took place, and the CIA's Office of General Counsel learned of them only in the 1970s.[23] This observation begs the question of how such a questionable set of objectives could have been approved in the first place. One explanation is that because the goals of the MK programs were known from the outset to be morally questionable, if not outright illegal, they were from the outset protected from broad discussion.

The highly compartmented nature of the MK programs permitted very little oversight. Evidence suggests that the participants strictly limited the number of people who were read into the program. As a result, few people, even senior CIA officials, knew about the LSD experiments, severely limiting the opportunities for dissent or alternative perspectives. Among those excluded from ongoing knowledge of the MK programs was the CIA's Medical Staff. Historical evidence also suggests that no members of Congress or officials in the Pentagon or the White House knew about the CIA's illegal use of drugs until the Senate hearings of the mid-1970s. Although CIA director Dulles had approved MK-Ultra, it is not clear if his successor, John McCone, was briefed on the existence of the program.[24]

Compounding the secrecy surrounding MK-Ultra was the fact that CIA programs involving the use of chemical and biological agents were granted a waiver from standard accounting practices, such as written contracts and periodic audits. This exemption seriously curtailed the documentary record on which oversight depends. Ironically, the waivers from standard practices meant that the LSD programs, which were among the most sensitive that the agency engaged in, received significantly less oversight than less controversial programs. This policy was clearly intentional, despite being contradictory to the fundamental purpose of oversight, namely, that the most sensitive programs should receive the most thorough scrutiny.[25]

Although the reforms of the 1970s curtailed the ability of U.S. intelligence agencies to act independently of Congress, the oversight process remains problematic because of the delicate balance between secrecy and transparency. Under the current system, the House and Senate leaders and the chairmen and ranking minority members of the House and Senate intelligence committees are briefed on significant covert programs. This approach represents a great improvement over the total lack of institutionalized oversight that characterized the pre-1975 era, but it is still inadequate. An inherent tension exists between the desire to minimize the risk of security breaches that could undermine the effectiveness of highly classified programs and the need to engage a diverse set of individuals and institutions in the oversight process to prevent covert programs from straying beyond acceptable bounds. Unfortunately, during times of perceived crisis or existential hazard—precisely when transgressions are most likely—national security concerns tend to trump accountability and oversight.[26]

The lack of routine independent legal review of all major projects played a critical role in allowing the CIA's LSD programs to avoid challenge. Such a legal review should be part of every agency's approval process. Furthermore, the granting of waivers from formal accounting and audit standards should not be allowed; no institution is well served by blinding itself to its own mistakes. Agency ombudsmen and institutional protections for whistle-blowers also play important roles because they give concerned individuals a place to go with their concerns and shield them from retaliation.

Another element of governance relates to the physicians who participated in the covert CIA programs and, in principle, were governed not only by U.S. law but by the ethics of their profession. Unfortunately, the medical community has been reluctant to investigate and discipline physicians who have participated in illegal or unethical military or intelligence programs involving the administration of psychoactive drugs and enhanced interrogation.[27] Secure reporting mechanisms should be established so that physicians who have misgivings about activities they observe or participate in can report them without risk of retribution.

Conclusions

This historical case study has examined two different efforts to turn LSD into a weapon: first by the U.S. Army to develop LSD as an incapacitating agent for battlefield use, and then by the CIA to use the drug as a vehicle for mind control and enhanced interrogation. The CIA program was clearly illegal and immoral, both in its fundamental goals and in many of its methods. In contrast, the army program was legal at the time, and most of the work adhered to established rules for the conduct of human experiments. Since the entry into force of the Chemical Weapons Convention in 1997, however, any such program would be prohibited under international law.

In contrast to the army program, the CIA's LSD program violated criminal laws and international agreements in force at the time, and any such program today would be subject to the same prohibitions. The combination of the CWC, existing domestic laws, and government regulation of prescription drugs and narcotics provides a fairly robust barrier against the future development of mind-control drugs, pharmaceutical interrogation aids, or drugs that induce psychosis—but only if secret programs receive adequate oversight and governance. Since many countries have little transparency and poor oversight mechanisms, it is possible that several are, or will soon be, developing psychoactive chemicals for covert use. Even in the United States, one of the most transparent and lawful countries in the world, persistent media reports that drugs have been used during the interrogation of detainees at Guantanamo and elsewhere have dogged the CIA.[28]

Perhaps the most important lesson from the CIA's experiments with LSD is that some dual-use threats go beyond arms control and counterterrorism and into the realm of fundamental human rights. Although international legal opinion generally considers the use of drugs for police interrogation to constitute torture, the issue appears unsettled. At least one country (India) admits to using sodium pentothal occasionally during police interrogations.[29] Such developments create a disturbing precedent for the development of other chemical agents that affect the human mind.

Given the rapid increase in the understanding of brain chemistry and the development of drug-delivery systems that are more precise and specific, the potential for misuse is great. Our thoughts, beliefs, emotions, memories, and sanity may be subject to manipulation by pharmacological means, without our permission or even our awareness. This danger is not as remote as it might seem. Accordingly, there is a need for much greater discussion of the ethical issues involved in nonconsensual manipulation of the human mind, and perhaps explicit recognition of a basic right to protection from such assault.[30]

Notes

1. Adrienne Mayor, *Greek Fire, Poison Arrows, and Scorpion Bombs: Biological and Chemical Warfare in the Ancient World* (New York: Overlook Duckworth, 2003).

2. Javier Gonzáles-Maeso, Noelia V. Weisstaub, Mingming Zhou, et al., "Hallucinogens Recruit Specific Cortical 5-HT$_{2A}$ Receptor-Mediated Signaling Pathways to Affect Behavior," *Neuron* 53 (2007): 439–452.

3. John Marks, *The Search for the "Manchurian Candidate": The CIA and Mind Control* (New York: W. W. Norton, 1991), 3–4.

4. D. E. Nichols, "Hallucinogens," *Pharmacology and Therapeutics* 101 (2004): 131–181; John Tierney, "Hallucinogens Have Doctors Tuning In Again," *New York Times*, April 11, 2010, A1; Roland R. Griffiths and Charles S. Grob, "Hallucinogens as Medicine," *Scientific American*, December 2010, 77–79.

5. Martin Furmanski and Malcolm Dando, "Midspectrum Incapacitant Programs," in *Deadly Cultures: Biological Weapons since 1945*, ed. Mark Wheelis, Lajos Rózsa, and Malcolm Dando (Cambridge, MA: Harvard University Press, 2006), 236–251.

6. Ibid.

7. U.S. Senate, *Final Report of the Select Committee to Study Governmental Operations with Respect to Intelligence Activities* (Washington, DC: U.S. Government Printing Office, 1976), 392, 412. Hereafter cited as Church Committee, *Final Report*.

8. Marks, *Search*, 6–9.

9. H. P. Albarelli Jr., *A Terrible Mistake: The Murder of Frank Olson and the CIA's Secret Cold War Experiments* (Walterville, OR: Trine Day, 2009), 187–206.

10. Marks, *Search*, 24–29, 31–47; Albarelli, *Terrible Mistake*, 207–250; Colin A. Ross, *The CIA Doctors: Human Rights Violations by American Psychiatrists* (Richardson, TX: Manitou Communications, 2006).

11. Marks, *Search*, 63–73; Ross, *CIA Doctors*, 81–83.

12. The MK prefix denotes projects run by the CIA's Technical Services Staff, a unit within the clandestine Directorate of Operations that was also responsible for developing new weapons, disguises, and false papers.

13. Marks, *Search*, 59–62.

14. Marks, *Search*, 76–78, 94–109; Albarelli, *Terrible Mistake*, 242–243, 280–281; John Ranelagh, *The Agency: The Rise and Decline of the CIA* (London: Weidenfeld and Nicolson, 1986), 202–216; Church Committee, *Final Report*, 385–422.

15. Albarelli, *Terrible Mistake*, 689–694. Albarelli argues that Olson was drugged not as part of an experiment on unwitting LSD intoxication but to interrogate him with Artichoke methods because he had been talking loosely about CIA/SOD activities.

16. Marks, *Search*, 79–93; Church Committee, *Final Report*, 394–399.

17. Marks, *Search*, 110–111; Church Committee, *Final Report*, 391–392.

18. Church Committee, *Final Report*, 417–418; James S. Ketchum, *Chemical Warfare Secrets Almost Forgotten: A Personal Story of Medical Testing of Army Volunteers with Incapacitating Chemical Agents during the Cold War (1955–1975)* (Santa Rosa, CA: ChemBooks, 2006), 29–34.

19. *Convention Relative to the Treatment of Prisoners of War*, Geneva, July 27, 1929 (entered into force on June 19, 1931); *Convention (III) Relative to the Treatment of Prisoners of War*, Geneva, August 12, 1949 (entered into force on October 21, 1950).

20. Germany (Territory under Allied Occupation, 1945–1955, U.S. Zone), *Trials of War Criminals before the Nuremberg Military Tribunals under Law No. 10, Nuremberg, October 1946–April 1949* (Washington, DC: U.S. Government Printing Office, 1949–1953), 181–182.

21. Marks, *Search*, 29–31.

22. Albarelli, *Terrible Mistake*, 231.

23. Church Committee, *Final Report*, 408.

24. Ibid., 406–407.

25. Ibid., 386, 403–406.

26. John M. Oseth, *Regulating U.S. Intelligence Operations: A Study in Definition of the National Interest* (Lexington: University Press of Kentucky, 1985).

27. Steven H. Miles, *Oath Betrayed: America's Torture Doctors*, 2nd ed. (Berkeley: University of California Press, 2009); Luis Justo, "Doctors, Interrogation, and Torture," *British Medical Journal* 332 (2006): 1462–1463; Joby Warrick and Peter Finn, "Psychologists Helped Guide Interrogations: Extent of Health Professionals' Role at CIA Prisons Draws Fresh Outrage from Ethicists," *Washington Post*, April 18, 2009.

28. J. Meeks, "People the Law Forgot," *Guardian*, December 3, 2003; P. Sleven and J. Stephens, "Detainees' Medical Files Shared: Guantanamo Interrogators' Access Criticized," *Washington Post*, July 10, 2004, A1; Neil A. Lewis, "Man Mistakenly Abducted by CIA Seeks Reinstatement of Suit," *New York Times*, November 29, 2006, A15; Deborah Sontag, "Videotape Offers a Window into a Terror Suspect's Isolation," *New York Times*, December 4, 2006, A1; Raymond Bonner, "Detainee Says He Was Abused While in U.S. Custody," *New York Times*, March 20, 2007; Alissa J. Rubin, "Bomber's Final Messages Exhort Fighters against U.S.," *New York Times*, May 9, 2008; William Glaberson, "Arraigned, 9/11 Defendants Talk of Martyrdom," *New York Times*, June 6, 2008, A1.

29. Jason R. Odeshoo, "Truth or Dare? Terrorism and 'Truth Serum' in the Post-9/11 World," *Stanford Law Review* 57 (2004): 209–255; Jeremy Page, "'Truth Serum' Row in India After Interrogators Fail to Find Killer," *Times* (London), July 16, 2008; Rhys Blakely, "Mumbai Police to Use Truth Serum on 'Baby-Faced' Terrorist Azam Amir Kasab," *Times* (London), December 3, 2008.

30. Mark Wheelis, "'Non-Lethal' Chemical Weapons: A Faustian Bargain," *Issues in Science and Technology* 19, no. 3 (spring 2003): 74–78; Françoise J. Hampson, "International Law and the Regulation of Weapons," 231–260, and William J. Aceves, "Human Rights Law and the Use of Incapacitating Biochemical Weapons," both in *Incapacitating Biochemical Weapons: Promise or Peril?* ed. Alan M. Pearson, Marie Isabelle Chevrier, and Mark Wheelis (New York: Rowman and Littlefield, 2007), 261–284.

IV
Findings and Conclusions

21

Governance of Emerging Dual-Use Technologies

Kirk C. Bansak and Jonathan B. Tucker

Chapter 4 proposed a decision framework for assessing and managing the risks of emerging dual-use technologies in the biological and chemical fields. In an effort to test and refine the decision framework, part II applied it to fourteen case studies of emerging dual-use technologies, which were chosen to be illustrative while capturing the range of variability across several key parameters. This concluding chapter performs a comparative analysis of the contemporary case studies and organizes them into a typology based on their risk of misuse and governability. The typology in turn provides the basis for selecting packages of governance measures that are tailored to each technology.

This chapter also discusses the two historical cases of civil-to-military technology transfer included in part III. Although the historical cases are not amenable to analysis with the decision framework, they provide lessons that complement the findings from the contemporary case studies. In particular, the historical perspective offers insights into the motivations and socioscientific networks that mediate the translation of a technology from a peaceful to a hostile application. Finally, the chapter discusses how the decision framework could be institutionalized and implemented effectively at both the national and international levels.

Analysis of the Contemporary Cases

Tables 21.1 and 21.2 summarize the fourteen contemporary case studies with respect to the parameters used to measure the risk of misuse and governability for each technology, and translate this information into aggregate ordinal scores (high, medium, or low). Although in part II the case studies are grouped into the four functional categories of dual-use technologies identified in the Lemon-Relman report, an analysis of the empirical data reveals that the risk and governability of each technology do not correlate closely with its function. Accordingly, developing suitable governance strategies requires the more tailored analysis provided by the decision framework.

Table 21.1
Assessing the risk of misuse

Case Studies	Risk of misuse based on accessibility	Risk of misuse based on ease of misuse	Risk of misuse based on magnitude of potential harm	Risk of misuse based on imminence of potential misuse	Overall risk of misuse
Combinatorial chemistry and high-throughput screening (HTS)	High—commercially available, within the financial means of a state or a sophisticated terrorist group	Medium—effective use of the technology requires Ph.D.-level expertise to design experimental system and mine the resulting compound library to identify highly toxic agents	High—could facilitate the development of new types of chemical warfare agents by a state program, without requiring the use of rational design	High—immediate, given the current availability of this technology to states	**High**
DNA shuffling and directed evolution	High—readily available either from patented or unpatented methods	Medium—performing the technique requires only basic molecular biology skills; even so, applying the results demands specialized knowledge; misuse also depends on developing screening methodologies to identify harmful organisms	High—technique could permit the development of highly dangerous microorganisms and toxins, without the need to understand the underlying biological mechanisms	Medium—risk of misuse depends on the development of a screening procedure for harmful pathogens, which would not arise from legitimate R&D	**High**

Table 21.1
(continued)

Case Studies	Risk of misuse based on accessibility	Risk of misuse based on ease of misuse	Risk of misuse based on magnitude of potential harm	Risk of misuse based on imminence of potential misuse	Overall risk of misuse
Protein engineering	Medium—depends on R&D infrastructure and expertise	Medium—rational design requires a high level of expertise, but directed evolution is moderately easy	Medium—could lead to development of novel toxin agents	Medium—the imminence of risk resulting from the current maturity of the technology is offset by the need for a high level of expertise	Medium
Synthesis of viral genomes	High—genetic sequences needed for the synthesis of many viruses can be ordered from suppliers over the Internet	Medium—synthesizing sequences greater than 180 base pairs remains somewhat of an art, and technical hurdles to virus construction exist	High—could re-create dangerous pathogens, including extinct or highly controlled viruses	High—technical barriers still exist but are likely to diminish as the technology advances	High
Synthetic biology with standard parts	Medium—the field is still embryonic but may become increasingly accessible due to emergence of do-it-yourself biology and commercially available BioBrick kits	Low—misuse is difficult because of the current absence of weaponizable parts and devices	Medium—over the long term could potentially permit the development of novel and highly dangerous synthetic organisms	Low—long-term, but could be medium-term in the event of unexpected scientific breakthroughs	Low

Table 21.1
(continued)

Case Studies	Risk of misuse based on accessibility	Risk of misuse based on ease of misuse	Risk of misuse based on magnitude of potential harm	Risk of misuse based on imminence of potential misuse	Overall risk of misuse
Development of psychoactive drugs	High—psychoactive drugs are available and employed therapeutically and experimentally; purchasing such drugs is within the means of an individual, group, or state	Medium—Ph.D.-level technical skills are required for drug development	High—could lead to new agents for military use, and possibly terrorist use, coercion, interrogation, or torture	High—near-term	**High**
Synthesis of peptide bioregulators	High—widely accessible on a gram to kilogram scale, and a few suppliers can produce ton quantities; customized peptides can be ordered from suppliers over the Internet or made in automated desktop synthesizers currently on the market	Medium—misuse not easy unless diverted from established medical or other applications; requires knowledge, experience, and technical skills	Medium—could lead to new agents for military use, and possibly terrorist use, coercion, interrogation, or torture	High—short- to medium-term risk given the current interest in "nonlethal" weapons and rapid advances in bioregulator research; if a peptide bioregulator was selected as a chemical warfare agent, production could start on relatively short notice	**Medium**

Table 21.1
(continued)

Case Studies	Risk of misuse based on accessibility	Risk of misuse based on ease of misuse	Risk of misuse based on magnitude of potential harm	Risk of misuse based on imminence of potential misuse	Overall risk of misuse
Immunological modulation	Medium—accessible to those with Ph.D.-level skills; dual-use knowledge produced by individual studies can be found in open literature	Low—moderate to difficult due to the requirement for tacit knowledge in a variety of fields, and the need for animal and clinical testing	Medium—potential for both covert and mass operations; knowledge produced could identify new vulnerabilities	Medium—medium-term	**Medium**
Personal genomics	Low—while DNA sequencing technology is becoming increasingly accessible, the aggregate data and scientific knowledge that would be necessary for misuse are not yet available	Low—the capacity for weaponization does not yet exist, nor do the aggregated databases that could be exploited for misuse (e.g., development of ethnic weapons)	Medium—if genetically targetable agents could be developed, consequences could be high, but this is still theoretical	Low—large genomics databases do not yet exist but will in the next five years; scientific knowledge relevant for weaponization will take much longer	Low

Table 21.1
(continued)

Case Studies	Risk of misuse based on accessibility	Risk of misuse based on ease of misuse	Risk of misuse based on magnitude of potential harm	Risk of misuse based on imminence of potential misuse	Overall risk of misuse
RNA interference	Medium—most molecular biology research labs can perform the technique, but it requires a substantial investment of resources	Low—specific applications of the technique require tacit knowledge from a variety of fields	High—by combining RNAi with viral pathogens, it might be possible to develop highly infectious and virulent infectious agents and even ethnically targetable weapons	Low—long-term because many years of R&D would be required to create and optimize an RNAi-based weapon	**Medium**
Transcranial magnetic stimulation (TMS)	High—TMS devices can be purchased on the commercial market and are within the means of an individual, group, or state	Medium—moderate deskilling, with substantial risk to subjects if operator is poorly trained	Low—no identifiable risk for mass casualties; for individual subjects, there is a risk of seizures and disruption of neural networks	Low—the technology is available but has no potential utility for causing mass casualties, a fact that should be given particular weight.	Low

Table 21.1
(continued)

Case Studies	Risk of misuse based on accessibility	Risk of misuse based on ease of misuse	Risk of misuse based on magnitude of potential harm	Risk of misuse based on imminence of potential misuse	Overall risk of misuse
Chemical micro process devices	High—commercially available; within the financial means of a state or a sophisticated terrorist group	Medium—deskilled for proven applications, but explicit and tacit knowledge are needed to adapt the process of chemical synthesis to micro devices	Medium—could facilitate the large-scale production of chemical warfare agents in small, easily concealed production plants; however, necessary support facilities would be harder to hide	High—immediate, given the commercial availability of the hardware and software	**High**
Gene therapy	Medium—viral-vector technology is accessible, especially to scientists in industrialized nations	Low—the field requires substantial tacit knowledge; until viral-vector technology is perfected, it would be difficult to introduce genes into populations without being detected	Medium—potentially high if "stealth" viral-vector technology can be created	Low—long-term risk of misuse of viral-vector technology	Low
Aerosol vaccines	Medium—classic aerosol vaccines are described extensively in the scientific literature	Low—requires technical skills and knowledge of aerosol behavior in the environment and respiratory systems, knowledge of formulations, and favorable atmospheric conditions	Low—potential for mass casualties but no particular advantage over existing delivery systems for biowarfare agents	Low—long-term	Low

Table 21.2
Assessing governability

Case Studies	Susceptibility to governance based on embodiment	Susceptibility to governance based on maturity	Susceptibility to governance based on convergence	Susceptibility to governance based on rate of advance	Susceptibility to governance based on international diffusion	Overall governability
Combinatorial chemistry and high-throughput screening (HTS)	Medium (*hybrid*)—hybrid of hardware and integrated software	Medium (*very mature*)—commercially available	Medium (*moderate convergence*)—miniaturization, laboratory robotics, drug screening	High (*low rate of advance*)—was exponential in 1988–1992 but has since plateaued	Medium (*moderate international diffusion*)—major vendors in U.S.A., Europe, and Japan	Medium
DNA shuffling and directed evolution	Low (*intangible*)—a technique that is an extension of existing biotechnology	Medium (*very mature*)—widely used in research and industry for protein engineering; not available for sale but can be created by those seeking its benefits	High (*low convergence*)—not convergent; squarely within mainstream molecular biology	Low (*high rate of advance*)—seminal advances were made in 1990s, but the technique is now being applied to new situations	Low (*high international diffusion*)—technology is largely global, coincident with modern molecular biological capabilities	Low
Protein engineering	Medium (*hybrid*)—mostly a technique, but with some proprietary software	High (*moderately mature*)—advanced development of technique; some fusion toxins are commercially available	Medium (*moderate convergence*)—chemical synthesis of DNA, protein biochemistry, molecular modeling software	Medium (*moderate rate of advance*)—was rapid from mid-1980s to mid-1990s but is now incremental	Low (*high international diffusion*)—largely global, coincident with modern molecular biological capabilities	Medium

Table 21.2
(continued)

Case Studies	Susceptibility to governance based on embodiment	Susceptibility to governance based on maturity	Susceptibility to governance based on convergence	Susceptibility to governance based on rate of advance	Susceptibility to governance based on international diffusion	Overall governability
Synthesis of viral genomes	Medium (*hybrid*)—hybrid of DNA synthesizers plus advanced know-how	Medium (*very mature*)—commercially available from private suppliers, many of which screen customers and orders	Medium (*moderate convergence*)—integrates biotech, engineering, nanotech, bioinformatics	Low (*high rate of advance*)—rapid advance in speed, accuracy, and cost	Medium (*moderate international diffusion*)—approximately 50 companies worldwide produce gene-length sequences	Medium
Synthetic biology with standard parts	Medium (*hybrid*)—materials based, with vital hardware (DNA synthesizers) and bioinformatics	Low (*not mature*)—still in development, now emerging from proof-of-principle phase	Low (*high convergence*)—chemical DNA synthesis, molecular biology, engineering, bioinformatics	Medium (*moderate rate of advance*)—rapidly increasing number of parts, but each application still requires extensive R&D and refinement	Medium (*moderate international diffusion*)—increasing numbers of iGEM teams from a growing number of countries around the world	Medium

Table 21.2
(continued)

Case Studies	Susceptibility to governance based on embodiment	Susceptibility to governance based on maturity	Susceptibility to governance based on convergence	Susceptibility to governance based on rate of advance	Susceptibility to governance based on international diffusion	Overall governability
Development of psychoactive drugs	Low (*intangible*)—an intangible technology with some role for brain imaging machines	Medium (*very mature*)—many psychoactive drugs are commercially available	Medium (*moderate convergence*)—neuroscience, neuro-imaging, pharmacology, and molecular genetics	Low (*high rate of advance*)—with continuing discovery of new neurotransmitter/receptor systems in the brain	Medium (*moderate international diffusion*)—accessible to any country with a pharmaceutical R&D capability	**Medium**
Synthesis of peptide bioregulators	Medium (*hybrid*)—applications are based largely on intangible knowledge, but large-scale production of peptides requires specialized automated synthesizers	Medium (*very mature*)—peptide synthesis technology is commercially available	Medium (*moderate convergence*)—brain research, receptor research, systems biology, automated peptide synthesis	Low (*high rate of advance*)—rapid growth of pharmaceutical R&D	Medium (*moderate international diffusion*)—available in Canada, China, Europe, India, Japan, South Korea, U.S.A.	**Medium**

Table 21.2
(continued)

Case Studies	Susceptibility to governance based on embodiment	Susceptibility to governance based on maturity	Susceptibility to governance based on convergence	Susceptibility to governance based on rate of advance	Susceptibility to governance based on international diffusion	Overall governability
Immunological modulation	Low (*intangible*)—involves a confluence of science and technology	High (*moderately mature*)—advanced development, some commercial applications available	Low (*high convergence*)—general biotechnology and immunology; animal and human testing; delivery technologies	Low (*high rate of advance*)—a range of rates depending on methods, but generally rapid	Medium (*moderate international diffusion*)—accessible to all countries with pharmaceutical R&D capability	Low
Personal genomics	Low (*intangible*)—based on information technology, but databases might be subject to regulation	Medium (*very mature*)—personal genomics services and DNA sequencing technology are commercially available	Medium (*moderate convergence*)—DNA sequencing, systems biology, bioinformatics, epidemiology	Low (*high rate of advance*)—field advancing rapidly since 2007; was slow before then	Medium (*moderate international diffusion*)—most personal genomics services based in North America and Europe but extend services to other parts of the world	Medium

Table 21.2
(continued)

Case Studies	Susceptibility to governance based on embodiment	Susceptibility to governance based on maturity	Susceptibility to governance based on convergence	Susceptibility to governance based on rate of advance	Susceptibility to governance based on international diffusion	Overall governability
RNA interference	Low (*intangible*)—an intangible technique that relies on a natural phenomenon	Medium (*very mature*)—reagents are commercially available; drug applications are in clinical trials	Medium (*moderate convergence*)—human genomics, small-molecule biochemistry, genetic engineering, RNA synthesis	Low (*high rate of advance*)—but each application still requires extensive R&D and refinement	Low (*high international diffusion*)—research is taking place in industrial and developing countries; a large number of companies provide reagents and research services	Low
Transcranial magnetic stimulation (TMS)	High (*hardware*)—mainly hardware with some need for functional know-how	Medium (*very mature*)—advanced development for various applications; FDA approved for major depression	High (*low convergence*)—limited extent of convergence, though it can be used in conjunction with functional MRI and other brain imaging technologies	Medium (*moderate rate of advance*)—incremental, but growing interest in clinical use of TMS among neurologists and psychiatrists	High (*low international diffusion*)—at least nine U.S. and European companies offer versions of TMS	High

Table 21.2
(continued)

Case Studies	Susceptibility to governance based on embodiment	Susceptibility to governance based on maturity	Susceptibility to governance based on convergence	Susceptibility to governance based on rate of advance	Susceptibility to governance based on international diffusion	Overall governability
Chemical micro process devices	High (*hardware*)—primarily hardware, with a need for specialized computer software	High (*moderately mature*)—commercially available but with a limited number of manufacturers	Medium (*moderate convergence*)—advanced machining, specialized materials, theory of nanoreactions	Medium (*moderate rate of advance*)—increasing number of applications and types of devices	High (*low international diffusion*)—about 20 vendors in Europe, U.S.A., Japan, and China	High
Gene therapy	Low (*intangible*)—a technique with some wetware	High (*moderately mature*)—still in clinical testing	Medium (*moderate convergence*)—recombinant DNA, virology, immunology	High (*low rate of advance*)—incremental, with some major setbacks	Medium (*moderate international diffusion*)—29 countries are conducting clinical trials (vast majority in North America and Europe)	Medium
Aerosol vaccines	Medium (*hybrid*)—based predominantly on know-how (intangible) but requires access to aerosol generators	High (*moderately mature*)—a few veterinary vaccines commercially available, plus one new human vaccine (FluMist); past government-sponsored development	Medium (*moderate convergence*)—limited extent of convergence (vaccinology, aerobiology, formulation techniques)	Medium (*moderate rate of advance*)—rapid from 1970 to 1980; the field plateaued until the early 2000s but has increased slowly since then	High (*low international diffusion*)—limited to a few advanced industrial countries	Medium

Another important aspect of the decision framework is that it enables policy makers to identify which emerging technologies warrant the development of governance measures on a priority basis. The first two steps in the framework, assessing the risk of misuse and assessing governability, provide a simple screening mechanism. Only if a given technology has at least a medium risk of misuse *and* at least a medium level of governability is it advisable to develop a tailored package of governance measures and evaluate their associated costs and benefits. Because all emerging technologies are assumed to offer at least some potential utility, whenever the risk of misuse is low, there is no need to conduct a cost-benefit analysis.

As was explained in chapter 4, the risk of misuse and governability of an emerging technology are assessed through four and five parameters, respectively. In the case of chemical micro process devices, for example, the parameters used to assess the risk of misuse were scored as follows: The accessibility of the technology is high because microreactors are commercially available and within the financial means of both state and nonstate actors, although they are more expensive than standard batch reactors. The ease of misuse of the technology is medium because microreactors are still a niche market and would require both explicit and tacit knowledge to adapt them to the production of chemical warfare (CW) agents. Moreover, operating a microreactor plant would require at least as much skill as a conventional chemical plant. The magnitude of potential harm from the misuse of the technology is medium because although microreactors could facilitate scale-up and permit the production of toxic chemicals in small, easily concealed facilities, significant quantities of precursor chemicals and end product would still have to be stored on-site, necessitating warehouses and materials-handling facilities that would be harder to conceal than the production plant itself. Finally, the imminence of potential misuse of the technology is high because the current generation of microreactors could be used to manufacture highly toxic chemicals and potentially adapted for the production of blister and nerve agents. Based on the four parameters, the overall risk of misuse associated with chemical micro process devices is judged to be high.

Typology of Dual-Use Technologies

The three-by-three matrix in table 21.3 provides a typology of the fourteen emerging dual-use technologies examined in this book, based on the two key variables of risk of misuse and governability. The six technologies clustered in the shaded area in the upper right-hand corner of the matrix have a high or medium risk of misuse *and* a high or medium level of governability. As a result, these technologies warrant the development of specific governance strategies to manage their associated security risks.

Only one technology, chemical micro process devices, ranks high in both risk of misuse and governability, suggesting that it should receive the highest priority. The

Table 21.3
Typology of the contemporary case studies

		Governability		
		Low	Medium	High
Risk of misuse	High	DNA shuffling and directed evolution	Synthesis of viral genomes Combinatorial chemistry Development of psychoactive drugs	Chemical micro process devices
	Medium	Immunological modulation RNA interference	Protein engineering Synthesis of peptide bioregulators	
	Low		Synthetic biology with standard parts Personal genomics Gene therapy Aerosol vaccines	Transcranial magnetic stimulation

five other technologies in the shaded portion of the matrix—synthesis of viral genomes, combinatorial chemistry, development of psychoactive drugs, protein engineering, and synthesis of peptide bioregulators—have a combination of high and medium scores and should therefore be next in line for the development of governance measures.

The sole technology in the upper left-hand corner of the matrix, DNA shuffling and directed evolution, has a high risk of misuse but a low governability score because it is based on intangible knowledge, which is hard to control, and uses materials, equipment, and know-how that are widely available in laboratories around the world. Similarly, immunological modulation and RNA interference have a medium risk of misuse and a low governability score. Although all three of these technologies pose a significant risk of misuse, that they are based on widely available intangible information and are advancing rapidly means that they are not amenable to formal regulation or even soft-law measures. Even so, these technologies should be monitored carefully and, to the extent possible, subjected to informal governance measures, such as awareness-building campaigns and professional codes of conduct.

The technology in the lower right-hand corner of the matrix, transcranial magnetic stimulation (TMS), poses a low risk of misuse because it can be applied to only one individual at a time rather than to a large group or population. For this reason, the potential misuse of TMS to support coercive interrogation or to erase memories is more of a human-rights concern than a potential threat to national security. At the same time, TMS has a high governability score because its clinical use is heavily regulated on safety and ethical grounds. That TMS consists primarily of hardware also makes it more susceptible to governance. Because the risk of misuse of TMS is low, the technology does not warrant the development of governance strategies.

The same reasoning applies to synthetic biology with standard parts, personal genomics, gene therapy, and aerosol vaccines, all of which have a low risk of misuse and a medium level of governability. Because these technologies are unlikely to pose a security risk anytime soon, policy makers should focus their limited resources on the governance of other, higher-risk technologies. Even so, it would be prudent to continue monitoring the development of these technologies in case unexpected technological developments cause them to pose a greater risk of deliberate misuse in the future.

The current assessment that synthetic biology with standard parts poses a low risk of misuse may be surprising in view of the large amount of public attention and concern that have been devoted to the technology. There are two reasons for this assessment. First, synthetic biology with standard parts is distinct from the closely related technology of genome synthesis (or "synthetic genomics"), which poses a higher risk of misuse because it can be used to construct pathogenic viruses from scratch in the laboratory. Second, whereas genome synthesis is a mature technology, parts-based synthetic biology is still embryonic. Given the limited number of standard biological parts that are known to function reliably, as well as the need for a high level of expertise and tacit knowledge to assemble such parts into functional genetic circuits and modules, it would be extremely difficult for a nonexpert to exploit this technology for harmful purposes.

The governability of parts-based synthetic biology has a medium score because many of its practitioners have embraced an open-access approach, in which all bona fide researchers are granted unrestricted access to the Registry of Standard Biological Parts. Nevertheless it would be fairly easy to increase control over this technology by restricting access to the registry through some type of personal vetting process.

It is also important to note that because the related fields of genome synthesis and parts-based synthetic biology are both evolving rapidly, the current assessments of risk and governability must be considered snapshots in time that are likely to change in response to further technological advances, as well as social innovations such as the rise of the amateur-science groups like Do-It-Yourself Biology (DIYbio).

Accordingly, both genome synthesis and parts-based synthetic biology should be reassessed periodically as their development continues. Over the next decade, the design and synthesis of artificial genomes may become increasingly deskilled through the availability of reagent kits and how-to manuals. In addition, if the deliberate deskilling agenda of parts-based synthetic biology bears fruit, the risk of misuse of this technology could increase. At the same time, if the open-access approach to parts-based synthetic biology is extended beyond recognized research laboratories to include amateur and unaffiliated scientists, the governability of the technology would decline.

Tailored Governance Measures

Depending on where the fourteen contemporary emerging technologies fall within the risk/governability typology, the decision framework suggests different packages of governance measures, which may include hard-law, soft-law, and informal elements. (A simplified version of the framework is provided in figure 21.1.)

The packages of governance measures derived from the use of the decision framework must be subjected to a cost-benefit analysis so that policy makers can determine which combination of measures is optimal. In general, the greater the governability of a technology, the more likely it is that formal regulations will be cost-effective. It is also desirable to pursue synergies among the different modes of governance. For example, formal regulations and informal guidelines should be consistent with any existing safety and security practices that have evolved spontaneously.

Although the decision framework focuses primarily on assessing the security risks of emerging technologies and devising appropriate measures for managing those risks, non-security-related governance measures should be considered whenever they can reduce the potential for deliberate misuse. For example, because gene therapy is already regulated extensively on health, safety, and ethical grounds, the existing governance structure could be modified fairly easily to address security concerns, were they to arise in the future. With respect to personal genomics, privacy-oriented governance measures, such as the Genetic Information Nondiscrimination Act (GINA) passed by the U.S. Congress in 2008, might be amended to prevent the deliberate exploitation of personal genetic data for the development of ethnically targeted weapons. Because the risk of misuse of both gene therapy and personal genomics is currently assessed to be low, however, there is no urgency in taking these steps.

Categories of Dual-Use Risk

The introduction to this book discussed four general scenarios of potential misuse of emerging dual-use technologies in the biological and chemical fields. In the light of the case studies, these scenarios can now be revisited and classified into distinct categories of risk.

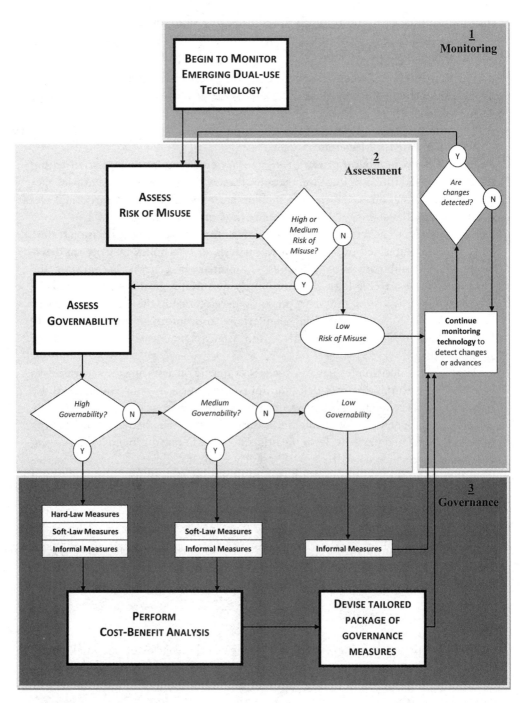

Figure 21.1
Decision framework for technology governance.

Accessibility Risk   Emerging dual-use technologies may facilitate or accelerate the production of standard biological or chemical warfare agents, thereby increasing their accessibility to state or nonstate actors seeking to cause harm. Examples of this type of risk include the application of genome synthesis to construct dangerous viruses from scratch and the use of chemical microreactors to produce standard chemical warfare agents, such as sarin nerve agent, with fewer safety and scale-up hurdles than is the case with batch reactors.

Novel Agent Risk   Emerging dual-use technologies could help to identify or develop novel biological or chemical warfare agents that are more effective for either lethal or incapacitating purposes than existing agents. This type of risk could arise from the use of combinatorial chemistry and high-throughput screening, DNA shuffling and directed evolution, or protein engineering. A small risk also exists that RNA interference and personal genomics might be used to develop so-called ethnic weapons that could be targeted selectively against particular ethnic or racial population groups.

Knowledge Risk   The growing understanding of human biology at the molecular and systems levels could reveal new vulnerabilities that might be exploited for offensive purposes. For example, the study of immunological modulation could provide a roadmap for developing engineered biological warfare agents that impede the immune response or are resistant to vaccines and other standard medical defenses.

Normative Risk   Emerging dual-use technologies may lead to harmful applications that undermine international legal norms. For example, the discovery of brain neuropeptides that affect cognition and emotion could lead to the development of a new generation of biochemical agents that are capable of inducing specific emotional states, such as fear, confusion, or euphoria, and can be delivered in aerosol form. Such agents might be used by police or paramilitary forces for riot-control and counterterrorism operations, or by authoritarian or totalitarian governments to repress domestic dissent. Such applications, were they to materialize, would seriously undermine the international legal norm against biological and chemical warfare, as well as international human rights law.

Analysis of the Historical Cases

Whereas the fourteen contemporary case studies in part II are all concerned with potential misuse by third parties, including outlaw states or nonstate actors such as terrorists, the historical cases in part III examine activities by the same government that sponsored the technology development. In the V-agents case, the British government decided during the 1950s to develop the pesticide Amiton into a chemical weapon, thereby rendering moot any national governance structure that might have prevented such an action. At that time, the development and production of chemical weapons, but not their use in war, were still permitted under international law. In the LSD case, a rogue element within the CIA pursued clearly illegal activities in an unaccountable manner. The two historical cases differ sharply from the contemporary cases, where the good faith of responsible governments is generally assumed.

Despite these differences, the historical case studies complement the analysis of the contemporary cases and help to inform the decision framework by permitting a nuanced, retrospective examination of the role of intent on the part of various actors and the importance of socioscientific networks. Drawing on the theory of the social construction of technology, which was briefly discussed in chapter 2, it is possible to analyze the two historical case studies through this lens. In both situations, a particular institution reinterpreted a civilian artifact or technique for a military purpose, promoted this interpretation, and obtained the consent of the other interested parties.

In the case of the V-agents, Porton Down took a failed pesticide (Amiton) and reinterpreted it as a chemical weapon; in the LSD case, the army and the CIA took a drug that was being used to study schizophrenia and reinterpreted it, respectively, as an incapacitating battlefield weapon and as a tool for covert intelligence operations. By tracing the socioscientific networks through which Amiton and LSD were each recast from a peaceful to a hostile purpose, the historical cases illuminate the social construction of technology and provide insights into the structural factors that facilitate misuse.

The V-Agents Case

The history of the V-series nerve agents suggests that the transformation of Amiton from a failed pesticide into a new class of chemical weapon was not inevitable. Contrary to determinist theories of technology development, Caitríona McLeish and Brian Balmer argue that the military application of Amiton did not arise automatically from the inherent properties of the compound—its high toxicity to humans and ability to penetrate the skin—but instead required the active intervention of defense officials. Accordingly, the V-agents case study focuses on the central role played in civil-to-military technology transfer by networks of governmental and private actors.

After World War II, officials at the British chemical warfare (CW) establishment at Porton Down decided to recruit the chemical industry to help identify new toxic compounds as candidate CW agents. Porton officials sent letters to private chemical companies, both directly and through the British chemical trade association, urging them to submit information on compounds highly toxic to mammals that they discovered inadvertently in the course of developing commercial drugs, dyes, and pesticides. In an attempt to legitimate this solicitation in the industry's eyes, Porton framed it disingenuously in "defensive" terms. A further effort by Porton to gain the cooperation of the British chemical industry involved the development of a commercial classification system for new chemicals that was designed to safeguard company trade secrets.

Despite these efforts, however, many chemical firms did not respond to the British government's outreach efforts, mainly because they lacked a financial incentive to study compounds that were too toxic for commercial use—and perhaps, in some cases, for ethical reasons as well. As a result, Porton's outreach to the chemical industry failed to generate the expected flood of research leads. However, the outreach effort did identify one product of military value: a commercial pesticide called Amiton, which had been developed and marketed by Imperial Chemical Industries but had proved too toxic for agricultural use. Within a matter of months, Porton recast the failed pesticide as VG, the first of a new class of chemical nerve agents that provided high potency, rapid action, stability, and persistence. (The "V" code reportedly stood for the word "venomous" because of the extreme lethality of these agents and their ability to penetrate the skin.)

The proactive role of British defense officials in turning a failed pesticide into a chemical weapon runs counter to the traditional determinist view of dual use as an inherent characteristic of a technology, resulting inevitably in its application for hostile purposes. Instead the transfer of Amiton from the civil to the military sector required the active intervention of a socioscientific network made up of government and private-sector actors, which served to "translate" a compound designed for peaceful purposes into a weapon. At the same time, the socioscientific network was characterized by strong internal tensions and required a concerted effort by Porton officials to keep it intact.

One can draw two important lessons from the V-agents case. First, intent is crucial to whether or not a particular technology will be misused. Second, civil-to-military technology transfer must contend with the social dynamics of the environment in which it occurs. Because technology translation takes place within the context of a sociotechnical network, the norms, customs, and interests of the participants (such as the market-based incentives of private companies and the ethical standards of civilian researchers) must be manipulated or suppressed for civil-to-military technology transfer to occur. In the V-agents case, the military framing of the technology ran counter to the financial interests—and in some cases the ethical norms—of the British chemical industry. Accordingly, this conflict had to be overcome through careful marketing and, at times, outright deception about the state's intended goals.

The LSD Case

The second historical case study chronicles the efforts by the U.S. Army and the CIA during the 1950s and 1960s to reinterpret a different civilian technology—the experimental drug LSD—as a tool for military and covert operations. The LSD case underlines the contextual nature of misuse by making a clear distinction between the army's attempted development of the drug as an incapacitating chemical weapon and the CIA's efforts to develop LSD as a tool for mind control, covert operations, and coercive interrogation. Because the army research program did not violate any treaties to which the United States was a party at the time, the development of LSD as a chemical incapacitant did not constitute "misuse" under international law. In contrast, the CIA's experimentation with LSD on unwitting human subjects during Project MK-Ultra violated U.S. criminal law and the Nuremberg Code, a set of ethical principles for human-subjects research that had been adopted in 1947 in response to the abusive medical experiments on prisoners and concentration-camp inmates conducted by Nazi doctors during World War II.

In seeking to develop LSD as a tool for mind control, the CIA pressured medical personnel to participate in unethical experiments. Whenever the goals of Project MK-Ultra came in conflict with the Hippocratic oath and other norms that guide the medical profession, agency officials sought to overcome these ethical barriers through appeals to patriotism and claims of Communist brainwashing. Moreover, although several physicians involved in the experiments developed serious moral qualms about their work, they had no recourse because of the intense secrecy shrouding the program and the lack of channels for principled dissent or whistle-blowing.

According to Mark Wheelis, the CIA's ethical abuses during MK-Ultra were made possible by a lack of organizational checks and balances, which empowered a "rogue element" within the CIA that was accountable to no one. Wheelis offers four explanations for this failure of governance: (1) the U.S. government's intense preoccupation with the Soviet military and ideological threat at the height of the Cold War eroded moral barriers; (2) no formal or informal oversight mechanisms were in place to monitor the activities of the CIA's clandestine service; (3) the program of human experimentation was tightly compartmented within the CIA in a deliberate effort to use secrecy to circumvent internal controls; and (4) the professional medical societies were reluctant to investigate or discipline their members who participated in unethical government activities.

Despite the apparent conflict between secrecy and governance, Wheelis contends that it is possible to have effective oversight even in a highly classified environment through measures such as independent legal analysis, ombudsmen, whistle-blower protections, and Institutional Review Board (IRB) review of human-subjects research. Such mechanisms can ensure respect for ethical norms in a way that is fully consistent with national security. Unfortunately governments have sometimes reinterpreted their own rules to allow such internal oversight systems to fail.

Wheelis also argues that a more equitable balance of power between CIA program managers and the medical professionals they supervised would have constrained the agency's ability to reinterpret LSD as an instrument of mind control and impose this interpretation on the medical staff who were ordered to conduct unethical human experiments. Because of this power imbalance, the ethical norms that imbue the practice of medicine generated resistance to—but were not sufficient to prevent—the translation of LSD into a chemical weapon.

Policy Implications of the Historical Cases

The two historical cases suggest that the contemporary policy discourse on emerging dual-use technologies lacks some important dimensions. First, both cases demonstrate that technological artifacts do not pose an inherent or inevitable risk of misuse. Instead the mediating influence of social processes is required for a technology to be misapplied for hostile purposes. As policy makers grapple with the difficult task of managing the dual-use risks of emerging technologies, they should not overlook the structural relationships between the civilian and military sectors or disregard the importance of socioscientific networks and ethical norms in influencing the behavior of physicians, psychiatrists, and scientists. In principle, such norms could be built on and reinforced through risk education and the active engagement of professional societies, providing a valuable complement to formal oversight mechanisms.

Second, the two historical case studies suggest that the socioscientific networks involved in civil-to-military technology transfer may be difficult to maintain and are potentially vulnerable to disruption. Accordingly, it may be possible to reduce the risk of misuse of an emerging technology by shaping the structural features of the socioscientific network in ways that promote a peaceful interpretation of the technology. For example, McLeish and Balmer call for fully implementing the purpose-based definition of a chemical weapon in the 1993 Chemical Weapons Convention (CWC). This treaty bans the development, production, stockpiling, and use for prohibited purposes of *all* toxic chemicals, including those that have yet to be invented, even though the treaty's routine verification regime is limited to lists of specific chemicals and their precursors that have been developed in the past as chemical weapons. Along similar lines, Wheelis proposes to regulate the socioscientific networks within the U.S. government by establishing stronger mechanisms for institutional review, creating safe reporting channels for whistle-blowers, and closing legal loopholes in the CWC that could be interpreted to allow the use of chemical incapacitating agents for paramilitary operations.

Third, although current U.S. policy focuses almost exclusively on the threat that nonstate actors could exploit advanced biotechnologies for hostile purposes, the risk that governments will misuse these capabilities, against either other countries or their own citizens, remains a serious concern. Historically states have been far more likely than nonstate actors to funnel emerging technologies into weapons development because they possess superior financial and technical resources. In addition to Project MK-Ultra, there are more recent cases in which rogue elements within a country's intelligence service or defense establishment have appropriated dual-use technologies for their own use, without the authorization of the government as a whole. In apartheid South Africa, for example, the South African Defense Force launched a secret chemical and biological warfare program code-named Project Coast, which developed toxic agents for the assassination of opposition leaders and tried unsuccessfully to invent ethnic weapons that could selectively target nonwhites.[1]

Fourth, the LSD case suggests that the potential misuse of emerging technologies goes beyond military or terrorist applications to include violations of human rights and international humanitarian law. Several governments are known to have used potent drugs to suppress the expression of opposing political views, including the Soviet and Chinese misuse of psychiatric medications to control dissidents[2] and Israel's use of the synthetic opiate fentanyl for the attempted assassination of Hamas leader Khaled Meshal.[3] Preventing a government from abusing its own citizens requires a high degree of transparency and accountability.

Implementation of the Decision Framework

In seeking to manage the risks of emerging dual-use technologies, it is not sufficient to use the decision framework to generate proposed sets of governance measures; effective implementation mechanisms must also be created. In general, implementation of the framework should start at the national level and, if the technology in question has spread to other countries, should include efforts at international harmonization. The following sections discuss possible approaches for implementing the decision framework at both the national and international levels.

Domestic Implementation

In seeking ways to manage the risks of emerging dual-use technologies, the contemporary case studies make various proposals for modifying existing governance structures and developing new measures when appropriate. Before implementing these strategies, however, it is necessary to determine who would be responsible for carrying them out. Are existing government agencies up to the task, or do the unique complexities of this area warrant the creation of a new, tailor-made institution?

Some analysts contend that existing U.S. government bodies are not adequate for regulating emerging technologies. The political economist Francis Fukuyama, for example, has proposed the creation of a new agency to manage the policy challenges associated with human biotechnologies, including reproductive cloning, preimplantation genetic diagnosis, and psychopharmacology.[4] Using a historical analogy, he contends that relying on existing institutions like the Food and Drug Administration and the National Institutes of Health to regulate the rapidly changing biotechnology landscape is akin to "trying to use the Interstate Commerce Commission, which was responsible for regulating trucks, to oversee civil aviation when that industry came into being, rather than creating a separate Federal Aviation Administration."[5]

Other scholars believe that addressing the challenges posed by emerging dual-use technologies does not require the founding of new institutions. The U.S. Presidential Commission for the Study of Bioethical Issues took this position with respect to synthetic biology. Responding to the announcement in May 2010 that scientists at the J. Craig Venter Institute had created the first self-replicating bacterial cell controlled by a synthetic genome, president Barack Obama asked the Bioethics Commission to examine the implications of this scientific milestone and to "develop recommendations about any actions the Federal government should take to ensure that America reaps the benefits of this developing field of science while identifying appropriate ethical boundaries and minimizing identified risks."[6] In its report, the commission recommended against creating new agencies, offices, or authorities to regulate synthetic biology and instead proposed leveraging the regulatory oversight provided by existing federal agencies with respect to various applications of synthetic biology.[7]

Supporting the position taken by the Bioethics Commission is the fact that emerging biotechnologies such as synthetic biology are not fundamentally new. Whereas civil aviation was a truly revolutionary mode of transport in 1926, when the Air Commerce Act created the Aeronautics Branch of the Department of Commerce (a predecessor of the Federal Aviation Administration), synthetic biology and the other dual-use technologies surveyed in this book have generally evolved from enabling technologies, such as automated DNA synthesis, or have emerged from the convergence of two or more preexisting disciplines. Nevertheless, whereas the Bioethics Commission focused narrowly on synthetic biology, the decision framework aims to address a wide variety of emerging dual-use technologies with different characteristics and may therefore warrant its own implementing body. Such an institution would serve two tasks: managing the risks of emerging dual-use technologies writ large and developing tailored governance strategies for individual technologies.

Penrose C. "Parney" Albright, who served as assistant secretary for science and technology in the Department of Homeland Security during the George W. Bush administration, has made a number of recommendations with respect to the U.S. biodefense program that are relevant to the national implementation of the decision framework. Albright points out that although several U.S. government departments and agencies perform biodefense work, no single entity oversees the entire effort, nor is biodefense the top priority of any one agency. In his view, this decentralized approach has resulted in numerous inefficiencies and flawed policies because of parochial agency interests and competition over roles and missions. As a corrective, Albright recommends creating a central coordinating body for biodefense, based in the White House, whose authority would supersede that of the various cabinet departments, so that the new body could adjudicate interagency disputes over roles, missions, and budgets.[8]

The Government Accountability Office (GAO), the investigative arm of the U.S. Congress, has also recommended the creation of "a focal point to coordinate federal biodefense activities." In making this recommendation, the GAO reiterated an earlier assessment that "interagency and intergovernmental activities can benefit from the leadership of a single entity with sufficient time, responsibility, authority, and resources needed to provide assurance that the federal programs are well coordinated, and that gaps and duplication in capabilities are avoided."[9]

An examination of two existing U.S. government advisory bodies, the National Science Advisory Board for Biosecurity (NSABB) and the Emerging Technologies Interagency Policy Coordination Committee (ETIPC), provides a starting point for thinking about how to implement the decision framework. The NSABB, created in 2004, provides guidance to federal policy makers with respect to the oversight of "dual-use research of concern" in the life sciences. The board consists of up to twenty-five nongovernmental experts with scientific, security, legal, and policy expertise, along with ex officio representatives from federal agencies that conduct or support biological or biomedical research.[10] Nevertheless, because the scope of the NSABB is limited to research in the life sciences rather than covering the full spectrum of emerging dual-use technologies in the biological and chemical fields, its pool of scientific expertise is not inclusive enough to implement the decision framework. The NSABB also serves in a strictly advisory capacity and has no executive authority.

The ETIPC, established in May 2010, is cochaired by three White House offices— the Office of Science and Technology Policy (OSTP), the Office of Information and Regulatory Affairs (OIRA), and the Office of the United States Trade Representative (USTR)—and includes representatives at the assistant-secretary level from about twenty federal agencies. The committee leads a coordinated, high-level interagency effort to assess the policy implications of emerging technologies, including the need for "risk-benefit-based oversight mechanisms that can ensure safety without stifling innovation, stigmatizing emerging technologies, or creating trade barriers."[11] This emphasis on the interactions among scientific, regulatory, and trade policies is relevant to the implementation of the decision framework. Nevertheless the ETIPC does not have a legal mandate from Congress and lacks the authority to serve an effective oversight role. In addition, the ETIPC was created primarily to promote scientific and technological development, and addressing security risks is not one of its priority tasks.

In sum, although the NSABB and ETIPC offer useful organizational models with respect to the types of expertise, institutional resources, and interagency representation they bring to bear, they do not have the legal authority, ability to oversee multiple agencies, or breadth of expertise required for the governance of emerging dual-use technologies in the biological and chemical fields. It therefore appears that effective national implementation of the decision framework would require a standing body within the executive branch that monitors emerging technologies, identifies those that pose potential dual-use risks, and oversees the risk-assessment process and the formulation of suitable governance options. Such an entity would require budgetary and adjudicative powers, as well as a specific focus on the security implications of emerging dual-use technologies.

Because the U.S. government currently has no such coordinating body, there are two main policy options. The first option is to grant new powers to an existing body, such as a restructured and strengthened ETIPC or a new unit within OSTP. The second option, which would be more resource intensive but probably more effective over the long term, is to establish a new entity within the Executive Office of the President that might be called the Dual-Use Technologies Oversight Board (DTOB). This entity would have the power to direct various U.S. government agencies to exercise their rule-making discretion and propose appropriate regulatory measures for emerging dual-use technologies, consistent with congressional guidelines. An appropriate place to house the new unit would be in the White House Office of Management and Budget (OMB), perhaps as a unit reporting to OIRA. The advantage of OMB is that it oversees financial management and regulatory policy making within the executive branch and has governance responsibilities across the U.S. government.

If a DTOB is created, its primary or sole mission should be to oversee the implementation of the decision framework. The board's staff should consist of officials who have experience in technology policy and security issues and have been seconded from other U.S. government agencies. These officials should consult on a regular basis with nongovernmental stakeholders, particularly the industry and scientific communities involved in developing dual-use technologies. The DTOB should also be structured to have a high level of technical expertise and institutional memory by limiting the number of political appointees in leadership roles. To the extent possible, the board should implement the decision framework by drawing on existing analytical and regulatory resources of the U.S. government to avoid the creation of redundant structures and make the best use of scarce resources.

As described in chapter 4, the decision framework consists of three interconnected processes: technology monitoring, technology assessment, and the selection of governance measures (including cost-benefit analysis). Because performing all these activities in-house would probably exceed the capabilities of the DTOB, the board could delegate some or all of these tasks to existing federal agencies or to congressionally chartered, independent research institutions.

With respect to the monitoring of emerging dual-use technologies, one possibility would be to create a standing interdisciplinary committee within the U.S. National Academies or, alternatively, the Science and Technology Policy Institute (STPI), a federally funded research and development center sponsored by the National Science Foundation. This technology-monitoring committee would have access to relevant classified information collected by the intelligence community and the U.S. National Laboratories, and it would submit annual or semiannual reports to the DTOB that identify emerging technologies of potential dual-use concern.

Although the board could delegate the two other elements of the decision framework—technology assessment and the development of governance options—to other government agencies, these tasks are closely interrelated and would need to be managed as such. Even if all three operational elements of the decision framework are carried out by other entities, the DTOB would oversee and coordinate all technology monitoring, assessment, and governance activities to ensure their seamless integration. Finally, the specific packages of hard-law, soft-law, and informal measures developed for each dual-use technology would be implemented by existing U.S. government agencies, nongovernmental institutions, and the scientific community.

Engaging Nongovernmental Actors and the General Public

In addition to developing and implementing hard-law regulations, the U.S. government has an important role to play with respect to soft-law and informal measures because it directly and indirectly influences the "parameters of the system within which people, institutions and other stakeholders behave so that self-regulation or the social ecosystem achieves the desired outcomes."[12] Soft-law and informal measures require active outreach to the scientific community and private industry to educate them about the risks of misuse and to secure their buy-in with respect to the need for oversight. Such consultations must be a two-way street, so that the government is aware of the interests and concerns of industry and the scientific community. Creating nongovernmental advisory committees to the DTOB would facilitate such communication.

Although engagement with the general public is essential to formulating policies that are acceptable to society as a whole, the implementation of the decision framework should be judicious in the way it incorporates public views. Whereas ethical and public-safety issues related to emerging technologies often involve subjective value judgments, assessing dual-use security risks and developing suitable governance options should proceed in a rigorous and objective fashion. Once subject-matter experts have properly identified the risks, priorities, and policy options, it is then appropriate to engage with the public—for example, by allowing interested citizens and civil-society groups to participate in the evaluation and selection of governance options. Public input is of particular importance when trying to find the right balance between mitigating security risks through potentially burdensome regulations and allowing technological innovation to progress freely. Of course, if the DTOB decided to pursue hard-law regulation as a way to manage dual-use risks, the responsible federal agencies would have to follow the standard rule-making process under the Administrative Procedures Act, which requires that the public be informed of, and given the chance to comment on, any proposed regulation before its adoption.[13]

International Implementation

If a dual-use technology has diffused internationally, policy makers should strive to achieve a consistent approach to governance across multiple countries through a process of harmonization or standardization. In such cases, the use of the decision framework to select governance measures at the domestic level should take into account whether or not similar measures can be implemented by all countries that already possess the technology in question. Because many soft-law and informal measures should be fairly easy to adopt, their promulgation on a national level would also promote their implementation internationally. Failure to achieve international harmonization would create gaps or inconsistencies in global technology governance that could prove troublesome, for two reasons. First, the failure by certain countries to regulate a particular dual-use technology would create safe havens where it might be misused for harmful purposes. Second, inconsistent governance measures from country to country would tend to increase the burden of regulatory compliance and hamper innovation.

Governance measures can be developed and promulgated at the international level through four different mechanisms, which are briefly summarized hereafter. Some of these approaches are better suited than others to the challenge of governing emerging dual-use technologies.

First, a group of states may come together to negotiate a multilateral treaty, or the UN Security Council may pass a binding resolution, without any prior regulatory activity at the national level. Whenever a multilateral treaty is not self-executing, it obligates each member state to adopt domestic implementing legislation that imposes obligations on its citizens and private companies. For example, the 1972 Biological Weapons Convention (BWC) and the 1993 Chemical Weapons Convention (CWC) both require the states that are parties to pass laws making the treaty provisions binding on their citizens, whether they live at home or abroad, and imposing penal sanctions for violations.

Although the CWC and the BWC outlaw offensive applications of chemistry and biology in general terms, they do not distinguish among specific technologies and hence are poorly suited for developing tailored approaches to governance. Another drawback of treaties as an instrument of technology governance is that they usually take many years to negotiate and, once concluded, are often too inflexible to keep pace with rapid technological change. For these reasons, more flexible approaches are needed, such as internationally harmonized guidelines that states can adopt on a voluntary basis.

Another possible approach is the negotiation of a "framework convention," a broadly worded international agreement that is designed to be augmented with specific protocols containing legally binding provisions and enforcement mechanisms. One example is the UN Framework Convention on Climate Change, which provided the foundation on which the international community negotiated the legally binding 1997 Kyoto Protocol limiting emissions of greenhouse gases. A framework convention for a particular dual-use technology of concern would be flexible because it could be updated over time as the technology matures and the risk of its misuse becomes more evident.[14]

Although the BWC and CWC are not suited for dealing with individual dual-use technologies, they could be leveraged to support the implementation of the decision framework at the international level by contributing to technology monitoring. At present, however, the BWC lacks a forum or mechanism to assess the implications of scientific and technological developments for the convention more frequently than at the five-year review conferences. In addition, although the CWC regime has a Scientific Advisory Board (SAB) that advises the director general of the treaty-implementing body, the Organization for the Prohibition of Chemical Weapons, the SAB does not have adequate resources to carry out its mandated functions. Strengthening the technology-monitoring capabilities of both treaty regimes would therefore be desirable.

Second, an informal coalition of like-minded states may adopt a set of common regulations, guidelines, or other measures to regulate a technology of concern. In 1985, responding to Iraq's extensive use in the Iran-Iraq War of chemical weapons, which were manufactured using precursor chemicals and production equipment imported from Western suppliers, a group of fifteen exporting countries established an informal forum called the Australia Group to harmonize their national export controls on precursors and equipment relevant to the production of chemical weapons. The group has since expanded to forty countries and the European Commission and increased its coverage to include dual-use biological items.[15]

In recent years, the Australia Group has established working groups on specific dual-use technologies, such as synthetic genomics and chemical microreactors. Even so, the participating countries would probably be resistant to adopting an entirely new set of measures designed to respond proactively to emerging dual-use risks rather than to obvious threats, such as Iraq's purchase of CW precursors during the

Iran-Iraq War. Another example of an ad hoc coalition of like-minded states is the Proliferation Security Initiative (PSI), established in 2003, through which more than ninety countries cooperate in interdicting illicit shipments of WMD-related materials.

Third, domestic nonproliferation and security legislation developed by one or more governments may serve as a model for other countries, leading over time to the emergence of a harmonized international regime. For example, a number of states have emulated the U.S. Select Agent Regulations, which impose strict access controls on a list of pathogens and toxins of bioterrorism concern. Using this approach, the U.S. government could develop a set of national governance measures for a particular dual-use technology and then engage with other countries on a bilateral or multilateral basis to promote an internationally harmonized approach. Although the harmonized standards would be modeled after the governance architecture of the leader government, the other countries could determine the specifics of their respective national implementation plans in accordance with their domestic institutions and requirements.

Just as nongovernmental stakeholders are key to the effective national implementation of the decision framework, so efforts at international harmonization should engage with foreign industry and scientific associations. International organizations can also play a useful role as vehicles for harmonization. In 2004, for example, the United States commissioned the World Health Organization (WHO) to develop a set of laboratory biosecurity guidelines, which have since been widely accepted internationally.[16] In addition, the Paris-based Organisation for Economic Co-operation and Development (OECD) would make an ideal partner for the development of harmonized approaches to technology governance, for several reasons. First, because the members of the OECD are the world's thirty-four most developed nations, they are likely to possess many emerging dual-use technologies. Second, the OECD routinely examines national regulatory policies, including those related to emerging technologies such as nanotechnology and synthetic biology. Third, the OECD's raison d'être is to stimulate economic development and international trade, lending credibility to international efforts to prevent the misuse of emerging technologies while preserving their benefits.

Fourth, an association of private companies or a coalition of nongovernmental organizations may establish a self-regulatory mechanism based on a set of voluntary guidelines or norms, which they then promote to similar entities from other countries. In the case of genome synthesis, for example, two groups of commercial gene-synthesis firms have developed their own guidelines to verify the bona fides of customers and to screen gene-synthesis requests to prevent the acquisition of pathogenic DNA sequences by criminals or terrorists. These industry initiatives in turn inspired the U.S. government to develop its own set of voluntary guidelines.[17]

A possible obstacle to the international harmonization of self-governance measures is that foreign industry groups may perceive economic disincentives to participating in such a regime.[18] To overcome these objections, national governments should highlight the mutual economic benefits arising from the creation of a level playing field.[19] Moreover, self-governance mechanisms can promote public confidence in an emerging technology, increasing its acceptability and the demand for its applications.[20]

It would also be desirable to foster synergies between government-led efforts to harmonize the governance of emerging dual-use technologies and a variety of non-governmental initiatives. For example, some scholars have proposed creating informal global-issue networks that link together scientists, government representatives, nongovernmental organizations, and private industries from different parts of the world.[21] This mechanism would facilitate the process of reaching an international consensus on strategies for the governance of specific technologies. Global networks of informed scientists could also assist national governments to monitor dual-use technologies by bringing concerns about misuse to the attention of the relevant authorities.[22] More generally, the creation of open-source databases and international forums for sharing information among experts in different countries would facilitate the process of harmonization.[23]

Figure 21.2 depicts the four approaches to international technology governance described in the foregoing paragraphs and shows how harmonization may proceed from the national level to the international level or vice versa. It may also be pos-

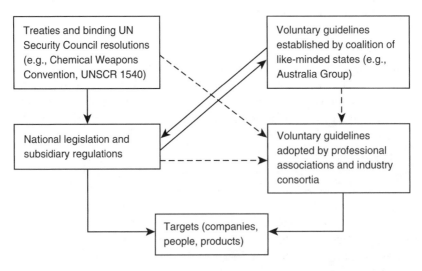

Figure 21.2
Relationships among national and international governance measures.

sible to integrate these different strategies. For example, the national implementation of a tailored governance strategy by a limited number of countries could serve to demonstrate its feasibility and make it easier to promote internationally.

Conclusions

This book has addressed the dual-use dilemma that characterizes many technological innovations in the biological and chemical fields. In addition to exploring the nature of the dual-use problem, it has proposed a decision framework that is designed to reduce the risk that emerging technologies will be misused for harmful purposes, while preserving their benefits. Although this book has focused on a particular subset of emerging dual-use technologies, namely, those in the biological and chemical fields, the decision framework is also relevant in a general sense to other areas of dynamic technological innovation, such as energy, robotics, and information technology. Moreover, George Poste of Arizona State University has argued that technological convergences will continue to build on themselves, cascading into new spaces for innovation, such as brain-machine interactions.[24]

As our complex technological civilization continues to accelerate into an uncertain future, governments alone cannot undertake the challenge of managing the risks of emerging dual-use technologies. Ultimately strategies of technology governance can only be as effective as the policy makers who select them, the government institutions that implement them, the scientists and corporate executives who comply with them, and the members of the public who, in a democratic system, must hold their political leaders accountable. Concerned citizens must do their part to monitor dual-use innovations and take steps to ensure that they enrich and enhance the human condition, rather than endangering or degrading it.

Notes

1. Chandré Gould and Alastair Hay, "The South African Biological Weapons Program," in *Deadly Cultures: Biological Weapons since 1945*, ed. Mark Wheelis, Lajos Rózsa, and Malcolm Dando (Cambridge, MA: Harvard University Press, 2006), 191–212.

2. Richard J. Bonnie, "Political Abuse of Psychiatry in the Soviet Union and in China: Complexities and Controversies," *Journal of the American Academy of Psychiatry and the Law* 30 (2002): 136–144.

3. Alan Cowell, "The Daring Attack That Blew Up in Israel's Face," *New York Times*, October 15, 1997, A1, A6.

4. Francis Fukuyama, *Our Posthuman Future: Consequences of the Biotechnology Revolution* (New York: Farrar, Straus and Giroux, 2002), 203–218.

5. Ibid., 212–213.

6. Presidential Commission for the Study of Bioethical Issues, *New Directions: The Ethics of Synthetic Biology and Emerging Technologies* (Washington, DC, December 2010), 8, 80–109, http://www.bioethics.gov/documents/synthetic-biology/PCSBI-Synthetic-Biology -Report-12.16.10.pdf.

7. The relevant agencies include the National Institutes of Health (NIH), the Centers for Disease Control and Prevention (CDC), the Food and Drug Administration (FDA), the U.S. Department of Agriculture (USDA), the Occupational Safety and Health Administration (OSHA), the Department of Transportation (DOT), and the Environmental Protection Agency (EPA).

8. Penrose C. "Parney" Albright, "Current Pressing Issues in Biodefense," presentation at University of California Washington Center in the seminar series "Current Issues in Public Policy and Biological Threats," January 28, 2011.

9. U.S. Government Accountability Office (GAO), "Opportunities to Reduce Potential Duplication in Government Programs, Save Tax Dollars, and Enhance Revenue," Report to Congressional Addressees, GAO-11-318SP (Washington, DC: U.S. Government Accountability Office, 2011), 92–95.

10. Secretary of Health and Human Services, Renewal of the Charter of the National Science Advisory Board for Biosecurity, March 10, 2010, http://oba.od.nih.gov/biosecurity/PDF/NSABB%202010%20Charter_Renewal%20.pdf.

11. Heather Evans, White House Office of Science and Technology Policy, "Emerging Technologies IPC Has Inaugural Meeting," May 15, 2010, http://www.whitehouse.gov/blog/2010/05/15/emerging-technologies-ipc-has-inaugural-meeting.

12. Mihail Roco, "Possibilities for Global Governance of Converging Technologies," *Journal of Nanoparticle Research* 10, no. 1 (January 2008): 3, 19.

13. Administrative Procedures Act, Pub. L. No. 79-404, 60 Stat. 238 (1946).

14. Alexander Kelle, Kathryn Nixdorff, and Malcolm Dando, *Controlling Biochemical Weapons: Adapting Multilateral Arms Control for the 21st Century* (Basingstoke: Palgrave Macmillan, 2006); Gary E. Merchant and Douglas J. Sylvester, "Transnational Models for Regulation of Nanotechnology," *Journal of Law, Medicine, and Ethics* (winter 2006): 720–721.

15. Australia Group Web site, http://www.australiagroup.net.

16. World Health Organization, *Biorisk Management: Laboratory Biosecurity Guidance*, WHO/CDS/EPR/2006.6 (Geneva: WHO, 2006), http://www.who.int/csr/resources/publications/biosafety/WHO_CDS_EPR_2006_6.pdf.

17. Jonathan B. Tucker, "Double-Edged DNA: Preventing the Misuse of Gene Synthesis," *Issues in Science and Technology* 26, no. 3 (spring 2010): 23–32.

18. Roco, "Possibilities for Global Governance of Converging Technologies," 13, 19.

19. Jonathan B. Tucker, "Preventing Terrorist Access to Dangerous Pathogens: The Need for International Biosecurity Standards," *Disarmament Diplomacy* 66 (September 2002), http://www.acronym.org.uk/dd/dd66/66op2.htm.

20. Gary E. Marchant and Douglas J. Sylvester, "Transnational Models for Regulation of Nanotechnology," *Journal of Law, Medicine, and Ethics* (winter 2006): 716.

21. Caroline S. Wagner, *The New Invisible College: Science for Development* (Washington, DC: Brookings Institution Press, 2008); Anne-Marie Slaughter, *A New World Order* (Princeton, NJ: Princeton University Press, 2005); Jean- François Rischard, "Global Issue Networks," *Washington Quarterly* 26, no. 1 (winter 2002–2003): 17.

22. National Research Council, *Globalization, Biosecurity, and the Future of the Life Sciences*, 251–256.

23. Roco, "Possibilities for Global Governance," 15–18, 19.

24. George Poste, "The Impact of the Life Sciences on National Security," presentation at the conference "Preserving National Security: The Growing Role of the Life Sciences," sponsored by the Center for Biosecurity of the University of Pittsburgh Medical Center, March 3, 2011.

Select Bibliography

Theoretical Works

Alic, John A., Lewis M. Branscomb, Harvey Brooks, Ashton B. Carter, and Gerald L. Epstein. *Beyond Spinoff: Military and Commercial Technologies in a Changing World*. Boston, MA: Harvard Business School Press, 1992.

Bijker, Wiebe E. *Of Bicycles, Bakelites, and Bulbs: Towards a Theory of Technological Change*. Cambridge, MA: MIT Press, 1995.

Bijker, Wiebe E., Thomas P. Hughes, and Trevor J. Pinch, eds. *The Social Construction of Technological Systems: New Directions in the Sociology and History of Technology*. Cambridge, MA: MIT Press, 1987.

Chubin, Daryl, and Ellen Chen. *Science off the Pedestal: Social Perspectives on Science and Technology*. Belmont, CA: Wadsworth Publishing, 1989.

Hackett, Edward J., Olga Amsterdamska, Michael Lynch, and Judy Wajcman, eds. *The Handbook of Science and Technology Studies*. 3rd ed. Cambridge, MA: MIT Press, 2007.

Knorr-Cetina, Karin. *The Manufacture of Knowledge: An Essay on the Constructivist and Contextual Nature of Science*. Oxford: Pergamon Press, 1981.

Latour, Bruno, and Steve Woolgar. *Laboratory Life: The Social Construction of Scientific Facts*. Beverly Hills, CA: Sage Publications, 1979.

Perrow, Charles. *Normal Accidents: Living with High-Risk Technologies*. Princeton, NJ: Princeton University Press, 1984.

Pool, Robert. *Beyond Engineering: How Society Shapes Technology*. Oxford: Oxford University Press, 1997.

Sclove, Richard. *Democracy and Technology*. New York: Guilford Press, 1995.

Star, Susan Leigh. *Ecologies of Knowledge: Work and Politics in Science and Technology*. Albany, NY: SUNY Press, 1995.

Ethics of Dual-Use Technologies

Bedau, Mark, and Emily C. Parke. *The Ethics of Protocells: Moral and Social Implications of Creating Life in the Laboratory*. Cambridge, MA: MIT Press, 2009.

Fukuyama, Francis. *Our Posthuman Future: Consequences of the Biotechnology Revolution*. New York: Macmillan, 2003.

Miller, Seumas, and Michael J. Selgelid. *Ethical and Philosophical Consideration of the Dual-Use Dilemma in the Biological Sciences*. New York: Springer, 2009.

Presidential Commission for the Study of Bioethical Issues. *New Directions: The Ethics of Synthetic Biology and Emerging Technologies*. Washington, DC, 2010.

Rappert, Brian, ed. *Education and Ethics in the Life Sciences: Strengthening the Prohibition of Biological Weapons*. Canberra, Australia: Australian National University Press, 2010.

Sollie, Paul, and Marcus Düwell. *Evaluating New Technologies: Methodological Problems for the Ethical Assessment of Technology Developments*. New York: Springer, 2009.

History of Dual-Use Technologies

Geissler, Erhard, and John Ellis van Courtland Moon, eds. *Biological and Toxin Weapons: Research, Development, and Use from the Middle Ages to 1945*. SIPRI Chemical and Biological Warfare Studies. Oxford: Oxford University Press for the Stockholm International Peace Research Institute, 1999.

McNeill, William H. *The Pursuit of Power: Technology, Armed Force, and Society since A.D. 1000*. Chicago: University of Chicago Press, 1982.

Tucker, Jonathan B., ed. *Toxic Terror: Assessing Terrorist Use of Chemical and Biological Weapons*. Cambridge, MA: MIT Press, 2000.

Wheelis, Mark, Lajos Rózsa, and Malcolm Dando, eds. *Deadly Cultures: Biological Weapons since 1945*. Cambridge, MA: Harvard University Press, 2006.

Risk Assessments of Dual-Use Technologies

Beck, Ulrich. *World Risk Society*. Cambridge: Polity Press, 1999.

British Medical Association. *Biotechnology, Weapons, and Humanity*. Amsterdam: Harwood Academic Publishers, 1999.

Carlson, Robert H. *Biology Is Technology: The Promise, Peril, and New Business of Engineering Life*. Cambridge, MA: Harvard University Press, 2010.

Commission on the Prevention of Weapons of Mass Destruction Proliferation and Terrorism. *World at Risk*. New York: Vintage Press, 2008.

Dando, Malcolm R., and Vivienne Nathanson. *Biotechnology, Weapons, and Humanity II*. London: British Medical Association, 2004.

Dumas, Lloyd J. *Lethal Arrogance: Human Fallibility and Dangerous Technologies*. New York: St. Martin's Press, 1999.

Dumas, Lloyd J. *The Technology Trap: Where Human Error and Malevolence Meet Powerful Technologies*. Santa Barbara, CA: ABC-CLIO, 2010.

Epstein, Gerald L. *Global Evolution of Dual-Use Biotechnology*. Washington, DC: Center for Strategic and International Studies, 2005.

Fischer, Julie E. *Dual-Use Technologies: Inexorable Progress, Inseparable Peril*. Washington, DC: Center for Strategic and International Studies, 2005.

Katona, Peter, ed. *Global Biosecurity: Threats and Responses*. London: Routledge Press, 2011.

Koblentz, Gregory D. *Living Weapons: Biological Warfare and International Security*. Ithaca, NY: Cornell University Press, 2009.

Levi, Michael A. *On Nuclear Terrorism*. Cambridge, MA: Harvard University Press, 2009.

Malik, Amitav. *Technology and Security in the 21st Century*. Oxford: Oxford University Press for the Stockholm International Peace Research Institute, 2004.

Moreno, Jonathan D. *Mind Wars: Brain Research and National Defense*. New York: Dana Press, 2006.

National Research Council. *Biotechnology Research in an Age of Terrorism*. Washington, DC: National Academies Press, 2004.

National Research Council. *Globalization, Biosecurity, and the Future of the Life Sciences*. Washington, DC: National Academies Press, 2008.

National Research Council. *National Security and Homeland Defense: Challenges for the Chemical Sciences in the 21st Century*. Washington, DC: National Academies Press, 2002.

National Research Council. *Understanding Risk: Informing Decisions in a Democratic Society*. Washington, DC: National Academies Press, 1996.

Schmidt, Markus, Alexander Kelle, Agomoni Ganguli-Mitra, and Huib de Vriend, eds. *Synthetic Biology: The Technoscience and Its Societal Consequences*. New York: Springer, 2009.

Technology Governance

Fidler, David P., and Lawrence O. Gostin. *Biosecurity in the Global Age: Biological Weapons, Public Health, and the Rule of Law*. Stanford, CA: Stanford University Press, 2008.

Garfinkel, Michele, Drew Endy, Gerald L. Epstein, and Robert M. Friedman. *Synthetic Genomics: Options for Governance*. Rockville, MD: J. Craig Venter Institute, with MIT and the Center for Strategic and International Studies, 2007.

Geissler, Erhard, and John P. Woodall, eds. *Control of Dual-Threat Agents: The Vaccines for Peace Programme*. Oxford: Oxford University Press, 1994.

Joyner, Daniel H. *International Law and the Proliferation of Weapons of Mass Destruction*. London: Oxford University Press, 2009.

Kelle, Alexander, Kathryn Nixdorff, and Malcolm Dando. *Controlling Biochemical Weapons: Adapting Multilateral Arms Control for the 21st Century*. Basingstoke: Palgrave Macmillan, 2006.

Kooiman, Jan, ed. *Modern Governance: New Government-Society Interactions*. London: Sage Publishers, 1993.

Lakoff, Andrew, and Stephen J. Collier. *Biosecurity Interventions: Global Health and Security in Question*. New York: Columbia University Press, 2008.

National Research Council. *An International Perspective on Advancing Technologies and Strategies for Managing Dual-Use Risks: Report of a Workshop*. Washington, DC: National Academies Press, 2005.

National Research Council. *Responsible Research with Biological Select Agents and Toxins*. Washington, DC: National Academies Press, 2009.

National Research Council. *Science and Security in a Post 9/11 World: A Report Based on Regional Discussions between the Science and Security Communities*. Washington, DC: National Academies Press, 2007.

National Research Council. *Seeking Security: Pathogens, Open Access, and Genome Databases*. Washington, DC: National Academies Press, 2004.

National Research Council. *Sequence-Based Classification of Select Agents: A Brighter Line*. Washington, DC: National Academies Press, 2010.

National Research Council. *A Survey of Attitudes and Actions on Dual-Use Research in the Life Sciences*. Washington, DC: National Academies Press, 2009.

National Science Advisory Board for Biosecurity. *Addressing Biosecurity Concerns Related to Synthetic Biology*. Bethesda, MD: National Institutes of Health, April 2010.

National Science Advisory Board for Biosecurity. *Addressing Biosecurity Concerns Related to the Synthesis of Select Agents*. Bethesda, MD: National Institutes of Health, December 2006.

National Science Advisory Board for Biosecurity. *Proposed Framework for the Oversight of Dual Use Life Sciences Research: Strategies for Minimizing the Potential Misuse of Research Information*. Bethesda, MD: National Institutes of Health, June 2007.

Rappert, Brian. *Biotechnology, Security, and the Search for Limits: An Inquiry into Research and Methods*. New York: Palgrave Macmillan, 2007.

Rappert, Brian. *Experimental Secrets: International Security, Codes, and the Future of Research*. New York: University Press of America, 2008.

Rappert, Brian, and Chandré Gould, eds. *Biosecurity: Origins, Transformations, and Practices*. New York: Palgrave Macmillan, 2009.

Rappert, Brian, and Caitríona McLeish, eds. *A Web of Prevention: Biological Weapons, Life Sciences, and the Governance of Research*. London: Earthscan, 2007.

Sims, Nicholas A. *The Evolution of Biological Disarmament*. SIPRI Chemical and Biological Warfare Studies. Oxford: Oxford University Press for the Stockholm International Peace Research Institute, 2001.

Sims, Nicholas A. *The Future of Biological Disarmament: Strengthening the Treaty Ban on Weapons*. London: Routledge, 2009.

Steinbruner, John, Elisa D. Harris, Nancy Gallagher, and Stacy M. Okutani. *Controlling Dangerous Pathogens*. College Park, MD: Center for International Security Studies at Maryland, 2007.

Contributors

Hussein Alramini is a graduate of Middlebury College with a bachelor's degree in molecular biology and biochemistry.

Brian Balmer is a senior lecturer in the Department of Science and Technology Studies at University College London.

Kirk C. Bansak is a graduate student at the Fletcher School of Law and Diplomacy, Tufts University, Medford, Massachusetts.

Nancy Connell is professor of infectious disease at the Center for Emerging Pathogens, University of Medicine and Dentistry of New Jersey, Newark, N.J.

Malcolm R. Dando is professor of international security at the University of Bradford, U.K.

Gerald L. Epstein is director of the Center for Science, Technology, and National Security Policy at the American Association for the Advancement of Science, Washington, D.C.

Gail Javitt is a research scientist at the Berman Institute of Bioethics at the Johns Hopkins University in Baltimore and Counsel at Sidley Austin LLP.

Catherine Jefferson is a research fellow at the Harvard Sussex Program on Chemical and Biological Weapons at the University of Sussex, U.K.

Alexander Kelle is a lecturer in politics and international relations at the University of Bath, U.K.

Lori P. Knowles is a research associate at the Health Law Institute of the University of Alberta, Canada.

Filippa Lentzos is a senior research fellow in the BIOS Centre at the London School of Economics and Political Science.

Caitríona McLeish is a research fellow at the Harvard Sussex Program on Chemical and Biological Weapons at the University of Sussex, U.K.

Matthew Metz is project officer with the Biomedical Advanced Research and Development Authority of the U.S. Department of Health and Human Services.

Nishal Mohan is project director of the Virtual Biosecurity Center at the Federation of American Scientists, Washington, D.C.

Jonathan D. Moreno is the David and Lyn Silfen University Professor and Professor of Medical Ethics and of History and Sociology of Science at the University of Pennsylvania, Philadelphia, PA.

Anya Prince is a joint-degree student in law and policy at Georgetown University Law Center and the Georgetown Public Policy Institute, Washington, D.C.

Pamela Silver is professor in the Department of Systems Biology at Harvard Medical School.

Amy E. Smithson is a senior fellow in the Washington, D.C., office of the James Martin Center for Nonproliferation Studies of the Monterey Institute of International Studies.

Ralf Trapp is an independent consultant specializing in chemical and biological weapons issues who is based in Geneva, Switzerland.

Jonathan B. Tucker[†] was a senior fellow at the Federation of American Scientists.

Mark Wheelis is a former lecturer in biology at the University of California, Davis.

Raymond A. Zilinskas is director of the Chemical and Biological Weapons Nonproliferation Program at the James Martin Center for Nonproliferation Studies, Monterey, Calif.

[†]deceased

Index